Communications
in Computer and Information Science   1905

Editorial Board Members

Joaquim Filipe ⓘ, *Polytechnic Institute of Setúbal, Setúbal, Portugal*
Ashish Ghosh ⓘ, *Indian Statistical Institute, Kolkata, India*
Lizhu Zhou, *Tsinghua University, Beijing, China*

AF148511

Rationale

The CCIS series is devoted to the publication of proceedings of computer science conferences. Its aim is to efficiently disseminate original research results in informatics in printed and electronic form. While the focus is on publication of peer-reviewed full papers presenting mature work, inclusion of reviewed short papers reporting on work in progress is welcome, too. Besides globally relevant meetings with internationally representative program committees guaranteeing a strict peer-reviewing and paper selection process, conferences run by societies or of high regional or national relevance are also considered for publication.

Topics

The topical scope of CCIS spans the entire spectrum of informatics ranging from foundational topics in the theory of computing to information and communications science and technology and a broad variety of interdisciplinary application fields.

Information for Volume Editors and Authors

Publication in CCIS is free of charge. No royalties are paid, however, we offer registered conference participants temporary free access to the online version of the conference proceedings on SpringerLink (http://link.springer.com) by means of an http referrer from the conference website and/or a number of complimentary printed copies, as specified in the official acceptance email of the event.

CCIS proceedings can be published in time for distribution at conferences or as post-proceedings, and delivered in the form of printed books and/or electronically as USBs and/or e-content licenses for accessing proceedings at SpringerLink. Furthermore, CCIS proceedings are included in the CCIS electronic book series hosted in the SpringerLink digital library at http://link.springer.com/bookseries/7899. Conferences publishing in CCIS are allowed to use Online Conference Service (OCS) for managing the whole proceedings lifecycle (from submission and reviewing to preparing for publication) free of charge.

Publication process

The language of publication is exclusively English. Authors publishing in CCIS have to sign the Springer CCIS copyright transfer form, however, they are free to use their material published in CCIS for substantially changed, more elaborate subsequent publications elsewhere. For the preparation of the camera-ready papers/files, authors have to strictly adhere to the Springer CCIS Authors' Instructions and are strongly encouraged to use the CCIS LaTeX style files or templates.

Abstracting/Indexing

CCIS is abstracted/indexed in DBLP, Google Scholar, EI-Compendex, Mathematical Reviews, SCImago, Scopus. CCIS volumes are also submitted for the inclusion in ISI Proceedings.

How to start

To start the evaluation of your proposal for inclusion in the CCIS series, please send an e-mail to ccis@springer.com.

Dmitry I. Ignatov · Michael Khachay ·
Andrey Kutuzov · Habet Madoyan ·
Ilya Makarov · Irina Nikishina ·
Alexander Panchenko · Maxim Panov ·
Panos M. Pardalos · Andrey V. Savchenko ·
Evgenii Tsymbalov · Elena Tutubalina ·
Sergey Zagoruyko
Editors

Recent Trends in Analysis of Images, Social Networks and Texts

11th International Conference, AIST 2023
Yerevan, Armenia, September 28–30
Revised Selected Papers

 Springer

Editors
Dmitry I. Ignatov 🆔
National Research University Higher School
of Economics
Moscow, Russia

Andrey Kutuzov 🆔
University of Oslo
Oslo, Norway

Ilya Makarov 🆔
Artificial Intelligence Research Institute
Moscow, Russia

Alexander Panchenko 🆔
Skolkovo Institute of Science and Technology
Moscow, Russia

Panos M. Pardalos 🆔
Industrial and Systems Engineering
University of Florida
Gainesville, FL, USA

Evgenii Tsymbalov 🆔
Apptek
Aachen, Nordrhein-Westfalen, Germany

Sergey Zagoruyko 🆔
MTS AI
Moscow, Russia

Michael Khachay 🆔
Krasovskii Institute of Mathematics
and Mechanics of Russian Academy of Sciences
Yekaterinburg, Russia

Habet Madoyan
American University of Armenia
Yerevan, Armenia

Irina Nikishina 🆔
Universität Hamburg
Hamburg, Germany

Maxim Panov 🆔
Mohamed bin Zayed University of Artificial
Intelligence and Technology Innovation Institute
Abu Dhabi, United Arab Emirates

Andrey V. Savchenko 🆔
National Research University Higher School
of Economics
Nizhny Novgorod, Russia

Elena Tutubalina 🆔
Kazan Federal University and HSE University
Moscow, Russia

ISSN 1865-0929   ISSN 1865-0937 (electronic)
Communications in Computer and Information Science
ISBN 978-3-031-67007-7   ISBN 978-3-031-67008-4 (eBook)
https://doi.org/10.1007/978-3-031-67008-4

© The Editor(s) (if applicable) and The Author(s), under exclusive license
to Springer Nature Switzerland AG 2024, corrected publication 2024

This work is subject to copyright. All rights are solely and exclusively licensed by the Publisher, whether the whole or part of the material is concerned, specifically the rights of translation, reprinting, reuse of illustrations, recitation, broadcasting, reproduction on microfilms or in any other physical way, and transmission or information storage and retrieval, electronic adaptation, computer software, or by similar or dissimilar methodology now known or hereafter developed.
The use of general descriptive names, registered names, trademarks, service marks, etc. in this publication does not imply, even in the absence of a specific statement, that such names are exempt from the relevant protective laws and regulations and therefore free for general use.
The publisher, the authors and the editors are safe to assume that the advice and information in this book are believed to be true and accurate at the date of publication. Neither the publisher nor the authors or the editors give a warranty, expressed or implied, with respect to the material contained herein or for any errors or omissions that may have been made. The publisher remains neutral with regard to jurisdictional claims in published maps and institutional affiliations.

This Springer imprint is published by the registered company Springer Nature Switzerland AG
The registered company address is: Gewerbestrasse 11, 6330 Cham, Switzerland

If disposing of this product, please recycle the paper.

Preface

This volume contains the refereed proceedings of the 11th International Conference on Analysis of Images, Social Networks and Texts (AIST 2023)[1]. The previous conferences (during 2012-2021) attracted a significant number of data scientists, students, researchers, academics, and engineers working on interdisciplinary data analysis of images, texts, and networks. The broad scope of AIST makes it an event where researchers from different domains, such as computer vision and natural language processing, exploiting various data analysis techniques, can meet and exchange ideas. As the test of time has shown, this leads to the cross-fertilisation of ideas between researchers relying on modern data analysis machinery.

Therefore, AIST 2023 brought together all kinds of applications of data mining and machine learning techniques. The conference allowed specialists from different fields to meet each other, present their work, and discuss both theoretical and practical aspects of their data analysis problems. Another important aim of the conference was to stimulate scientists and people from industry to benefit from knowledge exchange and identify possible grounds for fruitful collaboration.

The conference was held during September 28–30, 2023. The conference was organized with the support of Zaven and Sonia Akian College of Science and Engineering, American University of Armenia.

This year, the key topics of AIST were grouped into five tracks:

1. Data Analysis and Machine Learning chaired by Evgenii Tsymbalov (Apptek, Germany) and Maxim Panov (Technology Innovation Institute, UAE)
2. Natural Language Processing chaired by Andrey Kutuzov (University of Oslo, Norway) and Elena Tutubalina (Kazan Federal University, Russia)
3. Network Analysis chaired by Irina Nikishina (Universität Hamburg, Germany) and Ilya Makarov (HSE University and AIRI, Russia)
4. Computer Vision chaired by Sergei Zagoruyko (MTS AI, Russia), and Andrey Savchenko (HSE University, Russia)
5. Theoretical Machine Learning and Optimization chaired by Panos Pardalos (University of Florida, USA) and Michael Khachay (IMM UB RAS and Ural Federal University, Russia)

The Program Committee and the reviewers of the conference included 137 well-known experts in data mining and machine learning, natural language processing, image processing, social network analysis, and related areas from leading institutions of many countries including Armenia, Austria, the Czech Republic, Finland, France, Georgia, Germany, Greece, India, Ireland, Italy, Mexico, Montenegro, the Netherlands, Norway, Portugal, Russia, Slovenia, Spain, the UAE, the UK, and the USA. This year, we received 106 submissions, mostly from Russia but also from Armenia, China, Finland, France,

---

[1] https://aistconf.org.

Georgia, Germany, India, Iraq, Kyrgyzstan, Saudi Arabia, Singapore, Switzerland, the UAE, and the USA.

Out of the 106 technical submissions, 13 were poster submissions, and 76 submissions remained after desk rejects and withdrawals. For the remaining papers, only 24 were accepted into the main LNCS volume. In order to encourage young practitioners and researchers, we included 21 papers in this companion volume published in Springer's Communications in Computer and Information Science (CCIS) series. Thus, the acceptance rate of this CCIS volume is 40%. Each submission was double-blind reviewed by at least three reviewers, experts in their fields, in order to supply detailed and helpful comments. Based on careful selection and presence of the authors of posters at the conference, we also included 12 abstracts of posters in the back matter of the volume and one demo paper as a short paper.

The conference featured several invited talks and tutorials dedicated to current trends and challenges in the respective areas.

The invited talks from academia covered a wide range of machine learning and artificial intelligence areas:

- Narine Sarvazyan (George Washington University and American University of Armenia): "Decoding Hyperspectral Imaging: From Basic Principles to Medical Application"
- Hakim Hacid (Technology Innovation Institute): "Towards Edge AI: Principles, current state, and perspectives"
- Samuel Horvath (MBZUAI): "Towards Real-World Federated Learning: Addressing Client Heterogeneity and Model Size"
- Artem Shelmanov (MBZUAI): "Safety of Deploying NLP Models: Uncertainty Quantification of Generative LLMs"
- Muhammad Shahid Iqbal Malik (HSE University): "Threatening Content and Target Identification in low-resource languages using NLP Techniques"

We would like to thank the authors for submitting their papers and the members of the Program Committee for their efforts in providing exhaustive reviews.

According to the track chairs, and taking into account the reviews and presentation quality, the Best Paper Awards were granted to the following papers:

- Track 1. Data Analysis and Machine Learning: "Ensemble Clustering with Heterogeneous Transfer Learning" by Vladimir Berikov;
- Track 2. Natural Language Processing: "Benchmarking Multi-Label Topic Classification in Kyrgyz Language" by Anton Alekseev, Sergey Nikolenko, and Gulnara Kabaeva;
- Track 3. Social Network Analysis: "Limit Distributions of Friendship Index in Scale-Free Networks" by Sergei Sidorov, Sergei Mironov, and Alexey Grigoriev;
- Track 4. Computer Vision: "DeepLOC: Deep Learning-based Bone Pathology Localization and Classification in Wrist X-ray Images" by Razan Dibo, Andrey Galichin, Pavel Astashev, Dmitry V. Dylov, and Oleg Y. Rogov;
- Track 5. Theoretical Machine Learning and Optimization: "Is Canfield Right? On the Asymptotic Coefficients for the Maximum Antichain of Partitions and Related Counting Inequalities" by Dmitry Ignatov.

We would also like to express our special gratitude to all the invited speakers and industry representatives. We deeply thank all the partners and sponsors, especially the hosting organization: Zaven and Sonia Akian College of Science and Engineering, American University of Armenia, with special thanks to Habet Madoyan and Amalya Hambardzumyan, as local organizers. Our special thanks go to Springer for their help, starting from the first conference call to the final version of the proceedings. Last but not least, we are grateful to the volunteers, whose endless energy saved us at the most critical stages of the conference preparation.

Here, we would like to mention that the Russian word "aist" is more than just a simple abbreviation as (in Russian) it means "stork". Since it is a wonderful free bird, a symbol of happiness and peace, this stork gave us the inspiration to organize the AIST conference series. So we believe that this conference will likewise bring inspiration to data scientists around the world!

December 2023

Dmitry I. Ignatov
Michael Khachay
Andrey Kutuzov
Habet Madoyan
Ilya Makarov
Irina Nikishina
Alexander Panchenko
Maxim Panov
Panos M. Pardalos
Andrey V. Savchenko
Evgenii Tsymbalov
Elena Tutubalina
Sergey Zagoruyko

Organisation

Organizing Institutions

- Zaven and Sonia Akian College of Science and Engineering, American University of Armenia (Yerevan, Armenia)
- Skolkovo Institute of Science and Technology (Moscow, Russia)
- Krasovskii Institute of Mathematics and Mechanics, Ural Branch of the Russian Academy of Sciences (Yekaterinburg, Russia)
- School of Data Analysis and Artificial Intelligence, HSE University (Moscow, Russia)
- Laboratory of Algorithms and Technologies for Networks Analysis, HSE University (Nizhny Novgorod, Russia)
- Laboratory for Models and Methods of Computational Pragmatics, HSE University (Moscow, Russia)

Program Committee Chairs

Michael Khachay	Krasovskii Institute of Mathematics and Mechanics of Russian Academy of Sciences & Ural Federal University, Russia
Andrey Kutuzov	University of Oslo, Norway
Ilya Makarov	HSE University, Moscow and Artificial Intelligence Research Institute, Russia
Irina Nikishina	Universität Hamburg, Germany
Maxim Panov	Technology Innovation Institute, UAE
Panos Pardalos	University of Florida, USA
Andrey Savchenko	HSE University, Nizhny Novgorod, Russia
Elena Tutubalina	HSE University, Moscow and Kazan Federal University, Russia
Evgenii Tsymbalov	Apptek, Germany
Sergey Zagoruyko	MTS AI, Russia

Proceedings Chairs

Dmitry I. Ignatov	HSE University, Moscow, Russia
Evgenii Tsymbalov	Apptek, Germany

Steering Committee

Dmitry I. Ignatov	HSE University, Moscow, Russia
Michael Khachay	Krasovskii Institute of Mathematics and Mechanics of Russian Academy of Sciences, Russia & Ural Federal University, Yekaterinburg, Russia
Alexander Panchenko	Skolkovo Institute of Science and Technology, Russia
Andrey Savchenko	HSE University, Nizhny Novgorod, Russia

Program Committee

Anton Alekseev	St. Petersburg Department of V.A. Steklov Institute of Mathematics of the Russian Academy of Sciences, Russia
Ammar Ali	ITMO University, Russia
Vladimir Arlazarov	Smart Engines Service LLC, Federal Research Center "Computer Science and Control" of Russian Academy of Sciences, Russia
Ekaterina Artemova	HSE University, Moscow, Russia
Yulia Badryzlova	HSE University, Moscow, Russia
Jaume Baixeries	Universitat Politècnica de Catalunya, Spain
Amir Bakarov	HSE University, Moscow, Russia
Artem Baklanov	International Institute for Applied Systems Analysis, Austria
Jeremy Barnes	University of the Basque Country, Spain
Nikita Basov	University of Manchester, UK
Vladimir Batagelj	University of Ljubljana, Slovenia
Tatiana Batura	A.P. Ershov Institute of Informatics Systems SB RAS; Novosibirsk State University, Russia
Vladimir Berikov	Sobolev Institute of Mathematics SB RAS, Russia
Malay Bhattacharyya	Indian Statistical Institute, India
Mikhail Bogatyrev	Tula State University, Russia
Elena Bolshakova	Moscow State Lomonosov University, Russia
Pavel Braslavski	Ural Federal University, Russia and Nazarbaev University, Kazakhstan
Aram Butavyan	American University of Armenia, Armenia
Alexey Chernyavskiy	Philips Innovation Labs, Russia
Vera Davydova	Sber AI, Russia
Radhakrishnan Delhibabu	Vellore Institute of Technology, India
Oksana Dereza	University of Galway, Ireland

Michael Diskin	HSE University, Russia
Anna Dmitrieva	University of Helsinki, Finland
Ekaterina Dmitrieva	HSE University, Moscow, Russia
Boris Dobrov	Research Computing Center of Lomonosov Moscow State University, Russia
Ivan Drokin	Prequel App, USA
Pavel Efimov	ITMO University, Russia
Anton Eremeev	Omsk Branch of Sobolev Institute of Mathematics SB RAS, Russia
Elena Ericheva	botkin.ai, Russia
Anna Ermolayeva	Peoples' Friendship University of Russia, Russia
Kirill Fedyanin	Technology Innovation Institute, UAE
Alena Fenogenova	SberDevices, Sberbank, Russia
Alexander Fishkov	Skolkovo Institute of Science and Technology, Russia
Georgii Gaikov	MTS AI, Russia
Yuriy Gapanyuk	Bauman Moscow State Technical University, Russia
Olga Gerasimova	HSE University, Moscow, Russia
Edward Kh. Gimadi	Sobolev Institute of Mathematics, SB RAS, Russia
Petr Gladilin	Huawei and ITMO University, Russia
Maksim Glazkov	Neuro.net, Russia
Anna Glazkova	University of Tyumen, Russia
Alexander Gneushev	Moscow Institute of Physics and Technology, Russia
Elizaveta Goncharova	Artificial Intelligence Research Institute, Russia
Amalya Hambardzumyan	American University of Armenia, Armenia
Anastasia Ianina	Moscow Institute of Physics and Technology, Russia
Dmitry Ilvovsky	HSE University, Moscow, Russia
Vladimir Ivanov	Innopolis University, Russia
Ilia Karpov	HSE University, Moscow, Russia
Mikhail Khachay	IMM URAN, Russia
Vladimir Khandeev	Sobolev Institute of Mathematics, Siberian Branch of the Russian Academy of Sciences, Russia
Javad Khodadoust	Tecnológico de Monterrey, Mexico
Daniel Kireev	VisionLabs, Russia
Denis Kirjanov	HSE University, Russia
Dmitrii Kiselev	HSE University and Sber, Russia
Yury Kochetov	Sobolev Institute of Mathematics, Russia
Sergei Koltcov	HSE University, St. Petersburg, Russia

Jan Konecny	Palacký University Olomouc, Czech Republic
Anton Konushin	HSE University, Russia
Andrei Kopylov	Tula State University, Russia
Nikita Kotelevskii	Skolkovo Institute of Science and Technology, Russia
Evgeny Kotelnikov	Vyatka State University, Russia
Anastasia Kotelnikova	Vyatka State University, Russia
Angelina Kudriavtseva	MTS AI, Russia
Maria Kunilovskaya	University of Saarland, Germany
Anvar Kurmukov	IITP RAS, Russia
Andrey Kutuzov	University of Oslo, Norway
Dmitri Kvasov	University of Calabria, Italy
Florence Le Ber	icube, France
Alexander Lepskiy	HSE University, Moscow, Russia
Tatyana Levanova	Sobolev Institute of Mathematics SB RAS, Omsk Branch, Russia
Natalia Loukachevitch	Moscow State University, Russia
Olga Lyashevskaya	HSE University, Moscow, Russia
Habet Madoyan	American University of Armenia, Armenia
Jaafar Mahmoud	ITMO University, Russia
Ilya Makarov	Artificial Intelligence Research Institute, Russia
Alexey Malafeev	HSE University, Nizhny Novgorod, Russia
Muhammad Shahid Iqbal Malik	HSE University, Moscow, Russia
Vladislav Mikhailov	HSE University, Moscow, Russia
Olga Mitrofanova	St. Petersburg State University, Russia
Petter Mæhlum	University of Oslo, Norway
Amedeo Napoli	LORIA Nancy (CNRS - Inria - Université de Lorraine), France
Irina Nikishina	University of Hamburg, Germany
Damien Nouvel	INaLCO, France
Victor Ohanyan	American University of Armenia, Armenia
Walaa Othman	ITMO University, Russia
Evgeniy M. Ozhegov	HSE University, St. Petersburg, Russia
Alexander Panchenko	Artificial Intelligence Research Institute, Russia
Maxim Panov	Technology Innovation Institute, UAE
Panos Pardalos	University of Florida, USA
Stefan Pickl	University of the Bundeswehr Munich, Germany
Dina Pisarevskaya	Queen Mary University of London, UK
Lidia Pivovarova	University of Helsinki, Finland
Vladimir Pleshko	RCO, Russia

Arnak Poghosyan	Vmware, Inc.; Institute of Mathematics; American University of Armenia; Yerevan State University, Armenia
Aleksandr Rubashevskii	Skolkovo Institute of Science and Technology, Russia
Yuliya Rubtsova	University of Bonn, Germany
Alexey Ruchay	Chelyabinsk State University, Russia
Nicolay Rusnachenko	Bauman Moscow State Technical University, Russia
Alexander Sapin	ClickHouse B.V., Netherlands
Andrey Savchenko	HSE University, Nizhny Novgorod, Russia
Oleg Seredin	Tula State University, Russia
Tatiana Shavrina	HSE University, Moscow, Russia
Denis Sidorov	Energy Systems Institute SB RAS, Russia
Henry Soldano	Laboratoire d'Informatique de Paris Nord, France
Alexey Sorokin	Moscow State University, Russia
Andrey Sozykin	Krasovskii Institute of Mathematics and Mechanics, Russia
Dmitry Stepanov	Program System Institute of Russian Academy of Sciences, Russia
Tatiana Tchemisova	University of Aveiro, Portugal
Mikhail Tikhomirov	Lomonosov Moscow State University, Russia
Yulya Trofimova	HSE University, St. Petersburg, Russia
Christos Tryfonopoulos	University of the Peloponnese, Greece
Magda Tsintsadze	Iv.Javakhishvili Tbilisi State University, Georgia
Evgenii Tsymbalov	Apptek, Germany
Elena Tutubalina	Kazan Federal University, Russia
Alsu Vakhitova	MTS AI, Russia
Daiana Vavilova	Kalashnikov Izhevsk State Technical University, Russia
Petr Vytovtov	Kalashnikov Izhevsk State Technical University, Russia
Dmitry Yashunin	Harman International, USA
Varduhi Yeghiazaryan	American University of Armenia, Armenia
Sergey Zagoruyko	MTS AI, Russia
Alexey Zaytsev	Skolkovo Institute of Science and Technology, Russia
Nikolai Zolotykh	University of Nizhny Novgorod, Russia

Additional Reviewers

Ali, Ammar
Filatov, Andrei
Kireev, Daniel
Kovalenko, Aleksandr
Kurkin, Maxim
Kuvshinova, Ksenia
Musaev, Ruslan
Romanov, Vitaly
Severin, Nikita
Zhevnenko, Dmitry

Organizing Committee

Habet Madoyan (Organizing Chair)	American University of Armenia, Armenia
Amalya Hambardzumyan	American University of Armenia, Armenia
Dmitry Ignatov	HSE University, Moscow, Russia
Irina Nikishina	Universität Hamburg, Germany
Alexander Panchenko	Skolkovo Institute of Science and Technology and Artificial Intelligence Research Institute, Russia
Maxim Panov	Technology Innovation Institute, UAE
Evgenii Tsymbalov	Apptek, Germany

Sponsoring Institutions

American University of Armenia, Armenia
Artificial Intelligence Research Institute, Russia

Abstracts of Posters

Time Series Analysis and Complexity Theory

Armen Vagharshakyan

Institute of Mathematics, Armenia
avaghars@kent.edu

Abstract. Given a family of time series, we seek to determine how large a sample we need to train a model on to obtain a small forecasting error with high probability. We prove using the theory of distribution of lattice points in convex domains and the Vapnik-Chervonenkis approach to the fundamental theorem of statistical learning that after taking out the linear trend, the standard deviation of the resulting time series divided by the mean is an effective way of measuring the forecastability of a time series. We also improve the quantitative version of the Fundamental Theorem of Statistical Learning using Monte-Carlo type methods. (Joint work with A. Iosevich et al.)

Keywords: time series, Vapnik-Cervonenkis theory, Monte-Carlo methods

Theoretical Guarantees for Neural Control Variates in MCMC

Denis Belomestny[1], Artur Goldman[2], Naumov Alexey[2] and Sergey Samsonov[2]

[1] Duisburg-Essen University, Germany
[2] HSE University, Russia
denis.belomestny@uni-due.de, arturgoldman184@gmail.com,
{anaumov,svsamsonov}@hse.ru

Abstract. In this work, we propose a variance reduction approach for Markov chains based on additive control variates and the minimization of an appropriate estimate for the asymptotic variance. We focus on the particular case when control variates are represented as deep neural networks. We derive the optimal convergence rate of the asymptotic variance under various ergodicity assumptions on the underlying Markov chain. The proposed approach relies upon recent results on the stochastic errors of variance reduction algorithms and function approximation theory.

Keywords: Variance reduction, MCMC, deep neural networks, control variates, Stein operator

Guaranteed Optimal Generative Modeling with Maximum Deviation from the Empirical Distribution

Elen Vardanyan[1], Arshak Minasyan[2], Sona Hunanyan[1],
Tigran Galstyan[3] and Arnak Dalalyan[2]

[1] Yerevan State University, Armenia
[2] CREST, ENSAE, IP Paris, France
[3] RAU, YerevaNN, Armenia

evardanyan@aua.am, arshak.minasyan@ensae.fr,
hunan.sona@gmail.com, tigran@yerevann.com,
arnak.dalalyan@ensae.fr

Abstract. Generative modeling is a widely-used machine learning method with various applications in scientific and industrial fields. Its primary objective is to simulate new examples drawn from an unknown distribution given training data while ensuring diversity and avoiding replication of examples from the training data.

This paper presents theoretical insights into training a generative model with two properties: (i) the error of replacing the true data-generating distribution with the trained data-generating distribution should optimally converge to zero as the sample size approaches infinity, and (ii) the trained data-generating distribution should be far enough from any distribution replicating examples in the training data.

We provide non-asymptotic results in the form of finite sample risk bounds that quantify these properties and depend on relevant parameters such as sample size, the dimension of the ambient space, and the dimension of the latent space. Our results are applicable to general integral probability metrics used to quantify errors in probability distribution spaces, with the Wasserstein-1 distance being the central example.

Keywords: Generative models, Wasserstein GAN, minimax rate, sample complexity

Analyzing Local Representative Power of Vision Transformers

Ani Vanyan[1], Alvard Barseghyan[2,3], Vahan Huroyan[1], Hrant Khachatrian[1,3], Hakob Tamazyan[1,3] and Martin Danelljan[4]

[1] YerevaNN, Armenia
[2] CSIE Foundation, Armenia
[3] Yerevan State University, Armenia
[4] ETH Zurich, Switzerland
ani@yerevann.com, barseghyan.at@gmail.com

Abstract. In this paper, we present a comparative analysis of various self-supervised Vision Transformers (ViTs), focusing on their local representative power. Although self-supervised vision transformers are recognized for producing high-quality image representations, the efficacy of their patch representations remains less explored. To address this, we design experiments to analyze the quality of local, i.e. patch-level, representations in the context of few-shot semantic segmentation, instance identification, object retrieval and tracking. Our findings illustrate that DINO either matches or outperforms supervised ViTs, despite not utilizing image-level labels. Although DINOv2, which is pre-trained on a larger dataset, exhibits superior performance in most scenarios, it underperforms DINO in patch-based retrieval under specific image corruptions. We also discover that, while Masked Autoencoders (MAEs) perform similarly to DINO under linear probing, their representations are unsuitable for distance-based algorithms, such as k-NN. We perform a variance-based analysis to explain this phenomenon, and offer a counterintuitive but simple remedy to improve MAE embeddings. Furthermore, our results indicate that among the analyzed ViTs, MAEs demonstrate the least robustness to image degradation, and are less suitable for tracking.

Keywords: transformers, analysis, segmentation, tracking

Towards Robust Entity Matching: Cases Studies for the Pharmaceutical Databases

Aleksei Zhukov[1], Adelina Gamayunova[2],
Aleksander Novolotsky[1] and Denis Sidorov[3]

[1] IC Aptekar, Russia
[2] Irkutsk National Research Technical University, Russia
[3] Energy Systems Institute of Siberian Branch of RAS, Russia
zhukovalex13@gmail.com, theademari@gmail.com,
aptekar@irk.ru, contact.dns@gmail.com

Abstract. Entity matching is one of the principal challenges for success of the e-commerce development. It makes it possible to identify whether two entity records refer to the same real-world entity, i.e. matching the product offers from different vendors with the clients requests. The overview of the state of the art methods for attribute-based matching methods and systems is discussed. The novel approach to entity matching is proposed based on the fuzzy logic, machine learning and using the weighted cosine distance. The efficiency of the proposed system is demonstrated on the pharmaceutical databases provided by the pharmaceutical aggregator.

Keywords: Entity matching, NLP, Data integration, Knowledge fusion

Evaluating Naturalness of the Text Using Quantitative Features

Egor Saknikov and Anastasiya Bonch-Osmolovskaya

HSE University, Russia
{egorsalnikov1,abonch}@gmail.com

Abstract. From the NLP perspective, natural language generation (or NLG) is regarded as a set of specific practically significant tasks such as summarization, conditional text generation, etc. Each task has its formalized goal and a set of metrics to measure model's efficiency. This evaluation method has its own strengths. The most important one is its productivity, which noticeably boosted generative model's evolution. It also helped to prove practical value of such models, grasping the interest of not only academic scholars but also of industry players. However, this evaluation method has some serious flaws. In general, such practical orientation leads to oversimplification. Nowadays most of the evaluation systems tend to score only particular aspects of generative models. Aspects like general text quality, its "naturalness" and "humanlikeness" were overshadowed by more functional aspects.To analyze general quality of generated text and its "naturalness", the text should be seen as the system featuring formal and semantic integrity. Therefore, methods for such analysis should involve more abstract and task-agnostic metrics. In this work we explore some quantitative features, which are likely to represent the complex phenomenon "text naturalness". Moreover, these features can be used as a basis for hybrid metric to detect generated content.

Keywords: NLG, Generated content detection, Style evaluation

Smaller3D: Smaller Models for 3D Semantic Segmentation Using Minkowski Engine and Knowledge Distillation Methods

Alen Adamyan

American University of Armenia, Armenia
alen adamyan@edu.aua.am

Abstract. There are various optimization techniques in the realm of 3D, including point cloud-based approaches that use mesh, texture, and voxels which optimize how you store, and how do calculations in 3D. These techniques employ methods such as feed-forward networks, 3D convolutions, graph neural networks, transformers, and sparse tensors. However, the field of 3D is one of the most computationally expensive fields, and these methods have yet to achieve their full potential due to their large capacity, complexity, and computation limits. This poster proposes the application of knowledge distillation techniques, especially for sparse tensors in 3D deep learning, to reduce model sizes while maintaining performance. We analyze and propose different loss functions, including standard methods and combinations of various losses, to simulate the performance of state-of-the-art models of different Sparse Convolutional NNs. Our experiments are done on the standard ScanNet V2 dataset, and we achieved around 2.6% mIoU difference with a 4 times smaller model and around 8% with a 16 times smaller model on the latest state-of-the-art spatio-temporal convents based models.

Keywords: Knowledge Distillation, 3D semantic segmentation, Sparse Tensors

Machine Learning Pipelines Synthesis with Large Language Models

Ekaterina Trofimova and Emil Sataev

HSE University
etrofimova@hse.ru, ersataev@edu.hse.ru

Abstract. The implementation of data analysis pipelines using various machine learning models often comes down to building a sequence of repeating typical patterns. The complexity and domain-specific nature of these tasks often make them challenging for people who lack specialist knowledge in the field. Automation, thus, serves as a compelling solution to this problem. Our paper delves into analysis of the automatic generation of such sequences, guided by natural language descriptions of the task. We present a transformer-based model scheme for the Smart Generation of machine learning Pipelines (SGP). SGP is engineered to decipher a natural language description of a data analysis task, deduce the optimal machine learning solution pertinent to the task, and assess its score based on predefined metric. SGP eliminates the necessity for intricate understanding of machine learning paradigms, making data analysis a more approachable field.

Keywords: Data analysis pipelines, Large Language models, Machine learning pipelines synthesis, Natural language descriptions

Assessing Proficiency Levels of Russian Language Learners: Evaluation of Lexical Skills

Anton Vakhranev[1], Olesya Kisselev[2] and Mikhail Kopotev[3]

[1] HSE Univesity, Russia
[2] University of South Carolina, USA
[3] University of Helsinki, Finland
ag15bar@mail.ru, olesya.kisselev@utsa.edu,
mihail.kopotev@gmail.com

Abstract. This paper investigates how useful vocabulary lists can be for language proficiency assessment, specifically in the case of the Russian language. Our data include two types of lists: corpus-based frequency lists for learners and so-called 'lexical minimums' designed specifically for learners of Russian as a foreign language. Four commonly used lists for pedagogical purposes are considered: Visualize Russian Tool (VRT; Clancy, 2014–2023), SMARTool (Janda et al., 2018–2023), Lexical Minimum for Russian as a Foreign Language (LM; Golubeva, 2015), and KELLY (Kilgarriff et al., 2014). The correlations between these lists and written texts of Russian language learners, who were assigned proficiency levels ranging from A1 to B2 on the CEFR scale by professional assessors, were examined using in-house Python scripts and the clustering method known as Principal Component Analysis (PCA).

The results reveal limited overlap among the four vocabulary lists, with only one pair, LM and SMARTool, showing relatively strong overlap, specifically for B-levels (B2 $r = 0.97$, B1 $r = 0.92$). At this point, a conclusion can be made: the lists consist of rather random and ill-founded choices of lexemes that do not correlate either with each other or with the learners' texts. On the other hand, we found clear clustering of lexicons utilized in student texts at each proficiency level. The PCA method is especially effective for clustering beginners (level A on the CEFR level) versus intermediate students (level B on the CEFR level) while discriminating between A1 vs. A2 and B1 vs. B2 does not yield plausible results.

Based on these findings, we conclude that while vocabulary lists can guide classroom practices, they fall short as a comprehensive measure of proficiency or lexical sophistication and do not reflect actual language usage at any level from A1 to B2. Further research and refinement are necessary to establish parameters for the automated evaluation of lexical sophistication among Russian language learners.

Keywords: Russian learners, Proficiency levels, Learner vocabulary, Lexical assessment, Lexical clusterization, Principal Component Analysis

Genetic Variability and Disease Resistance Analysis of Grapevine from Armenia

Emma Hovhannisyan

ABI, Armenia
emma.hovhannisyan@abi.am

Abstract. The study analyzes population structure and genetic variability of Armenian cultivated and wild grapevines. Armenian grapevine genetics is shown to be different from the worldwide population. Further, the study considers powdery mildew disease, as some Armenian wild samples have shown to be resistant to the disease. By conducting GWAS on the wild Armenian samples, the study reveals new genetic markers that are significantly associated with powdery mildew disease resistance.

Keywords: grapevine, GWAS, SNP, machine learning

Raw Operational Data Visualization

Stanislav Glushkov and Rostislav Yavorskiy

Exactpro Systems LLC, UK
{stanislav.glushkov,rostislav.yavorski}
@exactprosystems.com

Abstract. Visualization turns out to be very effective for presenting essential information in vast amounts of data. In this presentation we suggest some approaches to the analysis of large amounts of data and the selection of information that is significant for a more detailed analysis. The dataset we work with consists of metrics measured from the operating system and from WebLogic Server monitoring beans. The data is distributed over 27900 CSV files, which all together have 4.51 million columns. That task require considerable pre-processing of the data. First, we treat each CSV file as a single data point in a higher dimension space and apply various dimension reduction techniques. Another approach is based on clustering the CSV files according to their properties and content.

Keywords: Big data, Visualization, Clustering

Using Physics-Informed Neural Networks to Predict Spatiotemporal Dynamical Processes Driven by Complex Ginzburg-Landau Equation

Aleksandr Hayrapetyan[1], Hakob Janesian[2],
Vahan Yeranosyan[2] and Mushegh Rafayelyan[1]

[1] Institute of Physics, Yerevan State University, Armenia
[2] American University of Armenia, Armenia
aleqsandr.hayrapetyan@ysu.am, hakobjanesian@gmail.com,
vahan_yeranosyan@edu.aua.am, mushegh.rafayelyan@ysu.am

Abstract. The Complex Ginzburg-Landau (CGL) equation plays a paramount role in describing the behavior of complex nonlinear physical systems, particularly in the modeling of light-matter interaction in liquid crystal (LC) systems. Namely, the CGL equation is used to model the spatiotemporal complexity of far-from-equilibrium dynamics in liquid crystals. This study presents a novel application of Physics- Informed Neural Networks (PINNs) to the CGL equation, thereby pioneering a model tailored to its structure for an enhanced prediction of state dynamics. The model, being rooted in the CGL equation represented as effectively forecasting dynamic states over space and time. In the context of our study, we specifically set the parameters b and a to be equal to zero. However, we deviate from the conventional approach where the bifurcation parameter adjacent to A is a constant. Instead, we introduce a spatiotemporally variable bifurcation parameter denoted by, thereby extending the complexity and applicability of our model. In our work, the CGL equation was integrated into the loss function of the PINN. The prediction of the spatiotemporal dynamics was obtained via a dense network and used to retrieve the bifurcation parameter. Notably, our model accurately predicted the bifurcation parameter, which is essential for system dynamics. This novel use of PINNs offers promising progress toward forecasting and controlling spatiotemporal dynamics of liquid crystals under the complex light-matter interaction governed by the CGL equation.

Keywords: Complex Ginzburg-Landau (CGL) equation, Liquid Crystal (LC) systems, Physics-Informed Neural Networks (PINNs), Spatiotemporal complexity, Far-from-equilibrium dynamics, Bifurcation parameter, Loss function, State dynamics

Contents

Natural Language Processing

Determination of the Number of Topics Intrinsically: Is It Possible?

Victor Bulatov[1], Vasiliy Alekseev[1(✉)], and Konstantin Vorontsov[2]

[1] Moscow Institute of Physics and Technology, Dolgoprudny, Russia
{viktor.bulatov,vasiliy.alekseyev}@phystech.edu
[2] Lomonosov Moscow State University, Moscow, Russia
k.vorontsov@iai.msu.ru

Abstract. The number of topics might be the most important parameter of a topic model. The topic modelling community has developed a set of various procedures to estimate the number of topics in a dataset, but there has not yet been a sufficiently complete comparison of existing practices. This study attempts to partially fill this gap by investigating the performance of various methods applied to several topic models on a number of publicly available corpora. Further analysis demonstrates that intrinsic methods are far from being reliable and accurate tools. The number of topics is shown to be a method- and a model-dependent quantity, as opposed to being an absolute property of a particular corpus. We conclude that other methods for dealing with this problem should be developed and suggest some promising directions for further research.

Keywords: Topic modelling · Number of topics · Coherence · Diversity · Perplexity · Stability · Entropy

1 Introduction

Topic models are statistical models which are usually employed for unsupervised text analysis. Topic modelling assumes that there are a number of latent topics which explain the collection. Following the convention, we will denote the number of documents by D, the number of topics by T and the size of vocabulary by W.

The topic model is trained by inferring two probability distributions: the "word-in-topic" distribution (colloquially referred to as $\phi_{wt} := p(w \mid t)$ or as an column of a stochastic matrix Φ with the shape $W \times T$) and the "topic-in-document" distribution (colloquially referred to as $\theta_{td} := p(t \mid d)$ or as a row of a stochastic matrix Θ with the shape $T \times D$). We limit our discussion to Φ, Θ parameter matrices since they are present in every topic model. Some topic models introduce additional parameters beside these; exploring intrinsic measures related to these parameters is beyond the scope of this work.

The number of topics T is the key hyperparameter of the most topic models. Naturally, there are a number of influential publications suggesting a way to select this hyperparameter [4,9,10,16,17,33]. However, there is no accepted

© The Author(s), under exclusive license to Springer Nature Switzerland AG 2024
D. I. Ignatov et al. (Eds.): AIST 2023, CCIS 1905, pp. 3–17, 2024.
https://doi.org/10.1007/978-3-031-67008-4_1

consensus on this matter. Namely, there is no agreement in the literature on the sequence of steps one must carry out to determine the best number of topics. Also, there appears to be a disagreement on which methods are appropriate [24].

We are most interested in *intrinsic* metrics which do not use any external resources, labels or human assessment. This approach is typically based on presenting different models with different numbers of topics, obtaining a measurement of a certain quality metric (possibly using cross-validation on held-out document sets) and selecting the number of topics corresponding to the best value. We do not explore extrinsic approaches aimed at optimizing some external criterion of interest, such as classification using a labelled validation dataset, because it is clear that explicitly optimizing for some secondary task provides a better result as measured by the model performance on that task.

Notably, we exclude models which learn the required number of topics automatically (such as hierarchical Dirichlet process [7]) for two reasons. First, they tend to add a new set of hyperparameters that require optimization. Second, they are not universal: any topic model containing Φ and Θ distributions could be scored according to any internal metric, while complex Bayesian topic models optimize some loss function specific to their parameters.[1]

This work investigates the number of metrics proposed in literature over the range of different corpora and different topic models and attempts to formulate a set of useful guidelines for practitioners.

The paper is structured as follows. In Section 2, we review papers related to our methodology on the whole, while Section 3 reviews and assesses various methods proposed to choose the "right" number of topics. Section 4 describes the design of the experiment; namely, topic models and corpora used. In Section 4, we give experimental evidence related to the issue of T determination. Finally, we conclude in Section 5 with a few points of discussion.

2   Related Work

Many researchers proposed a way to determine T (see Sec. 3), but they tend to focus their experiments to certain topic model families or by the range and/or size of datasets involved, as well as perform a very limited survey of other existing approaches. In this section, we are discussing previous work that reviews a range of methods.

We note the work [14] that proposes a new process for estimating the number of clusters in a dataset and compares it to a large number of traditional metrics as implemented in NbClust package. The comparison is evaluated on 20NG corpus and several subsets of WikiRef220 dataset; most of traditional metrics fail to obtain good results. Another comparable work [24] examines and enumerates a number of methods.

Known to us, two software packages implement a number of metrics and operate on a similar idea: they allow the user to explore the values of several

---

[1] That being said, such models sometimes contain model-agnostic metrics which they are implicitly optimizing. We will use these quality metrics if possible.

quality metrics among the range of different T and select a point that appears most advantageous.

The `ldatuning` package[2] for R [26] can apply the following methods to LDA model: D-Spectral [4], D-avg-COS [9], D-avg-JS [10], and holdPerp [17]. The recent work of [20] examines the performance of these methods on several generated datasets with a known value of T.

The TOM library[3] for Python [18] implements several methods for estimation of T. Such as D-Spectral [4] and toptokens-ssample-stab [16]. It supports LDA models and models based on Non-negative Matrix Factorization.

The OCTIS library[4] also implements a number of quality metrics but does not use them to select T.

3 Intrinsic Quality Metrics

This paper mainly focuses on intrinsic metrics, which will be discussed in this section. The Python source code for metrics' computation is publicly available.[5] The section is organized into several broad categories, each category containing a number of somewhat related ideas found in the literature.

Perplexity. The classic intrinsic approach is based on hold-out perplexity [17] (*holdPerp*). The work of [33] enhances this method by considering rates of perplexity change (*RPC*) instead of raw perplexities.

Stability. The authors of [16] employ stability analysis to judge the quality of modelling choices (most notably, the number of topics). This approach is often used when analysing clustering models such as k-means or Non-negative Matrix Factorization. Intuitively, solutions with the "incorrect" number of clusters are unstable since they are forced to either merge clusters in an arbitrary way or to create arbitrary partitions of data.

The stability (*toptokens-ssample-stab*) is measured by repeatedly creating a shuffled subsample of data, fitting a topic model to it, and then comparing the top-tokens of models created that way to the top-tokens of reference model build on the entire dataset. The numeric value is calculated as the Jaccard Similarity Index of the assignment obtained by the Hungarian algorithm.

Diversity and Sufficiency. Intuitively, when the number of topics is too large, the topic model produces a lot of small topics similar to each other.

The most influential work along these lines [9] proposes the usage of average cosine distance between topics (*D-avg-COS*) as a criterion for model selection. This idea is expanded in [10] by considering Jensen-Shannon divergence (*D-avg-JS*) instead of cosine distance. The work [29] employs another variant of diversity, average Euclidean distance (*D-avg-L2*).

In this work, we expand on existing methodology by considering the average distance to the closest topic (instead of average pairwise distance); as a result, *D-*

---

[2] https://github.com/nikita-moor/ldatuning.

[3] https://github.com/AdrienGuille/TOM.

[4] https://github.com/mind-Lab/octis.

[5] https://github.com/machine-intelligence-laboratory/OptimalNumberOfTopics.

cls-COS, *D-cls-JS*, and *D-cls-L2* are obtained. In addition, we employ *D-avg-H* and *D-cls-H*, which are based on Hellinger distance.

Another important development (*D-Spectral*) was proposed in [4] which integrates information in Φ and Θ matrices by considering spectral values of Φ and rows of unnormalized Θ. The proposed *Spectral Divergence Measure* reflects the degree of orthogonality between topic vectors.

The approach proposed in [31] starts with an excessively large number of topics and then uses regularization to set most of them to zero. Notably, the proposed regularizer is able to remove linear combinations of existing topics from the model. The function being maximized by the regularizer is a Kullback–Leibler divergence between uniform $u(t) = T^{-1}$ distribution and $p(t)$ distribution implicitly defined by the topic model, calculated as $\sum_d \theta_{td} \frac{n_d}{n}$. We will interpret this quantity as a quality criterion named *uni-theta-divergence*.

Clustering. One can also employ a number of metrics usually associated with network analysis and clustering analysis, notably Silhouette Coefficient (*SilhC*) and Calinski-Harabasz Index (*CHI*) [25]. Interestingly, Krasnov et al. [24] fail to reproduce Silhouette and CHI results on a particular dataset.

Information-Theoretic Criteria. Another method is the usage of Bayesian Information Criterion (*BIC*) which balances the goodness of fit and model complexity. Minimum Description Length (*MDL*) formalism was also used [13]. The work of [30] explores Akaike Information Criterion (*AIC*) and BIC as a function of the number of topics.

To calculate these metrics, we proceed as follows. First, we obtain $\mathfrak{L}(\Phi, \Theta)$, a model likelihood. Second, we need to find out the number of free parameters N_p, which could be calculated in two different ways: the dimensions of Φ ($(W-1)*T$) or the number of non-zero entries of Φ which we denote by $\#\Phi$ (note [30] argue that a number of free parameters in LDA and sparse models should be treated differently). The following expressions allow to calculate both sparse and non-sparse variations of metrics:

$$AIC = 2N_p - 2\mathfrak{L}, \qquad BIC = N_p \log(D) - 2\mathfrak{L}, \qquad MDL = N_p \log(TD) - 2\mathfrak{L}.$$

Entropy. The work of [22] develops an analogy between topic models and non-equilibrium complex systems where the number of topics is equivalent to the number of states each particle (word) can occupy. It is suggested that the "correct" number of topics should correspond to the equilibrium state, which is characterised by the minimum of entropy. That way the problem is reduced to finding the minimum of a particular function.

To compute entropy, one needs to determine the set $S = \{(w,t) \mid \phi_{wt} > \varepsilon_0\} \subset \Phi$ for some fixed ε_0. Afterwards, the energy is defined as $E = -\log \sum_{(w,t) \in S} \phi_{wt}$, free energy as $E_f = E - T \log(|S|/(WT))$. Finally, Renyi entropy is calculated as $-E_f/(T-1)$.

The work of [22] uses $\varepsilon_0 = (W)^{-1}$, but we found that it did not perform well in some cases. We elected to consider the cases of $\varepsilon_0 = 2(W)^{-1}$ and $\varepsilon_0 = 0.5(W)^{-1}$ as well. All these criteria are denoted by *renyi-0.5*, *renyi-1*, and *renyi-2*.

Lift. This quality measure (*lift-score*) was introduced in recent work [11], where it was observed that LDA models with more "advanced" informative priors correspond to higher lift-scores. Hence, lift-score could be helpful for tuning model hyperparameters. This poses an interesting question of whether lift-score could be used for T determination.

Top-Tokens Analysis. Although not reflected in scientific literature, another reasonable approach is to build many models with different T and pick the one that gives the highest coherence value [15]. *Coherence* is widely used quality metric for topic models, which is computed by using co-occurrence counts of top 10 most probable words of each topic (top-tokens).

4 Methodology

The question we aim to answer: is the notion of "optimal number of topics" well-defined? In other words, is there an agreement among different approaches proposed in literature? Also, are proposed approaches sufficiently robust to the parameters change?

Our methodology is as follows. We train a number of different topic models (such as PLSA, LDA, ARTM, and TARTM; see Sec. 4.1) with the T hyperparameter ranging from T_{min} to T_{max} for a number of iterations sufficient for convergence. The training process starts from three different random initializations, and several quality metrics are measured for each run. If a method requires a user-defined parameter, we explore various candidate values.

We repeat this process for many different corpora. To allow the calculation of held-out perplexity, each corpus is randomly split into train and test (80% train, 20% test). We do not shuffle documents, since the inference algorithm we use (BigARTM [12]) does not depend on the document order.

As a separate experiment, we use stability analysis for determination of T. Our implementation is adapted from TOM library [18], which is inspired by [1,16]. The procedure revolves about assessing topic diversity over subsampled datasets. We create 5 subsamples of the original dataset without replacement. The size of all dataset subsamples $D_i \subseteq D$, $i \in [0, S)$ is fixed and equal to 0.5 of the size of the original dataset. We train topic models for a range of topic numbers $t \in [T_{min}, T_{max}]$ on each of the obtained subsamples. A seed which determines initial weights in Φ_t matrix is fixed and set equal to 0 for all topic models. After data subsampling and model training, the last step is comparison of the models with the same numbers of topics t but trained on the different data subsamples $D_i, D_j, i \neq j$. To get a value of distance between topic models, we do a pairwise comparison using the Jaccard distance function. Then, having a $|t| \times |t|$ matrix of topic distances and by getting a solution of this linear sum assignment problem, we compute the distance between topic models $\rho_{stab}\big(\Phi_t(D_i), \Phi_t(D_j)\big)$, $i, j \in [0, S)$. These distances are then averaged over the number of comparisons of topic models ($\binom{5}{2} = 10$ in the experiments):

$$\frac{1}{\binom{S}{2}} \sum_{0 \leq i < j < S} \rho_{stab}\big(\Phi_t(D_i), \Phi_t(D_j)\big)$$

The lower this number, the better. This logic differs from [16], where the proposed stability score, on the other hand, is an estimate of the similarity of the models and therefore the higher the better. Another notable departure is the calculation involving $\binom{S}{2}$ pairwise comparisons instead of training an initial reference model S_0 and doing S comparisons using it as a benchmark. We note that such comparisons are implicitly based on the assumption that the reference model is good, which could be difficult to guarantee (especially considering that the optimal T is unknown, hence each candidate number t should be checked).

Ideally, the value "recommended" by some particular method should correspond to the pronounced minimum/maximum on the plot. This fact motivates our further analysis where we attempt to locate and classify global optima algorithmically as follows.

First, we take note of the highest and lowest points (h and l). We select all points which values fall into $[h - \alpha(h - l), h]$ interval (or $[l, l + \alpha(h - l)]$ if the score should be minimized). We hold $\alpha = 0.07$ in our analysis. Second, we test whether those points are adjacent to each other (if they are, the optimum is single and robust; otherwise the curve is either jumping or has several significant local optima). Additionally, we check if the optimum was achieved on the boundary of the explored range.

The source code of the methods for finding the optimal number of topics, as well as the experiments conducted, is publicly available.[6]

4.1   Topic Models Studied

In this section, we describe in detail the topic models used in the experiments.

PLSA. Probabilistic Latent Semantic Analysis (PLSA) [19] is a simple topic model without any additional hyperparameters aside from T.

LDA. Latent Dirichlet Allocation (LDA) is a well-known topic model, having prior η for Φ distribution and a prior α for Θ distribution (priors could be numbers or vectors). We implemented three variants of LDA model inside BigARTM/TopicNet technology stack: *double-symmetric* ($\eta = \alpha = \frac{1}{T}$), *asymmetric* (following the recommendation from [32] we use symmetric prior over Φ and asymmetric over Θ: $\eta = \frac{1}{T}$, $\alpha_{td} = \frac{1}{\sqrt{t+T}}$, $0 \le t \le T$) and *heuristic* (the values $\alpha = \frac{50}{T}$ and $\eta = 0.01$ which were used in [6]).

Decorrelated models. It has been shown that LDA tends to produce correlated topics when T is too high or too low [9]. Therefore, it is interesting to explore models that explicitly attempt to reduce pairwise topic correlations. The simplest example is TWC-LDA [28] which is already implemented in BigARTM library [31]. We consider three possible τ coefficients for decorrelation: 0.02, 0.05, 0.1, while holding the γ coefficient equal to 0 (i.e., employing the "relative regularization" feature).

Sparse models. Another property of LDA is the difficulty producing sparse models due to the smoothing priors [30]. Some information-based quality metrics treat sparse and smooth topic models differently [30], therefore it is important

---

[6] https://github.com/machine-intelligence-laboratory/OptimalNumberOfTopics.

Table 1. Ground truth on number of topics.

Dataset	D	W	$T_{expected}$	T_{min}	T_{max}
WikiRef220	220	4839	5	2	20
20NG	18846	2174	15−20	3	40
Reuters	10788	5074	90	5	150
Brown	500	7409	10−20	5	25
StackOverflow	895621	3430	40	5	60
PostNauka	3404	8417	15−30	5	50
ruwiki-good	8603	236018	10/90	5	100

to include sparse models in our analysis. The simplest sparse model divides its topics into two categories: background (general, stopword, uninformative, "slab") topics and specific (domain, foreground, focused, "spike") topics that are sparse compared to the background ones [11]. The BigARTM supports such models [27] using the combination of four regularizers: two regularizers increasing ϕ_{wt}, θ_{td} for general topics t, and two regularizers decreasing ϕ_{wt}, θ_{td} for domain topics t. We explore 0.05 and 0.1 as two dataset-adjusted values for smoothing/sparsing prior. In addition, we include a thetaless topic model (TARTM) designed in [21] as an example of a model where sparsity is an emergent property.

Sparse decorrelated models. For the sake of completeness, we combined different restrictions to obtain a model that is sparse and decorrelated simultaneously. Additive regularization of topic models (ARTM) allows us to combine several requirements in that manner using a set of regularizers [8]. Here, we combine both sets of regularizers and let them influence the model together.

4.2   Corpora Used

The following corpora were utilized (see Tab. 1). Corpora were lemmatized (using `pymorphy2` package [23] for Russian-language text and `NLTK` for English). Pre-defined stop word lists were employed to remove common words.

WikiRef220 dataset,[7] which firstly appeared in [14], consists of 220 news articles hyperlinked to a specific Wikipedia article. The documents are divided into 16 different groups depending on the article linked, but only 5 groups contain more than 5 entries. Following this line of reasoning, authors describe this dataset as having 5 topics and noise.

PostNauka is a corpus of 3,404 articles from the popular Russian online science magazine "PostNauka".[8] Each document is labelled with a number of tags, which make it possible to estimate the reasonable number of topics as laying in the range [10, 30]. The prior research on this dataset resulted in a topic model consisting of 19 topics [2].

---

[7] https://www.multisensorproject.eu/achievements/datasets.
[8] https://postnauka.ru.

20NewsGroups, **Brown Corpus** and **Reuters Corpus** are well-known datasets in NLP. The general consensus is that 20NG consists of 15-20 topics, Brown consists of 10-20 topics and Reuters consists of 50-100 topics.

StackOverflow is a well-known question and answer site that focuses on programming. There have been various studies done to find good topics on Stack-Overflow [5]. We use the already preprocessed version of this corpus from [1], which consists of $895,621$ documents.

Russian Wikipedia. We present a new dataset named "ruwiki-good" accessible through the TopicNet library [8]. To obtain it, we downloaded a Russian Wikipedia database dump and extracted 8603 articles falling into either of "featured" (избранные), "good" (хорошие) or "solid" (добротные) assessment grades. This corpus boasts a curated hierarchy of labels: each article falls into one of 11 main categories[9], with each of them being further subdivided into a various number of subcategories (e.g. "History" $\rightarrow$ "History of UK" $\rightarrow$ "Murders in the United Kingdom"). We believe it to be a valuable testing ground for issues related to topic granularity.

5   Results and Discussion

We organized the results of the conducted experiments in Table 2. In order to provide meaningful insight into the performance of the considered metrics, we designed three features to characterize their behaviour. We wanted to assess metric ability to provide topic number estimation independent from model random initialization, the "readability" of obtained plots and precision of the metric providing an expected number of topics.

The first column is Jaccard metric calculated the following way: for each random initialization, we extract the optimal value or range of values according to metric specifics. Then we calculate the Jaccard distance between intersection and union of those sets excluding cases when metric points at the boundaries of the experiment interval.

The second column gives a proportion of how many times the metric results were "readable" meaning that they fall in one of the categories: 1) have a pronounced min/max value/values, 2) have an interval/s around min/max value, or 3) have a region of alternating peaks. All other types of encountered metric behaviour can be described as either independent from the number of topics or not having any of the described above behaviour (having optimal value outside the range of the experiment).

The last metric is an average of a boolean value: was an expected number of topics in the range of optimal values provided by the metric for this model. The results in Tab. 2 cast doubt on the notion that the number of topics is a well-defined property of a particular corpus (or, at least, that current methods are suitable for deducing it).

---

[9]   Biology, Geography, Science, Arts, History, Culture and Society, Personalities, Religion and Philosophy, Sports and Entertainment, Technology, Economics

Table 2. Metric comparison by applicability averaged over datasets. Some of the diversity-based metrics are removed for brevity; overall their performance appears to be underwhelming.

Score	Jaccard	Informativity	Expected
AIC	0.280	0.542	0.578
AIC sparse	0.219	0.111	0.100
BIC	0.128	0.444	0.461
BIC sparse	0.274	0.164	0.128
MDL	0.096	0.488	0.414
MDL sparse	0.282	0.428	0.256
renyi-0.5	0.470	0.507	0.425
renyi-1	0.356	0.475	0.394
renyi-2	0.230	0.299	0.183
D-Spectral	0.456	0.144	0.083
D-avg-L2	0.682	0.250	0.119
D-cls-H	0.595	0.245	0.189
D-avg-JH	0.302	0.053	0.022
lift	0.383	0.123	0.033
holdout-perplexity	0.228	0.025	0.019
perplexity	0.218	0.023	0.014
CHI	0.277	0.157	0.008
SilhC	0.233	0.079	0.028
average coherence	0.780	0.472	0.208
uni-theta-divergence	0.470	0.197	0.047

5.1   Common Issues Encountered

Here, we present some general observations identified during the experiments.

Model dependency. Our first observation is that the "optimal number of topics" depends on the model used. It could be influenced by a particular choice of the topic modelling scheme or even by the hyperparameter configuration within the model scheme. Fig. 1 demonstrates this with WikiRef220, but this is common with other corpora as well.

Randomization. Another thing to note is the variance caused by random seed (the issue is further complicated if the model depends on the document order). As was mentioned earlier, we ran experiments using three random initializations. If one to look at each curve separately (instead of averaging all three together), it is often the case that their behaviours do not match. The most frequent case is having partially overlapping and adjacent peaks (e.g. `seed_0` gives maximum at 15 topics, `seed_1` gives 15 topics as well, but `seed_2` gives 14 topics). The common case of different but adjacent peaks is also easy to analyze, but

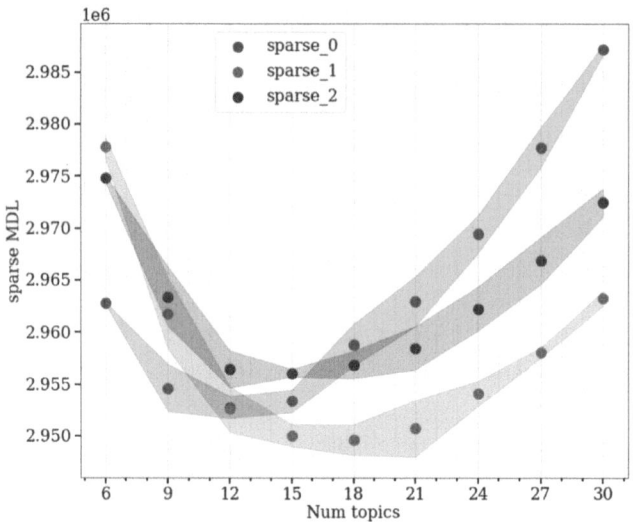

Fig. 1. Sparse MDL criterion for models with different sparsity hyperparameter values.

problematic cases of peaks being significantly separated or only some trajectories having noticeable peaks occur as well.

The direction indicated, for example, in the study [3,25] seems to be a more promising approach for tackling this problem. First, one needs to create a number of different topic models. Second, one needs to extract from these topic models a set of topics that are considered "good" (coherent, interpretable) or "strong" (robust, reproducible). These topics are saved for the later analysis; some care should be taken to ensure that all saved topics are unique enough. When this process converges, the set of different topics found that way produces the topic model we are looking for. The value of T is a by-product of the process.

Methods' disagreement. The question of agreement could be divided into two: 1) Do different criteria agree with each other? 2) Are variations of the same method consistent?

The answer to the first question is negative. Generally, the value determined by diversity-based methods is several times larger than the value determined by other methods. The differences between other methods are less drastic, but they are frequently significant. The sole exception is WikiRef220 dataset, where regions overlap, as seen on Fig. 2.

The answer to the second question is inconclusive. The values given by similar methods tend to differ only slightly. However, it is very common that some criterion gives an answer but a number of variations of it fail. Therefore, we recommend examining several related measures. The change of metric in diversity-based methods does not affect the location of peaks. That being said, the Euclidean metric appears to be the least informative of all.

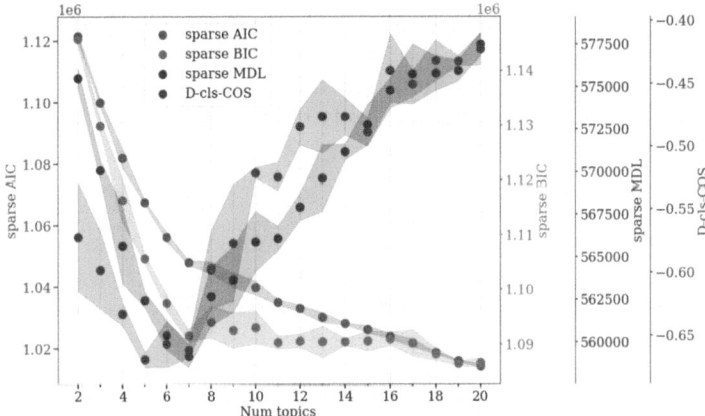

Fig. 2. A set of quality metrics exploring various T for PLSA, $1 < T < 21$. The metrics depicted are AIC, MDL that accounts for model sparsity, and cosine-based diversity (taken with a negative sign, so the minimum corresponds to the "best" value). We see all metrics agreeing with 7 being a reasonable value for T.

Objectivity concerns. Our approach of checking for sole pronounced minimum/maximum on the plot misses less reliable features of data. The features such as the location of the first peak, the location where the curve have flattened out, elbow points, and inflection points might be useful in practice. Unfortunately, using these features introduces too much subjectivity and noise; therefore it is not suitable for selecting the optimal T as an objective, absolute property of the corpus. We decided to stick to simpler approach for analyzing the T dependency curves.

It might be possible to improve some poor-performing methods by giving explicit guidelines on such cases. In cases where the curve flattens out, just subtracting some function linear in T is enough to produce a pronounced extremum. However, this issue is outside the scope of the current study.

5.2 Properties of the Studied Metrics

Here, we present some observations on the intrinsic quality criteria used for finding the optimal number of topics.

Diversity. The value of T provided by this method appears to significantly overestimate the number of topics needed. For the most datasets, the optimum appears to be located outside of the observed range of numbers of topics. The curve tends to flatten out instead of showing maximum (which is sometimes remedied by considering the average distance to closest instead of average pairwise distance between all topics). The location of optimum could change across different random initializations.

Information-theoretic. These methods are better employed in conjunction, since a single method often fails to produce an estimate of T for some models and datasets. Taken together, however, they usually give reasonable values of T (which differ from the "golden standard", however). The location of optimum appears remarkably stable across different random initializations.

Entropic. These metrics are most likely to give pronounced optima that are robust to random initializations. However, the location of the minimum significantly depends on the value of ε_0, and the "default" value of $\varepsilon_0 = (W)^{-1}$ fails to give expected results.

Clustering. Silhoutte Coefficient and Calinski–Harabasz index almost always fail to provide any estimate of T. We suspect that the feature space induced by Θ is not particularly suitable for cluster analysis. This is further supported by work of [24].

Spectral divergence. This method is very noisy and difficult to employ. It is unable to provide any estimate when applied to sparse models (the curve monotonously decreases).

Coherence. This method is very noisy and difficult to employ. We observe that average coherence is a declining function of T with frequent fluctuations. Hence, it is common to find global maximum reached on the small T, making this method ill-suitable for T selection.

Perplexity. The perplexity appears to be monotonous in almost every case, without any notable features helpful for selecting T. This behaviour contradicts earlier works where held-out perplexity had pronounced local minimum. We conjecture that it depends on precise implementation details, such as treatment of out-of-vocabulary words. Rate of perplexity change often produces stable peaks and plateaus, but they do not coincide with the expected value. The numeric result is highly model-dependent.

Lift. This metric indicates the optimal topic number as a maximum value from the plot. Across most datasets, the maximal value reached on the biggest topic model in the experiment well outside the expected optimal number for the datasets. However, for the Stack Overflow datasets, we observed pronounced maxima for different model families. We suspect this was achieved by aggressive token filtration of the original dataset.

Stability. Authors of this approach suggest looking for a minimal value or one of the multiple local minima as an indicator for optimal number of topics. However, in our experiments we did not always observed the desired behaviour on all datasets and had to settle for a plateau, or just a decline in the rate of instability increase. We also observed the following problems with the metric. First, it becomes too noisy for the models with small number of topics (less than 10−15). Second, this metric is not well-suited for sparse models: the obtained results are too noisy to conclude the number of topics. Third, an estimate of the optimal number of topics given by this method usually was lower than the expected number of topics.

6   Conclusion

Further analysis demonstrates that intrinsic methods are far from being reliable and accurate tools in the search of optimal number of topics.

From our experiments, we see that the best performance was achieved by the simplest approaches: AIC, BIC, MDL, renyi. Those metrics provide their judgement based on rough estimation of the model state unlike their counterparts deriving their value from finer topic-level structure of the models. The more intricate methods (lift, coherence, diversity) that attempt to directly measure some desirable qualities of topic model fail to give satisfactory results.

We see that many approaches provide a set of solutions or even a range of optimal number of topics. This contradicts the naive notion of optimal T being a single fixed value attached to a corpus. This behaviour can point at the other problem in the field: a concept of topic is ill-defined from granularity point. Every topic could be divided into subtopics, possibly, without worsening intrinsic metric such as topics' distinctness or clustering quality. Noting that we failed to find methods agreeing with each other on almost all datasets we conclude that the notion of "optimal number of topics" may have a few myths in it requiring a deeper consideration.

A text-book approach for the topic model to learn "latent" word-topic distributions leads to perception that topics exist within a collection and can be found no matter the approach for finding them. According to this point of view a dataset is defining the true number of distributions generating it. This number should not change from type of model extracting those distributions or a type of intrinsic metric finding this number to be "optimal". However, in our experiments, we see that this is not true. If anything the number of topics is mainly a model-dependent quantity, and partially defined by the implemented optimal topic number approach. If there is a lesson to be learned from the data that topic number has been misinterpreted by the community and its just another machine learning model hyperparameter to be tuned.

In light of this consideration we see a few ways how the community already deals with that problem. First, selecting a model according to secondary task. Second, building a hierarchy of topics and pruning it afterwards. Third, improving the process of human (semi-) supervision. An example of this approach would be the suggestion from [29]: employing a weak supervision from users to fine-tune threshold hyperparameter that determines the number of topics.

We also suggest looking in the following directions as they might turn out to be fruitful in dealing with the question. First, eliminating the T hyperparameter. Second, developing topic models more robust to the change of T. For example, consider a hypothetical procedure allowing one to build a topic model for a given number of topics in which all topics would be interpretable. Such procedure would render the question of T determination largely irrelevant. Third, new event detection algorithm with subsequent automatic change of T.

The ways we considered in this paper to evaluate this "optimal number of topics" using intrinsic model quality criteria do not inspire confidence. We conclude that the optimal number of topics depends not so much on the dataset as on the very method of determining the number of topics itself and the topic model being used, or even on the purpose for which topic modeling is applied.

While previously proposed methods remain work of fiction rather than real algorithms we suggest our readers to treat T merely as another model hyperparameter. Our findings suggest that practitioners should not try to estimate "natural" number of topics inherent in corpus; instead, one should focus on the questions similar to the following. How many documents should a topic consist of, on average? Which degree of granularity is desired? If external labels are available: how to incorporate this additional knowledge into the model? Are inferred topics unique enough?

In the end, we conclude that the main purpose of topic modeling should not be the search for the optimal number of topics, but the search for such a method of model training which, given the number of topics, results in a model whose topics in the absence of external criterion are all interpretable.

References

1. Agrawal, A., Fu, W., Menzies, T.: What is wrong with topic modeling? and how to fix it using search-based software engineering. Information and Software Technology (2018)
2. Alekseev, V., Bulatov, V., Vorontsov, K.: Intra-text coherence as a measure of topic models' interpretability. In: Computational Linguistics and Intellectual Technologies: Papers from the Annual International Conference Dialogue (2018)
3. Alekseev, V., Egorov, E., Vorontsov, K., Goncharov, A., Nurumov, K., Buldybayev, T.: Topicbank: Collection of coherent topics using multiple model training with their further use for topic model validation. Data & Knowledge Engineering (2021)
4. Arun, R., Suresh, V., Madhavan, C.V., Murthy, M.N.: On finding the natural number of topics with latent dirichlet allocation: Some observations. In: Pacific-Asia conference on knowledge discovery and data mining (2010)
5. Barua, A., Thomas, S.W., Hassan, A.E.: What are developers talking about? an analysis of topics and trends in stack overflow. Empirical Software Engineering (2014)
6. Biggers, L.R., Bocovich, C., Capshaw, R., Eddy, B.P., Etzkorn, L.H., Kraft, N.A.: Configuring latent dirichlet allocation based feature location. Empirical Software Engineering (2014)
7. Bryant, M., Sudderth, E.B.: Truly nonparametric online variational inference for hierarchical dirichlet processes. In: Advances in Neural Information Processing Systems (2012)
8. Bulatov, V., Alekseev, V., Vorontsov, K., Polyudova, D., Veselova, E., Goncharov, A., Egorov, E.: Topicnet: Making additive regularisation for topic modelling accessible. In: Proceedings of The 12th Language Resources and Evaluation Conference (2020)
9. Cao, J., Xia, T., Li, J., Zhang, Y., Tang, S.: A density-based method for adaptive lda model selection. Neurocomputing (2009)
10. Deveaud, R., SanJuan, E., Bellot, P.: Accurate and effective latent concept modeling for ad hoc information retrieval. Document numérique (2014)
11. Fan, A., Doshi-Velez, F., Miratrix, L.: Assessing topic model relevance: Evaluation and informative priors. The ASA Data Science Journal, Statistical Analysis and Data Mining (2019)

12. Frei, O., Apishev, M.: Parallel non-blocking deterministic algorithm for online topic modeling. In: International Conference on Analysis of Images, Social Networks and Texts (2016)
13. Gerlach, M., Peixoto, T.P., Altmann, E.G.: A network approach to topic models. Science advances (2018)
14. Gialampoukidis, I., Vrochidis, S., Kompatsiaris, I.: A hybrid framework for news clustering based on the dbscan-martingale and lda. In: International Conference on Machine Learning and Data Mining in Pattern Recognition (2016)
15. del Gobbo, E., Fontanella, S., Sarra, A., Fontanella, L.: Emerging topics in brexit debate on twitter around the deadlines. Social Indicators Research (2020)
16. Greene, D., O'Callaghan, D., Cunningham, P.: How many topics? stability analysis for topic models. In: Joint European Conference on Machine Learning and Knowledge Discovery in Databases (2014)
17. Griffiths, T.L., Steyvers, M.: Finding scientific topics. Proceedings of the National academy of Sciences (2004)
18. Guille, A., Soriano-Morales, E.P.: Tom: A library for topic modeling and browsing
19. Hofmann, T.: Probabilistic latent semantic analysis. In: Proceedings of the Fifteenth conference on Uncertainty in artificial intelligence (1999)
20. Hou-Liu, J.: Benchmarking and improving recovery of number of topics in latent dirichlet allocation models (2018)
21. Irkhin, I., Bulatov, V., Vorontsov, K.: Additive regularizarion of topic models with fast text vectorizartion. Computer research and modeling (2020)
22. Koltcov, S.: Application of rényi and tsallis entropies to topic modeling optimization. Statistical Mechanics and its Applications, Physica A (2018)
23. Korobov, M.: Morphological analyzer and generator for russian and ukrainian languages. In: Analysis of Images, Social Networks and Texts (2015)
24. Krasnov, F., Sen, A.: The number of topics optimization: clustering approach. Machine Learning and Knowledge Extraction (2019)
25. Mehta, V., Caceres, R.S., Carter, K.M.: Evaluating topic quality using model clustering. In: 2014 IEEE Symposium on Computational Intelligence and Data Mining
26. Murzintcev Nikita, N.C.: ldatuning: Tuning of the Latent Dirichlet Allocation Models Parameters (2020), https://CRAN.R-project.org/package=ldatuning
27. Potapenko, A., Vorontsov, K.: Robust plsa performs better than lda. In: European Conference on Information Retrieval (2013)
28. Tan, Y., Ou, Z.: Topic-weak-correlated latent dirichlet allocation. In: 2010 7th International Symposium on Chinese Spoken Language Processing. IEEE (2010)
29. Tang, J., Zhang, M., Mei, Q.: " look ma, no hands!" a parameter-free topic model. arXiv preprint arXiv:1409.2993 (2014)
30. Than, K., Ho, T.B.: Fully sparse topic models. In: Joint European Conference on Machine Learning and Knowledge Discovery in Databases (2012)
31. Vorontsov, K., Potapenko, A., Plavin, A.: Additive regularization of topic models for topic selection and sparse factorization. In: International Symposium on Statistical Learning and Data Sciences (2015)
32. Wallach, H.M., Mimno, D.M., McCallum, A.: Rethinking lda: Why priors matter. In: Advances in neural information processing systems (2009)
33. Zhao, W., Chen, J.J., Perkins, R., Liu, Z., Ge, W., Ding, Y., Zou, W.: A heuristic approach to determine an appropriate number of topics in topic modeling. In: BMC bioinformatics (2015)

Sentence Difficulty in Three Languages: Russian Dataset Compared to Italian and English

Vladimir Ivanov[1,2]([✉]) [iD] and Elbayoumi Mohamed Gamal[2]

[1] Innopolis University, Innopolis, Russia
v.ivanov@innopolis.ru
[2] Kazan Federal University, Kazan, Russia
m.elbayoumi@innopolis.university

Abstract. The task of predicting text complexity has been extensively studied in various languages. However, there has been relatively less research interest in predicting complexity at the sentence level specifically in the Russian language. In this paper, we conduct experiments using a novel dataset that includes sentence-level annotations for complexity. Our study focuses on examining simple syntactical features and baseline models, such as graph neural networks and pre-trained language models. Furthermore, we compare our findings with existing datasets in Italian and English.

Keywords: sentence complexity · text difficulty prediction · Russian language · English language · Italian language

1 Introduction

Text complexity evaluation and prediction have garnered significant attention in the field of natural language processing. Researchers have explored this task at various linguistic levels, including at the level of text fragments, individual words, and phrases [4,5,7,19,20]. However, the evaluation of complexity at the sentence level, known as sentence-level complexity evaluation (SCE), has received relatively less research interest.

SCE serves as an intermediate step between evaluating complexity at the text fragment level (such as several coherent sentences or a paragraph) and assessing the complexity of individual words or phrases. Understanding and predicting sentence-level complexity is crucial for various applications, including educational technology, automated writing assistance, and readability assessment.

SCE focuses specifically on the complexity of entire sentences, encompassing both lexical and syntactical features within a single linguistic unit. In contrast, evaluations at other levels may focus on individual words, phrases, or entire texts. By examining sentences, we aim to capture the nuances of sentence structure, coherence, and readability.

Sentence complexity in our study is defined as a multidimensional concept that includes factors such as sentence length, syntactic complexity, and vocabulary diversity. We will elaborate on this definition in subsequent sections.

© The Author(s), under exclusive license to Springer Nature Switzerland AG 2024
D. I. Ignatov et al. (Eds.): AIST 2023, CCIS 1905, pp. 18–28, 2024.
https://doi.org/10.1007/978-3-031-67008-4_2

While text readability often involves evaluating the overall ease of comprehending a text, sentence complexity delves into the factors that contribute to the complexity of individual sentences within that text. Understanding sentence complexity is a crucial aspect of text readability but involves a more granular analysis.

Recent works have explored sets of features that can be utilized for SCE, encompassing lexical and syntactical features derived from the target sentence, as well as contextual features obtained from surrounding sentences [8, 18]. However, SCE presents challenges in terms of interpreting results and selecting appropriate features for modeling. The advancement of deep neural networks has provided promising solutions, as they can effectively capture and represent a wide range of linguistic features.

Transformer-based neural architectures, such as the one introduced by Vaswani et al. [22], have revolutionized natural language processing. These architectures have demonstrated exceptional performance in various tasks, including machine translation, sentiment analysis, and question answering. Furthermore, pre-trained language models, such as BERT [6], have shown remarkable capabilities in capturing contextual information and improving the performance of downstream tasks.

Deep neural networks have been successfully applied to SCE in multiple languages, including Italian, English, Arabic, and Russian. For instance, Schicchi et al. in [17] employed a long short-term memory (LSTM) model with an attention mechanism for binary classification of Italian sentences based on their complexity. In the Arabic language, Khallaf et al. investigated different features and models, including pre-trained language models, to classify sentence difficulty using both binary classification and regression approaches [12]. Ivanov et al. [10] utilized Deep Graph Neural Networks to model sentence-level complexity in the Russian language.

To facilitate research in SCE, several datasets with manual annotations of sentence-level complexity have been developed using crowd-sourcing. For example, Brunato et al. [2] conducted a comprehensive analysis of features that affect human perception of sentence complexity, including lexical, morphosyntactic, and syntactic features. Their dataset, comprising over 1,000 sentences, revealed the importance of factors such as sentence length, maximum dependency length in a dependency syntax tree, average word length, and lexical density. Similarly, a comparable dataset consisting of 1,200 Russian sentences with over 23,000 individual complexity judgments has been annotated [11], and this dataset is the one used in our current study.

In this paper, we present a thorough review of the Russian dataset with sentence-level complexity annotations and compare it to existing datasets in English and Italian. We then evaluate simple models, such as Decision Tree, Support Vector Machines (SVM), Graph Convolution Networks, and the pre-trained language model BERT, for predicting sentence-level complexity. Additionally, we train and evaluate models using datasets in English and Italian to provide a comprehensive comparison.

The remainder of the paper is organized as follows. Section 2 provides an overview of related datasets and approaches in the field of text complexity evaluation and prediction. Section 3 describes datasets, as well as the features and modeling approaches employed in our study. Section 4 presents the experimental results of sentence complexity prediction, and finally, Sect. 5 concludes the paper with discussion and future directions.

2   Related Work

2.1   Datasets and Models

The text complexity is a well-studied and wide area, encompassing various aspects of language and readability. In this section, we survey research works closely related to the present study, focusing on the prediction and evaluation of sentence complexity. In [9], the relative complexity of sentences in the context of readability for deaf people was examined. The authors collected a corpus consisting of pairs of sentences with paraphrases and focused on classifying the paraphrases into three levels (left, right, same). To model complexity, Inui and Yamamoto developed a rule-based method and compared its performance to that of an SVM classifier. Their findings provided insights into the readability of sentences for the deaf community. Vajjala and Meurers evaluated an SVM classifier to predict the relative complexity of pairs of complex and simplified sentences [21]. Their study aimed to assess the difference in complexity between the two sentence versions, shedding light on the potential simplification of sentences for improved readability. A comparative analysis of different algorithms for sentence complexity evaluation using an English dataset with seven complexity categories conducted by Maqsood et al. in [16]. Their study explored the effectiveness of various models in capturing sentence complexity variations across different linguistic contexts.

In 2016, Schumacher et al. [18] investigated models for predicting the relative reading difficulty of sentences, both with and without surrounding context. To capture the impact of context, the authors binned sentences based on grade levels and examined lexical and grammatical features. They trained a logistic regression classifier and a Bayesian ranker, demonstrating that considering the context improves the prediction of sentence readability. This work highlighted the importance of contextual information in understanding sentence complexity.

In [2] Brunato et al. employed crowdsourcing to model human perception of single-sentence difficulty in Italian and English. Their study analyzed a diverse set of linguistic features and their contributions to the perception of sentence complexity. Features such as average number of characters per word and average number of words per sentence were found to be important factors in determining sentence complexity. The authors explicitly controlled for sentence length by binning the dataset into groups of sentences with the same length, further emphasizing the impact of sentence length on complexity perception.

Building upon the dataset used in the previous work, Iavarone et al. [8] conducted a study on modeling sentence complexity in context in English. They reported results for complexity prediction using SVM and BERT models. Despite the high performance of the BERT-based model, the authors concluded that it did not seem to leverage syntactic features for predicting sentence complexity. Motivated by this finding, our study explicitly compares RuBERT with a GNN-based model that directly operates on the dependency tree structure. While a direct comparison of MAE values is not possible due to different scales, the average MAE values reported for the BERT-based model (around 0.05) slightly differ from our MAE of 0.08. Finally, Lobosco et al. [13] proposed deep neural networks for sentence complexity classification. Their model utilized the Tree-Tagger to extract syntactic features, two LSTM layers, and a linear layer for classification. Experimental results demonstrated the effectiveness of the approach compared to baselines such as Support Vector Machines, Gradient Boosting, and Random Forest, for both Italian and English datasets.

3 Datasets

3.1 Sentence Complexity Dataset for Russian

Here we briefly describe the dataset for Russian [11], which is used in our study. The authors selected 1,200 sentences from the SynTagRus corpus and then used Yandex Toloka to annotate the collection. Therefore, each sentence has a dependency parse tree, morphological annotation, which makes it possible to extract various types of linguistic features. Authors grouped the sentences by the sentence length (10, 15, 20, 25, 30, 35 tokens), such that each group contained 200 sentences. In the crowdsourcing annotation they applied a 7-point scale.

Authors explored basic correlations between features, concluding that features with the highest correlations are those related to maximum and average path length, proportions of nouns and phrases, and frequency as well as sentence/token lengths. These findings suggest that certain linguistic features play a significant role in predicting sentence complexity in Russian. Then they trained a simple linear regression model with three features: sentence length in characters (SLC), average path length (APL), and the number of clauses (NCL). The model has MSE = 0.32 ($\pm$0.03), MAE = 0.45 ($\pm$0.02), while a model with a single parameter (SLC) has MSE = 0.33 and MAE = 0.46.

$$Compl.Score = -1.61 + 0.014 * SLC + 0.146 * APL + 0.057 * NCL$$

In this equation, the coefficients have the following interpretations:

- The coefficient -1.61 represents the intercept or the baseline complexity score when all feature values are zero. In other words, it represents the base complexity level of a sentence.
- The coefficient $0.014 * SLC$ indicates how sentence complexity increases or decreases with an increase in sentence length in characters. A positive value suggests that longer sentences tend to be more complex.
- The coefficient $0.146 * APL$ represents the impact of the average path length on sentence complexity. A positive value suggests that as the average path length increases, the sentence complexity also tends to increase.
- The coefficient $0.057 * NCL$ indicates how the number of clauses affects sentence complexity. A positive value suggests that sentences with more clauses tend to be more complex.

Finally, they fine-tuned the pre-trained RuBERT model on 80% of the dataset. The performance of the fine-tuned RuBERT on the rest 20% was 0.54 (MSE) and 0.57 (MAE).

3.2   Comparing Russian Data to English, Italian and Arabic

In [12] a larger dataset consisting of over 5,000 Arabic sentences was proposed. However, this dataset only had two labels and primarily focused on binary classification. Notably, the parameter of sentence length was intentionally excluded from the list of features. In contrast, for the Russian dataset, the scale ranges from 1 to 7, allowing for both classification and regression experimental setups.

English and Italian data were presented and examined in [2]. They calculated the correlation between various linguistic features and human judgments of sentence difficulty using Spearman's rank correlation coefficient. For Russian dataset similar experiment has been conducted [11]. Authors explored the relationship between linguistic features and sentence complexity using Pearson correlation. When comparing the findings carried out for different languages, we can observe some similarities. Most studies found a strong positive correlation between sentence length (either in characters, or in words) and the sentence complexity. However, in each of the analyzed datasets, one can find short sentences with a high complexity score. Additionally, these studies identified similar sets of linguistic features that significantly correlate with sentence complexity, such as average token length and the number of clauses, etc. These commonalities highlight the importance of these features in determining sentence complexity across different languages and datasets.

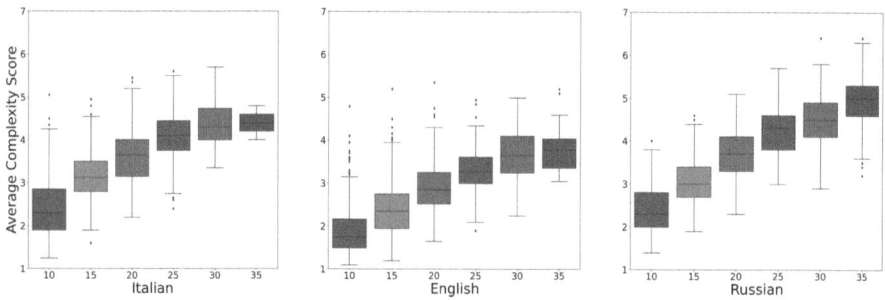

Fig. 1. Dependency between the average sentence complexity and the sentence length in Italian, English and Russian.

However, there are differences as well. One can see from Fig. 1 that Italian and English data have slightly different pattern comparing to the Russian. First, sentences longer that 30–35 tokens are more rare, hence Italian and English have relatively less samples in the '35' bins and very few extremely complex sentences (with score greater than six. In Russian all bins have the same number of samples. Further, the mean sentence complexity scores are also different (Fig. 2). The *'order'* is the following: *'English ≤ Italian ≤ Russian'* with corresponding means and standard deviations: 2.79 (±0.87); 3.28 (±0.92); 3.83 (±1.03). This discrepancy between three estimates can be explained as follows. In general, the average sentence length in English (and in Italian) is bigger than in Russian language. However, analysis of quantiles distribution for the sentence length in the collected dataset shows that Russian subset has relatively more sentences with extremely high length, than the English (and the Italian) subsets. Therefore, we have more assessments with higher complexity in Russian subset. These sampling and annotation biases can be the reason behind the difference between the histograms. One could adjust data by oversampling (for English and Italian) or undersampling (for Russian), but, in general, we do not consider this a major issue.

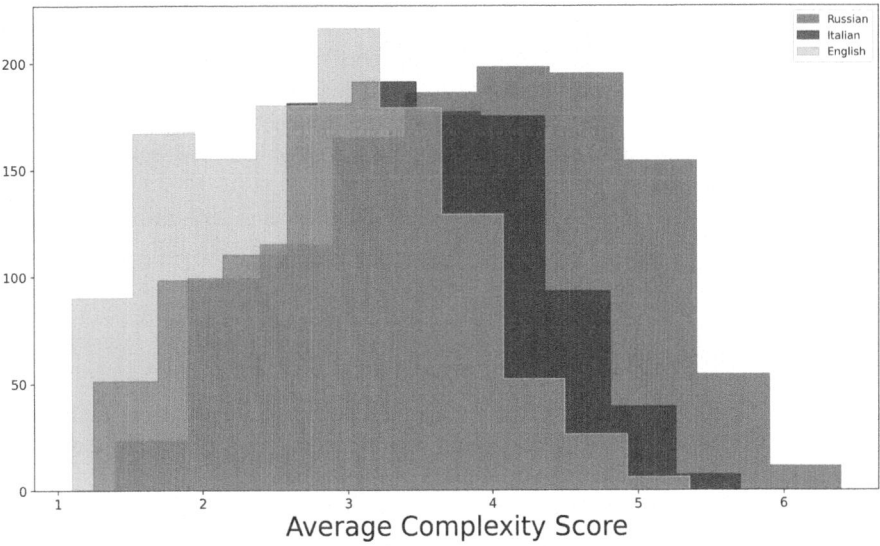

Fig. 2. The distribution of the mean sentence complexity scores in Italian, English and Russian.

4   Modeling Sentence Complexity

We began by loading the three datasets and calculating the average judgment score for each sentence. To facilitate further analysis, we also computed the sentence length and lemma frequencies for each sentence in the datasets. We used the Russian frequency list from [14], the English frequency list [3], and the Italian frequency list [15]. Next, we pre-processed the sentences for each dataset using the appropriate language model from the spaCy library. The pre-processing steps included tokenization, lemmatization, and the removal of stopwords. We extracted the Term Frequency-Inverse Document Frequency (TF-IDF) features for the preprocessed sentences using the TfidfVectorizer from the Scikit-learn library. Finally, we combined the TF-IDF features with the additional features, 'sentence_length' and 'lemma_freq', to create the final feature matrices for the Russian, English, and Italian datasets. These feature matrices were then used for training and evaluating various machine learning models in subsequent sections of our study.

By pre-processing and extracting features such as TF-IDF, sentence length, and lemma frequencies, we aimed to obtain a comprehensive representation of the sentences in the datasets. This would enable the effective comparison of sentence complexity across Russian, English, and Italian texts and facilitate the development of robust models for sentence complexity prediction.

4.1 Decision Trees, SVMs, and Linear Regression

To investigate the utility of various machine learning models for sentence complexity prediction, we considered Decision Trees and Support Vector Machines (SVM) in addition to linear regression. We pre-processed the text data by tokenizing, lemmatizing, and removing stopwords, then calculated TF-IDF features for each language. The final feature set for each dataset included the TF-IDF features combined with sentence_length and lemma_freq attributes.

We divided the data into training and testing sets (80% and 20% respectively) and trained Decision Tree, and SVM models on each language dataset (results are presented in the Table 1).

4.2 Graph Neural Networks

As a Graph Neural Network baseline we applied the Graph convolution model proposed in [10]. This model combines lexical representation from the fastText with the syntactical representations learned from the dependency tree.

Typically, GNNs are based on a message-passing mechanism when each node in a graph has an internal state (node embedding) that can be transferred to its neighbors. During message-passing, the node 'sends' its state to adjacent nodes (neighbors). Thus, after each message-passing, all neighbors aggregate the messages from adjacent nodes. Usually, these messages are passed over the network for a fixed number of iterations. These message-passing steps are treated as network layers. Node embeddings aggregated after the whole process can be viewed as contextual representations. The context size depends on the number of layers.

This paper employed a Multi-layer Graph Convolutional Network (GCN) with self-attention. The GCN is a model that has the following layer propagation rule to calculate the hidden representation of the i-th node $h_i^{(l+1)}$ using hidden representations of all its neighbors $h_j^{(l)}$ from the previous layer (l).

$$h_i^{(l+1)} = \sigma \left(\sum_{j \in \mathcal{N}(i)} \frac{1}{c_{ij}} h_j^{(l)} W^{(l)} \right)$$

Here $\mathcal{N}(i)$ represents a set of neighbors of the i-th node. The ReLU activation function is used as a default setting for the non-linear transformation function $\sigma(\cdot)$. The matrix $W^{(l)}$ contains trainable parameters of a convolution filter and the c_{ij} represents a normalization constant. At the last layer, each node of a graph has its vector representation; these representations are averaged and passed through a fully connected layer to train the model for classification or regression.

The first and the second layers apply the multi-head attention with six heads, and the third layer is a convolutional layer that outputs 64-dimensional vector for each node in a graph. The result is projected via a linear layer to the desired output.

In a sentence-level complexity prediction, each sentence is treated as a graph derived from a dependency syntax tree of the sentence. Precisely, the graph edges are constructed from the edges of a dependency tree for a given sentence, plus backward edges that help message-passing. Each tree node represents a token from the source sentence. When the GNN is initialized, node representation in the input layer are loaded from the fastText embedding for Russian (fastText, [1]).

In Russian and Italian we had access to both fastText embeddings and TreeBanks. Unfortunately, for English dataset, dependency trees were not available. The GCN model contains two graph convolution layers; it was trained and evaluated on the same 80/20 split.

4.3   Pre-trained Language Models

To assess the performance of pre-trained models for sentence complexity prediction, we used the BERT model for each language. Specifically, we used the 'DeepPavlov/rubert-base-cased' model for Russian, the 'bert-base-uncased' model for English, and the 'dbmdz/bert-base-italian-uncased' model for Italian[1] We employed the Hugging Face Transformers library to fine-tune these models on our respective datasets. We used a maximum sequence length of 128 and a batch size of 16. We fine-tuned each model for 100 epochs using the AdamW optimizer with a learning rate of 2×10^{-5} and a linear scheduler with warmup.

5   Results and Discussion

Table 1 presents the evaluation results of the different models on the datasets: English, Italian, and Russian. The models being compared are BERT, Decision Tree, Linear

Table 1. Comparison of model performance metrics (MAE, MSE, and R^2) for the datasets ('N/A' corresponds to cases when linear regression has failed to converge; '–' corresponds to cases where dependency trees of sentences were not available.).

	English			Italian			Russian		
Model	MAE	MSE	R^2	MAE	MSE	R^2	MAE	MSE	R^2
BERT-based	.080	.009	.987	.076	.008	.990	.055	.004	.995
Decision Tree	.434	.329	.519	.607	.587	.382	.545	.507	.590
Linear Regression	N/A	N/A	N/A	.533	.484	.490	.483	.357	.711
SVM	.336	.182	.734	.744	.845	.110	.668	.701	.433
GCN-based	–	–	–	.623	.624	.332	.657	.629	.294

[1] Here, we refer to model names from the Huggingface, https://huggingface.co.

Regression, and SVM. Three metrics are used to evaluate the performance of these models: Mean Squared Error (MSE), Mean Absolute Error (MAE), and R^2 score.

These results demonstrate that the fine-tuned BERT models perform well in predicting sentence complexity for all three languages, with relatively low MSE and MAE values and high R^2 scores. This indicates that pre-trained models, such as BERT, can be effectively leveraged for sentence complexity prediction tasks in various languages. Here is the interpretation of the results:

1. **BERT**: The BERT model outperforms the other models across all three languages, with the lowest MSE, MAE, and the highest R^2 score. This indicates that the BERT model is the most accurate and has the best overall performance.
2. **Decision Tree**: The Decision Tree model performs better than Linear Regression and SVM in terms of R^2 score for the English dataset. However, its performance is lower than BERT across all metrics and languages.
3. **Linear Regression**: The Linear Regression model performs well on the Russian dataset, with a high R^2 score (0.71) and low errors (MSE and MAE). However, its fit on the English dataset is poor. For the Italian dataset, its performance is moderate.
4. **SVM**: The SVM model performs moderately on the English dataset but has lower R^2 scores compared to BERT and Decision Tree. For the Italian and Russian datasets, its performance is lower than the other models, with higher MSE and MAE values and lower R^2 scores.
5. **GCN-based**: GCN-based model (like the SVM with rich set of features) demonstrates an inferior performance.

The BERT model is the best-performing model for predicting sentence complexity across all three languages. The other models show varying degrees of performance depending on the dataset, with the Decision Tree and Linear Regression models demonstrating some potential, while the SVM and GCN models show overall weaker performance.

6   Conclusion

In this paper, we use a dataset of 1,200 Russian sentences in order to evaluate different models for sentence level complexity prediction. We compared the models performance on two other languages. The analysis of the dataset shows that it is slightly unbalanced towards difficult sentences, with a correlation between sentence length and complexity score. The results of the study can be applied in a hybrid model for text complexity prediction which combines signals from different levels including complexity scores of individual words, sentences and passages. Another research direction is development and evaluation of a cross-lingual and multilingual models for sentence complexity prediction.

Acknowledgments. This work is funded by Russian Science Foundation, grant # 22-21-00334.

References

1. Bojanowski, P., Grave, E., Joulin, A., Mikolov, T.: Enriching word vectors with subword information. CoRR abs/1607.04606 (2016). http://arxiv.org/abs/1607.04606
2. Brunato, D., De Mattei, L., Dell'Orletta, F., Iavarone, B., Venturi, G.: Is this sentence difficult? Do you agree? In: Proceedings of the 2018 Conference on Empirical Methods in Natural Language Processing, Brussels, Belgium, pp. 2690–2699. Association for Computational Linguistics (2018). https://doi.org/10.18653/v1/D18-1289, https://aclanthology.org/D18-1289
3. Brysbaert, M., New, B.: Subtlex-us frequency list with POS information final text version (2013)
4. Collins-Thompson, K., Callan, J.: Predicting reading difficulty with statistical language models. J. Am. Soc. Inform. Sci. Technol. **56**(13), 1448–1462 (2005)
5. Crossley, S.A., Greenfield, J., McNamara, D.S.: Assessing text readability using cognitively based indices. TESOL Q. **42**(3), 475–493 (2008)
6. Devlin, J., Chang, M.W., Lee, K., Toutanova, K.: BERT: pre-training of deep bidirectional transformers for language understanding. arXiv preprint arXiv:1810.04805 (2018)
7. Heilman, M., Collins-Thompson, K., Eskenazi, M.: An analysis of statistical models and features for reading difficulty prediction. In: Proceedings of the Third Workshop on Innovative Use of NLP for Building Educational Applications, pp. 71–79 (2008)
8. Iavarone, B., Brunato, D., Dell'Orletta, F.: Sentence complexity in context. In: Proceedings of the Workshop on Cognitive Modeling and Computational Linguistics, pp. 186–199. Association for Computational Linguistics, Online (2021). https://doi.org/10.18653/v1/2021.cmcl-1.23
9. Inui, K., Yamamoto, S.: Corpus-based acquisition of sentence readability ranking models for deaf people. In: Proceedings of the Sixth Natural Language Processing Pacific Rim Symposium, 27–30 November 2001, Tokyo, Japan, pp. 159–166, Hitotsubashi Memorial Hall, National Center of Sciences (2001). http://www.afnlp.org/nlprs2001/pdf/0035-01.pdf
10. Ivanov, V.: Sentence-level complexity in Russian: an evaluation of BERT and graph neural networks. Front. Artif. Intell. **5** (2022)
11. Ivanov, V., Elbayoumi, M.G.: A new dataset for sentence-level complexity in Russian. In: Computational Linguistics and Intellectual Technologies: Papers from the Annual International Conference "Dialogue" (2023)
12. Khallaf, N., Sharoff, S.: Automatic difficulty classification of Arabic sentences. In: Workshop on Arabic Natural Language Processing (2021)
13. Lo Bosco, G., Pilato, G., Schicchi, D.: Deepeva: a deep neural network architecture for assessing sentence complexity in Italian and English languages. Array **12**, 100097 (2021) https://doi.org/10.1016/j.array.2021.100097, https://www.sciencedirect.com/science/article/pii/S2590005621000424
14. Lyashevskaya, O., Sharov, S.A.: Frequency dictionary of the modern Russian language (the Russian National Corpus) (2009)
15. Lyding, V., et al.: The paisa' corpus of Italian web texts (2014). https://doi.org/10.3115/v1/W14-0406
16. Maqsood, S., et al.: Assessing English language sentences readability using machine learning models. PeerJ Comput. Sci. **7**, e818 (2022)
17. Schicchi, D., Pilato, G., Bosco, G.L.: Deep neural attention-based model for the evaluation of Italian sentences complexity. In: 2020 IEEE 14th International Conference on Semantic Computing (ICSC), pp. 253–256. IEEE (2020)

18. Schumacher, E., Eskenazi, M., Frishkoff, G., Collins-Thompson, K.: Predicting the relative difficulty of single sentences with and without surrounding context. In: Proceedings of the 2016 Conference on Empirical Methods in Natural Language Processing, pp. 1871–1881, Austin, Texas. Association for Computational Linguistics (2016). https://doi.org/10.18653/ v1/D16-1192, https://aclanthology.org/D16-1192

19. Shardlow, M., Cooper, M., Zampieri, M.: CompLex — a new corpus for lexical complexity prediction from Likert Scale data. In: Proceedings of the 1st Workshop on Tools and Resources to Empower People with REAding DIfficulties (READI), Marseille, France, pp. 57–62. European Language Resources Association (2020). https://aclanthology.org/2020. readi-1.9

20. Shardlow, M., Evans, R., Paetzold, G.H., Zampieri, M.: SemEval-2021 task 1: lexical complexity prediction. In: Proceedings of the 15th International Workshop on Semantic Evaluation (SemEval-2021). pp. 1–16. Association for Computational Linguistics, Online (2021). https://doi.org/10.18653/v1/2021.semeval-1.1, https://aclanthology.org/2021.semeval-1.1

21. Vajjala, S., Meurers, D.: Assessing the relative reading level of sentence pairs for text simplification. In: Proceedings of the 14th Conference of the European Chapter of the Association for Computational Linguistics, Gothenburg, Sweden, pp. 288–297. Association for Computational Linguistics (2014). https://doi.org/10.3115/v1/E14-1031, https://aclanthology.org/ E14-1031

22. Vaswani, A., et al.: Attention is all you need. In: Advances in Neural Information Processing Systems, vol. 30 (2017)

Machine Translation Models Stand Strong in the Face of Adversarial Attacks

Pavel Burnyshev[1]([✉]), Elizaveta Kostenok[1,2,3], and Alexey Zaytsev[1]

[1] Skoltech, Moscow, Russia
pavel.Burnyshev@skoltech.ru
[2] MIPT, Dolgoprudny, Russia
[3] IITP RAS, Moscow, Russia

Abstract. Adversarial attacks expose vulnerabilities of deep learning models by introducing minor perturbations to the input, which lead to substantial alterations in the output. Our research focuses on the impact of such adversarial attacks on sequence-to-sequence (seq2seq) models, specifically machine translation models. We introduce algorithms that incorporate basic text perturbation heuristics and more advanced strategies, such as the gradient-based attack, which utilizes a differentiable approximation of the inherently non-differentiable translation metric. Through our investigation, we provide evidence that machine translation models display robustness displayed robustness against best performed known adversarial attacks, as the degree of perturbation in the output is directly proportional to the perturbation in the input. However, among underdogs, our attacks outperform alternatives, providing the best relative performance. Another strong candidate is an attack based on mixing of individual characters.

Keywords: Adversarial attack · Robustness · Neural machine translation

1 Introduction

Modern neural machine translation models demonstrate high-quality generated translations, and they are widely used in real-world applications as part of automatic translation systems. For this reason, the robustness and reliability of such models become crucial factors.

Adversarial attacks, as detailed [3,20], encompass a broad range of techniques aimed at exposing and probing the vulnerabilities of these models. These attacks introduce slight perturbations to the input data, which can, in turn, lead to significant misinterpretations or errors in the output. The aim is to understand the model's weak points and stability under these "attacks".

The core concept of an adversarial attack is not conditioned on the data nature: an attacker tries to significantly change the model output by modifying the input object. Nevertheless, constructing adversaries for NLP models is complicated due to the discrete structure of the text data [1,15,23]. As we can not

© The Author(s), under exclusive license to Springer Nature Switzerland AG 2024
D. I. Ignatov et al. (Eds.): AIST 2023, CCIS 1905, pp. 29–41, 2024.
https://doi.org/10.1007/978-3-031-67008-4_3

straightly use derivatives of the loss function, we compute differentiable approximations of metrics [25] and derivatives of the adversarial loss with respect to non-discrete token embeddings. We can use this idea to generate adversarial examples from the embedding space [8]. The work [7] goes further in this way by proposing the use of a generative model to make the adversarial attack work.

However, one can spot a common point in significant part of all these articles: they mostly pay attention to models with an output that consists of a single number. Nowadays, many use cases for natural language processing models focus on sequence2sequence problems, where both input and output for a model are sequences. One particular example of such a problem is a classic machine translation. The input, in this case, is a sequence in one language, and the output is a sequence in another language. Our research can help not only investigate vulnerabilities of these models to adversarial perturbations but also provide new insights on the possibility of detecting anomalies and estimating uncertainty for these models.

Our main contributions to adversarial attacks on machine translation models are:

- We propose new techniques to construct adversaries on the machine translation task. The first algorithm replaces input tokens based on the gradient of the target function with respect to the model's embeddings. Another approach exploits approximations of non-differentiable metrics.
- We conduct a fair comparison of different attacks based on a diverse set of metrics for a machine translation problem.
- Our experiments demonstrate that modern machine translation models are only slightly vulnerable to adversarial inputs. They do degrade for carefully created adversarial examples via a range of techniques, while the effect is less evident compared to drastic performance drops for computer vision and NLP classification models [11].
- The biggest vulnerability comes from attacks that work at character levels suggesting that in this case the adversarial examples fall out of the domain of data used for training.

2   Related Work

Various types of adversarial attacks on machine translation models have detected their sensitivity to disrupted inputs [2,19]. The first family of attack strategies finds the most loss-increasing perturbations of the source sentence using a gradient in the embedding space. The HotFlip attack [6] vectorizes simple char-level operations such as replacement, deletion, and insertion and uses directional derivatives to select the change of input sample. Targeted attack [14] uses gradient projections in the latent space to make perturbations. It preserves the similarity between initial and adversarial translations by inserting a target keyword into adversarial output. AdvGen algorithm [4] works on word level and craft adversarial examples based on the similarity between the loss gradient and distance between initial word and adversarial candidates.

The second group of attacks exploits differentiable estimations of standard NLP metrics to control text perturbations. Authors of [25] propose such approximation to BLEU, and authors of [9] use a deep learning model to estimate Levenshtein distance between sentences. Dependence on metrics allows selecting perturbations in discrete space more naturally.

The third type of attack can successfully fool a machine translation model by imitating typos or letter emission. Authors of [1] add synthetic noise to attacked sentences which includes replacement of letters and varying their order. In addition to swapped characters, distorted inputs can contain emojis and profanity [17].

Several approaches can produce high-quality adversarial examples but require more complicated training and generation processes. GAN-based framework [23] operates on a sentence level, and its training process is adapted for the discrete data structure. Authors of [26] propose a reinforcement learning paradigm to generate meaning-preserving examples.

There are certain methods to evaluate adversarial attacks on NLP data. Attack Success Rate measures the proportion of successful attacks, which reduces twice the BLEU score of adversarial translation compared to initial translation [5]. Authors of [11] propose an evaluation framework for attacks on seq2seq models that focuses on the semantic equivalence of the pre- and post-perturbation input.

In this study, we provide a comparison of principal attack types: gradient-based, synthetic, and metric approximation. Our modifications to existing methods allow both saving the semantic and grammar correctness of adversaries and altering the attacked translation.

3   Methods

3.1   General Description of a Machine Translation Model

The backbone of the majority of modern research and production Translation models is a Transformer model [18]. It consists of Encoder and Decoder parts, each of which includes sequential application of a multi-head attention mechanism that forces latent representation of tokens to interact with each other. The encoder of the model maps the input sentence $X = \{x_1, x_2 \dots x_n\}$ into latent representation $Z = \{z_1, z_2 \dots z_k\}$. Decoder likewise translates it into output embedding representation. The decoder output goes into the classification head, which chooses the next token y_j, the process repeats until the model generates a special end token. Choice of the next output token $y_{<j}$ depends on input text X, hidden representations z and already generated text $y_{<j}$:

$$p_\theta(Y|X) = \prod_{j=1}^{m} p_\theta(y_j|y_{<j}, X, Z),$$

where θ are model parameters and $Y = \{y_1, \dots, y_{k'}\}$. The loss function can be defined as $J(\theta, X, Y) = \frac{1}{n} \sum_{i=1}^{n} -\log P(y_i|X, \theta)$.

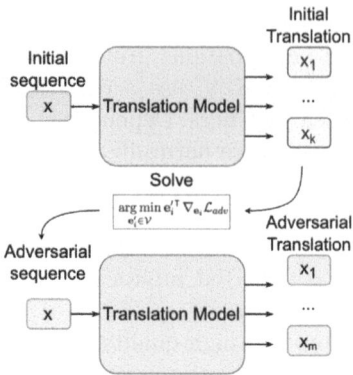

Fig. 1. Gradient Attack on Machine Translation models

3.2 Gradient Machine Translation Attack

The proposed gradient attack algorithm has a white-box full access to the model's parameters θ, adversarial loss $\mathcal{L}_{adv}$ we want to minimize, and an input sequence of tokens X that corresponds to a text. We suppose that for a set of tokens, we have a dictionary of embeddings $\mathcal{V}$. The model works on the token level, and the number of tokens in the alphabet is $|\mathcal{V}|$.

The core idea of the attack is inspired by Hotflip [6] attack: we iteratively replace input tokens according to the adversarial loss, calculated with respect to the model's input embeddings $\mathbf{e}$. The new token's embedding would minimize the first-order Taylor approximation of adversarial loss:

$$\arg\min_{\mathbf{e}'_i \in \mathcal{V}} [\mathbf{e}'_i - \mathbf{e}_i]^\top \nabla_{\mathbf{e}_i} \mathcal{L}_{adv}.$$

$\nabla_{\mathbf{e}_i} \mathcal{L}_{adv}$ means computing gradient with respect to token at position i. The subtracted part of the expression does not depend on substitute embeddings, so the optimization problem reduces to

$$\arg\min_{\mathbf{e}'_i \in \mathcal{V}} [\mathbf{e}'_i]^\top \nabla_{\mathbf{e}_i} \mathcal{L}_{adv}.$$

To select the replacement token, we try all possible indexes i and compare their respective difference in loss.

The overall approach is illustrated in Fig. 1.

It's essential to preserve the semantics and grammar of the initial text. Otherwise, the attack discriminators [24] would always detect the attack. So, following [6], we use several constraints in our experiments. They aim to save the initial meaning of the sentence and prevent an attacker from turning a sentence into a meaningless string of characters that doesn't resemble the initial meaning of a sentence:

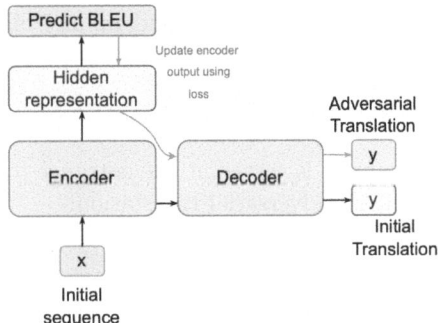

Fig. 2. BLEUER Attack, based on predicting initial BLEU score

1. The cosine distance between new and replaced embeddings must not be smaller than the threshold.
2. Attacker can replace each token position only once.
3. Attacker can separate all tokens in the vocabulary into two parts. The first part of the subset of tokens always stays at the beginning of the word. Another part stays at the second and next positions. We discourage the algorithm from replacing tokens from one part with tokens from another.
4. We disallow replacement of tokens denoting punctuation, first and last tokens of the sentence, and stop words.

3.3 BLEUER Attack

The gradient attack described before does not guarantee any estimations on the primary translation metrics change: BLEU, METEOR, etc. Instead of optimizing adversarial loss, which does not directly depend on text metric, we can incorporate dependence from the differentiable approximation of the metric. An illustration of the approach for BLEU score is presented in Fig. 2.

Before applying an attack, an adversary needs to train extra layers to predict BLEU. We translated a subset from the text corpus using the initial translation model and computed initial BLEU scores for pairs of sequences. Those scores are used as targets during the training of additional layers on the top of the encoder part of the model. During the training, we minimize MSE loss between the predicted value of the BLEU score and the original one:

$$J = \mathrm{MSE}(f(z), \mathrm{BLEU}(Y_{orig}, Y_{trans})),$$

where z is the encoder output, f means applying additional layers, Y_{orig} is expected translation from data corpus, Y_{trans} is the model's translation. The attack algorithm executes the following steps:

1. Get encoder outputs z;
2. Calculate prediction of BLEU score $f(z)$ using differentiable layers and loss value J;

3. Calculate gradients with respect to loss and update encoder outputs z, so that approximate BLEU score decreases:

$$z := z + \varepsilon \cdot \nabla_{\mathbf{e}_i} \mathrm{MSE}(f(z), 1)).$$

The updated encoder output is served at the entrance to the decoder part of the model, which generates adversarial translation.

3.4   MBART Attack

The proposed gradient attack and BLEUER attack approaches can be combined in the attack, which we call the MBART attack. Depending on gradients, obtained after predicting BLEU score on encoder outputs, we iteratively replace input tokens. After several iterations we get adversarial input X_{adv} and use the model to get adversarial translation Y_{adv}.

3.5   Synthetic Attacks

We propose an extremely naive synthetic attack as an alternative method to attack machine translation tasks. Synthetic attack [1] initially simulates errors or mistakes that can happen during daily usage of translation systems: keyboard typos, accidental omission/addition, or swapping chars. Additionally, we tried some uncommon sentence perturbations, such as randomly swapping a subset of words or randomly swapping a subset of chars in one word. As the main hyperparameter of such an attack, we used a portion of perturbed words or chars in the sentence.

4   Experiments

We perform the experiments with Marian and MBART Transformer models. For them, the comparison is between the three approaches described above: gradient, BLEUER, and synthetic attacks, since each of them represents a principal type of attack method. We also conduct a comparison of existing and novel attack algorithms. We pay special attention to balancing the trade-off between preserving the original sentence and altering the attacked translation. The attack should not be easily recognized by adversarial detectors, so the text should save logical and semantic literacy and grammar structure. We use a wide range of automatic linguistic metrics to evaluate attack approaches from this point of view. The code of our experiments will be available on a public online repository in the case of acceptance.

4.1   Metrics

Machine translation attacks aim to decrease the quality of translation metrics. We used 6 metrics in our experiments. **BLEU** is the most famous metric for evaluating the similarity of two sentences, and it is highly correlated with the human concept of text similarity. **chrF** metric is calculated as an F-score between character n-grams [12]. **METEOR** is another n-gram metric. It is calculated as an F-score for unigrams. **WER** metric considers a number of basic text operations: adding, deleting, and swapping characters for transforming one text into another. **Paraphrase similarity** metric is built upon pre-trained Sentence-Trans-former [13], model, which matches texts into 768-dimensional vectors. Cosine distance between such vectors correlates greatly with a human opinion of text similarity. **BertScore** [21] leverages vectors, obtained from pre-trained models. Bert Score has been found to correspond with human judgment at the sentence level meaning. It calculates each token's precision, recall, and F1 measures in assessed sentences.

4.2   Baselines

In addition to the proposed methods, we evaluated naive approaches and state-of-the-art approaches.

In particular, we consider our variant of **Gradient** attack itself and a modification of it **Gradient attack and ML constraint**. For the later attack, we utilize constraints on how much we can change the initial sentence. These constraints are described above in the methods section and aim at keeping the meaning and structure of the attacked sentence similar to the initial one. The attack uses Marian [10] Transformers, pre-trained on English-Russian corpora.

We consider two types of attacks that consider an approximation of the target metric during an attack: **BLEUER** and **MBART** attacks. For these attacks, we train an additional head that takes encoder outputs as an input. These heads are predicting BLEU or BertScore correspondingly. As these heads are differentiable, we incorporate these scores into the loss function to maximize the difference between the initial translation Y and the attacked sentence translation $Y_{attacked}$ and minimize the difference between the initial sentence X and its adversarial perturbation $X_{attacked}$. For training of BLEUER and MBART, we use the validation data of **wmt-14** dataset.

To make sure that all main types of attacks are considered, we evaluate methods from the literature. **Prefix attack** inserts tokens at the beginning as we try to select tokens that serve as a prompt. **SWLS** is the attack from the article [22]. This attack tries to leverage a bidirectional translation model and looks for perturbations that maximize the difference between adversarial sequence $X_{attacked}$ and its back-translation.

The last two methods consider attacks at separate character levels. **Char swap** tries to randomly swap characters to make the attack stronger. **Char + grad swap** is a version of our gradient attack at the character level.

Table 1. Attack samples for the Machine Translation task for different types of attacks

Attack type	Sentence type	Sentence
Gradient	Orig. sentence	Cars get many more miles to the gallon
	Attacked sentence	Cars get many more miles to the ormoneon
	Orig. translation	Автомобили проезжают больше миль на один галлон
	Translation	Машины проехали еще много миль до галлона
	Attacked translation	Машины проехали гораздо больше миль до гормона
BLEUER	Orig. sentence	Cars get many more miles to the gallon.
	Attacked sentence	Cars get many more miles to the gall on
	Orig. translation	Автомобили проезжают больше миль на один галлон
	Translation	Автомобили проезжают гораздо больше миль до галлона
	Attacked translation	Автомобили получают гораздо больше миль до галлона
Synthetic	Orig. sentence	Cars get many more miles to the gallon
	Attacked sentence	arCs egt myna emro ielsm to het gllnoa
	Orig. translation	Автомобили проезжают больше миль на один галлон
	Translation	Машины проехали еще много миль до галлона
	Attacked translation	arCs eggt myna empro ielsm to het glnoa

4.3 Attack Examples

At first, we visually examined the results of the conducted attacks by comparing examples of adversarially perturbed sentences. While sometimes the results are imperfect, in general, we see the desired effect. Examples of such sentences are provided in Table 1.

4.4 Experiment Setup

For gradient attack 3.2 and "BLEUER" attack 3.3 we used Marian [10] Transformers, pre-trained on English-Russian text corpora. For MBART 3.4 attack we used MBart-50 [16]). For "BLEUER" we additionally trained layers for approximating actual BLEU metric. For training we used validation data of **wmt-14** dataset.

4.5 Main Results

There is an important factor to be considered while evaluating machine translation adversarial attacks: perturbations should preserve the lexicon and grammar structure of the initial sentence. Authors of [11] proposed a new definition,

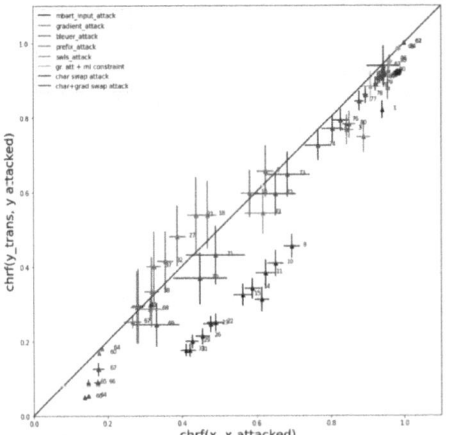

Fig. 3. Pareto frontiers for ChRF metric for considered attack methods. We aim at the lower right corner with high change of the translated sentence, but small change of the sentence to translate

meaning-preserving perturbations, which underline the importance of the correct assessment of an attack. We decided to take care of the balance of perturbing initial sentences and translations. Computing two similarities between the source sentence X and its perturbed sentence X_{attacked} and the similarity between the initial translation Y and translation of the attacked sentence Y_{attacked} is key to holding such a balance. Suppose the violation of the initial sentence approximately coincides with the violation of the translation. In that case, we cannot talk about the attack's success: the model honestly works out on a distorted sentence. An ideal attack would slightly change the initial similarity metric but significantly decrease the similarity between translations.

We vary the ability to introduce modifications into the initial sentence by modifying hyperparameters for all attacks. For each attack setting, they form a Pareto frontier, which can help us analyze the attack's impact. Numbers near the dots indicate hyperparameters of the attack. For the gradient attack and BLEUER attack, we provided a threshold for minimum cosine distance between vectors of original and substitute tokens for each dot. For synthetic attacks, we provided a maximum number of basic transformations for each dot.

Pareto frontiers for the full set of considered attacks are presented in Fig. 3. In general, the considered attacks could not show high values of attack success rates, supporting the evidence that modern translation models are robust due to the architectural features of models, computational expenses on training, and the colossal size of datasets. The top-performing attack is based on a swap at the level of characters. Both modifications show a significant improvement over others jointly providing a desired Pareto frontier.

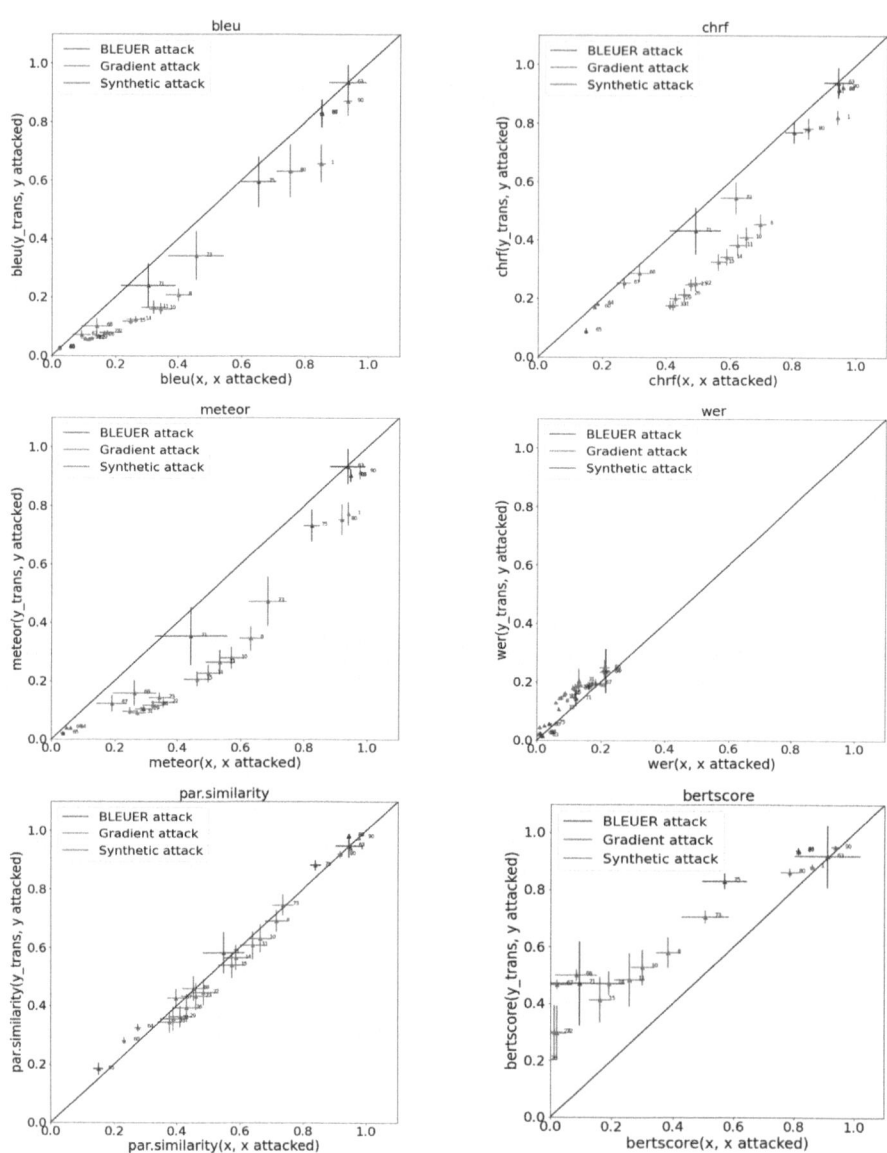

Fig. 4. Pareto frontiers for BLEU, ChRF, METEOR, WER, Par. Similarity, BertScore metrics for different attacks. Better attacks should aim lower right corner with a big similarity between input sequences before (x) and after an attack (x_{attacked}) and low similarity between translated sequences before (y) and after (y_{attacked})

4.6   Performance with Respect to Different Metrics

We provide Pareto frontiers for 6 automatic text metrics for 3 types of attacks: gradient attack, BLEUER, and a synthetic attack. Experimental results are presented in Fig. 4. Hitting as low and to the right as possible is the most successful attack, showing min distance between original and adversarial sentences and maximum distance between original and adversarial translations. It is rather evident from the graphics that most dots correspond to the same distortion of the initial and translation sequences. We can not ignore that the dots of the most straightforward method, synthetic attack on average, lie lower than the dots of more complicated approaches. That fact is especially noticeable for chrF metric due to a char-level of that attack. Simple char operations break the structure of tokens, heavily damaging deep models for machine translation, which usually work on the token level. The numerical summary is given in Table 2. It also supports the evidence that Synthetic attack provides superior metrics compared to an embedding-based approach that leverages the gradients of a model.

Table 2. Numerical comparison of the best attack settings based on the differences between the initial similarity metric and the similarity of translations

Metric type	BLEUER	Gradient	Synthetic
BLEU ↑	0.08	0.11	**0.20**
chrF ↑	0.09	0.13	**0.25**
METEOR ↑	0.14	0.24	**0.27**
WER ↓	−0.02	−0.04	**−0.07**
Paraphrase similarity ↑	−0.01	0.00	**0.06**
BertScore ↑	**0.00**	−0.01	−0.01

5   Conclusion

Adversarial attacks face limitations in the NLP domain. Especially for the machine translation task, both creating adversarial sequences and evaluating attacks become non-trivial. Most of the existing approaches have a high attack success rate, but they still suffer from lacking semantics and losing lexicon and grammar correctness. In our investigations, we focus on how we can make attacks more meaningful and valuable in analyzing the translation model's vulnerabilities. We tried to control translation metrics directly by using differentiable approximations.

The primary outcome of MT experiments is that we still did not find a method that guaranteed that translation would be changed stronger than a source sentence. We compared a range of metrics between initial and corrupted sentences and between initial and attacked translations. We made many additional rules and constraints which forced the attack algorithm not to collapse the initial sentence and save its semantic meaning totally, but they did not significantly change the situation.

Acknowledgements. The research was supported by the Russian Science Foundation grant 20-71-10135.

References

1. Belinkov, Y., Bisk, Y.: Synthetic and natural noise both break neural machine translation (2017)
2. Chakraborty, A., Alam, M., Dey, V., Chattopadhyay, A., Mukhopadhyay, D.: Adversarial attacks and defences: a survey. CoRR abs/1810.00069 (2018). http://arxiv.org/abs/1810.00069
3. Chen, J., Tam, D., Raffel, C., Bansal, M., Yang, D.: An empirical survey of data augmentation for limited data learning in NLP (2021)
4. Cheng, Y., Jiang, L., Macherey, W.: Robust neural machine translation with doubly adversarial inputs. In: Proceedings of the 57th Annual Meeting of the Association for Computational Linguistics, pp. 4324–4333 (2019)
5. Ebrahimi, J., Lowd, D., Dou, D.: On adversarial examples for character-level neural machine translation (2018)
6. Ebrahimi, J., Rao, A., Lowd, D., Dou, D.: HotFlip: white-box adversarial examples for text classification. In: Proceedings of the 56th Annual Meeting of the Association for Computational Linguistics (Volume 2: Short Papers), pp. 31–36 (2018)
7. Fursov, I., et al.: A differentiable language model adversarial attack on text classifiers. IEEE Access **10**, 17966–17976 (2022)
8. Fursov, I., Zaytsev, A., Kluchnikov, N., Kravchenko, A., Burnaev, E.: Gradient-based adversarial attacks on categorical sequence models via traversing an embedded world. In: van der Aalst, W.M.P., et al. (eds.) AIST 2020. LNCS, vol. 12602, pp. 356–368. Springer, Cham (2021). https://doi.org/10.1007/978-3-030-72610-2_27
9. Gomez, L., Rusinol, M., Karatzas, D.: LSDE: levenshtein space deep embedding for query-by-string word spotting, pp. 499–504 (2017). https://doi.org/10.1109/ICDAR.2017.88
10. Junczys-Dowmunt, M., et al.: Marian: fast neural machine translation in C++. In: Proceedings of ACL 2018, System Demonstrations. pp. 116–121. Association for Computational Linguistics, Melbourne, Australia, July 2018. https://doi.org/10.18653/v1/P18-4020. https://aclanthology.org/P18-4020
11. Michel, P., Li, X., Neubig, G., Pino, J.: On evaluation of adversarial perturbations for sequence-to-sequence models. In: Proceedings of the 2019 Conference of the North (2019)
12. Popović, M.: chrF: character N-gram F-score for automatic MT evaluation. In: Proceedings of the Tenth Workshop on Statistical Machine Translation, pp. 392–395. Association for Computational Linguistics, Lisbon, Portugal, September 2015. https://doi.org/10.18653/v1/W15-3049. https://aclanthology.org/W15-3049
13. Reimers, N., Gurevych, I.: Sentence-BERT: sentence embeddings using Siamese BERT-networks. In: Proceedings of the 2019 Conference on Empirical Methods in Natural Language Processing. Association for Computational Linguistics, November 2019. http://arxiv.org/abs/1908.10084

14. Sadrizadeh, S., Aghdam, A.D., Dolamic, L., Frossard, P.: Targeted adversarial attacks against neural machine translation. In: ICASSP 2023-2023 IEEE International Conference on Acoustics, Speech and Signal Processing (ICASSP), pp. 1–5. IEEE (2023)
15. Samanta, S., Mehta, S.: Towards crafting text adversarial samples (2017)
16. Tang, Y., et al.: Multilingual translation with extensible multilingual pretraining and finetuning. CoRR abs/2008.00401 (2020). https://arxiv.org/abs/2008.00401
17. Vaibhav, Singh, S., Stewart, C., Neubig, G.: Improving robustness of machine translation with synthetic noise. In: Meeting of the North American Chapter of the Association for Computational Linguistics (NAACL), Minneapolis, USA, June 2019
18. Vaswani, A., et al.: Attention is all you need (2017)
19. Wang, W., Tang, B., Wang, R., Wang, L., Ye, A.: A survey on adversarial attacks and defenses in text. CoRR abs/1902.07285 (2019). http://arxiv.org/abs/1902.07285
20. Xu, H., et al.: Adversarial attacks and defenses in images, graphs and text: a review. Int. J. Autom. Comput. **17**, 151–178 (2020)
21. Zhang, T., Kishore, V., Wu, F., Weinberger, K.Q., Artzi, Y.: BERTscore: evaluating text generation with BERT (2019)
22. Zhang, X., Zhang, J., Chen, Z., He, K.: Crafting adversarial examples for neural machine translation. In: Proceedings of the 59th Annual Meeting of the Association for Computational Linguistics and the 11th International Joint Conference on Natural Language Processing (Volume 1: Long Papers), pp. 1967–1977 (2021)
23. Zhao, Z., Dua, D., Singh, S.: Generating natural adversarial examples (2017)
24. Zhou, Y., Jiang, J.Y., Chang, K.W., Wang, W.: Learning to discriminate perturbations for blocking adversarial attacks in text classification. In: Proceedings of the 2019 Conference on Empirical Methods in Natural Language Processing and the 9th International Joint Conference on Natural Language Processing (EMNLP-IJCNLP), pp. 4904–4913. Association for Computational Linguistics, Hong Kong, China, November 2019. https://doi.org/10.18653/v1/D19-1496. https://aclanthology.org/D19-1496
25. Zhukov, V., Golikov, E., Kretov, M.: Differentiable lower bound for expected BLEU score (2017)
26. Zou, W., Huang, S., Xie, J., Dai, X., Chen, J.: A reinforced generation of adversarial examples for neural machine translation. In: Annual Meeting of the Association for Computational Linguistics (2019)

Whether Large Language Models Learn at the Inference Stage? Effects of Active Learning and Labelling with LLMs on their Reasoning

Vladlen Kulikov[1], Radoslav Neychev[1], and Ilya Makarov[2,3(✉)]

[1] MIPT, Moscow, Russia
{kulikov.vg,neychev}@phystech.edu
[2] AIRI, Moscow, Russia
makarov@airi.net
[3] NUST MISiS, Moscow, Russia

Abstract. In-context learning (zero, one and few-shot learning), chain-of-thoughts and prompt engineering have been more and more widely researched and discussed recently, in connection with the discovered possibility of Large Language Models (LLMs) based on examples (and in particular demonstrations of instructions, some actions with examples, including demonstration of reasoning) to solve certain types of problems at the Inference stage, without the need for fine-tuning, and in general without any additional training of LLMs on these tasks.

In our work, due to the lack of a general term, we refer to this emergent ability, observed in LLMs, as the L&R effect (Learning and Reasoning at the Inference Stage effect). This effect, along with the opportunities it creates, leaves open the question of what exactly is the reason for the emergence of such a surprising ability in LLMs.

In this paper we formulate a hypothesis, consistent with both the literature and our experiments, that explains the cause of the L&R effect and answers the main question: Do LLMs learn at the Inference or not?

Keywords: Learning · Reasoning · Inference · Large Language Models

1 Introduction

Interestingly, in the field of natural language processing (NLP), no single term adequately characterizes the emergent abilities under study. The terms "learning" and "reasoning" can be somewhat misleading, as will be discussed below.

The work was Financially supported by the Ministry of Science and Higher Education of the RF, contract 075-11-2022-036.

The original version of the chapter has been revised. An acknowledgement had not been displayed correctly. This has been corrected. A correction to this chapter can be found at https://doi.org/10.1007/978-3-031-67008-4_23

© The Author(s), under exclusive license to Springer Nature Switzerland AG 2024, corrected publication 2024
D. I. Ignatov et al. (Eds.): AIST 2023, CCIS 1905, pp. 42–53, 2024.
https://doi.org/10.1007/978-3-031-67008-4_4

However, the terms 'learning' and 'reasoning' are well established in the field. Therefore, we will use the term 'L&R effect', which refers to 'Learning and Reasoning at the Inference Stage'. This term describes the ability of Language Learning Models (LLMs) at the Inference Stage based on:

1) demonstration of examples, instructions describing the task, or
2) demonstrating both examples and instructions, and any actions with them (for example, logical or arithmetic operations, including chains of reasoning) give the correct answer, without any additional training of the model.

The L&R effect is emergent in nature [1] (emergence is a qualitative property that occurs spontaneously in a system when it reaches a certain threshold of complexity). The terminology used in this paper systematically follows the terminology used in the [2–4] for the terms *in-context, few-shot, one-shot, zero-shot learning* and the term *chain-of-thought(CoT)* is understood according to [5] as a series of intermediate reasoning steps.

Main questions in this work:

I. L&R effect - What is it?
II. Do LLMs learn and reason at the Inference?

To answer these questions, the following hypothesis was formulated. The validity of this hypothesis is consistent with the facts and observations in Part 2 and was tested experimentally in Part 3. Hypothesis:

Teaching LLMs based on language data leads to the creation of inner language spaces that contain not only the language itself linguistically but also some of the patterns of rules of reasoning implicitly embedded in the structures of human languages.

LLMs "learn" at the Inference stage without actually learning:

1. LLMs use language spaces of reasoning that they have previously created at the Train stage.
2. Demonstrating examples of reasoning with data (or just instructions with and without data) focuses the LLM's "attention" on areas of the language space of reasoning where similar "rules" already exist, allowing models to work with this and similar reasoning patterns or data.

2 Related Works

First, consider a special case of few-shot learning for machine translation where a reasonable explanation of the phenomena is internal language spaces.

In the paper – [6], the authors demonstrate the potential of few-shot translation systems trained on unpaired language data, and it turns out that with only 5 few-shot examples of high-quality translations, the quality of the resulting translations can match state-of-the-art models as well as more general commercial translation systems. That is, 5 few-shot examples of high-quality inference translation, for initially unpaired language data for the model, is enough to match the two languages very well.

And this special case, which we emphasized, is particularly revealing, because it makes it easy to see that without pre-formed language constructs at the training stage, the model cannot learn how to make high-quality translations in initially unpaired languages, just on the basis of 5 inference examples.

We think that LLMs create, at the training stage, an internal language spaces for each language. Since we are dealing with human languages, these internal language spaces have a common basic structure and can be matched by several n-dimensional points. These "calibration" points are fed into the model at the Inference as a few-shot examples [7–9].

At the Inference stage, these language spaces created at the Train stage, are matched, allowing LLMs to translate between languages that were not originally paired.

Now let's look at the facts and observation about reasoning abilities of LLMs.

The first thing to note is what the authors in [10] call rule-like generalization. According the article, LLMs receive in-context examples and draw conclusions based on "rules", i.e., using some "reasoning", the ability to create and apply which appears when they learn languages and according to [11] - if the LLM was trained on more "weak" data, "weaker" than the language - then the L&R effect is not observed.

Consider [10] in more detail. The researchers observed distinct patterns of generalization in transformers when they processed information in-context versus in-weight. The results indicate that generalization based on in-weight information is entirely rule-based when the model is trained on synthetic data, whereas generalization based on in-context information is primarily exemplar-based. Furthermore, research has shown that it is feasible to prompt the transformer to make generalizations based on rules by pre-training it on a classification problem that relies solely on rules. Remarkably, as the size of the language model increases [12–15], it becomes more proficient at making rule-based generalizations from contextual information. These findings suggest that natural language data may have a significant impact on the development of rule-like generalization, with the effectiveness being determined by the scale of the model.

The second point will be about chain-of-thought. A series of intermediate reasoning steps improves the ability of LLMs to reproduce complex reasoning at the Inference stage [5] and from same authors - paper [16] about self-consistency, to replace the naive greedy decoding used in chain-of-thought prompting.

In details, after previous article, about rule-like generalization, we have an understanding that LLMs pre-trained in languages, receiving examples at the Inference, are surprisingly rule-based when generalizing on these examples. Based on this understanding, we can quite logically assume that the addition of an example-only prompt with rules for working with these examples would be more "understandable" by LLMs and could probably give better results. Indeed, what we might assume from our analysis corresponds to what was called in [5] chain-of-thoughts – a series of intermediate reasoning steps, which, as expected, as we now understand, improves the ability of LLMs to reproduce complex reasoning at the Inference stage.

In [16], the authors proposed to modify the chain-of-thoughts method, which led to a further improvement in the results. The self-consistency technique is composed of three main steps. Firstly, a language model is prompted through the chain-of-thought (CoT) method. Next, instead of using the "greedy decode" approach, the CoT prompting is modified to randomly sample from the language model's decoder, resulting in the generation of various reasoning paths. Finally, these reasoning paths are marginalized, and the most consistent answer is selected from the resulting answer set.

It first samples a diverse set of reasoning paths instead of only taking the greedy one, and then selects the most consistent answer by marginalizing out the sampled reasoning paths. Self-consistency leverages the intuition that a complex reasoning problem typically admits multiple different ways of thinking leading to its unique correct answer.

However, in addition to the last two articles, which seem to indicate that LLMs learn, there are following facts and observations that contradict this assumption.

In research [17] was found that validity of reasoning matters only a small portion to the performance (it remains 80–90% of the original). While relevance to the input query and following the order along the reasoning steps are the key to the effectiveness of chain-of-thought prompting

In articles [18] explored a chain-of-thoughts with formal logic. LLMs are quite capable of doing correct single steps of deduction and are generally able to reason even in fictional contexts. However, they have difficulty planning proofs - when there are multiple valid inference steps available, they cannot systematically explore different options .

In [19] was studied 12 LLMs including GPT-3 and it was found that the models can use prior from pretraining and at the same time ignore the task defined by the demonstrations. Genuine demo values are not really required - randomly swapping labels in demos has almost no performance impact on a number of classification and multiple-choice tasks. While other aspects of the demonstration are key - namely, providing some examples: "the label space, the distribution of the input text, the overall format of the sequence".

In the work [20] was observed that models often learn just as well with misleading and irrelevant templates as well as instructive ones, and the choice of target words is more important than the meaning of the overall prompts. In general, the results obtained contradict the widely accepted hypothesis in the literature that prompts serve as semantically meaningful instructions for performing a task and that writing highly effective prompts requires domain knowledge.

A lot of work today is devoted to ways to improve the prompt [21–23]. This is called prompt engineering, for example: [24, 25] - it doesn't matter what the prompt is, only the result matters. By successfully setting up the prompt, including training, supplementing with web search, etc. - possible to improve the quality (prompt engineering). That is, it is no longer about what is right and in what form it seems right to us to feed at the Inference stage into the model as a demonstration of examples, reasoning and instructions, but what exactly will work with the best quality.

3   Methodology

The main question for experiments was:

Do the LLMs learn to reason at the Inference stage or only use the reasoning abilities learned at the Train stage?

Since we were primarily interested in identifying patterns rather than detailed assessments, we manually ran a small number of the same tests on very different models (a total of 400 tests were performed, which were also visually checked manually).

The work, the methodology and tests of which we follow [17], was performed using InstructGPT-175B [26] (text-davinci-002 and addition text-davinci-003). Therefore, it was especially interesting to compare the results of this work with:

1. The OpenSource line of LLMs Bloom [27] and Bloomz [28]: bloomz-560M, bloomz-1 B1, bloomz-1.7B, bloomz-3B, bloomz-7B1, bloom 176B + mt0-xxl 13B;
2. Quanaco 33B [29] which is LLAMA33B [30] fine-tuned with Quantized LoRA(Low Rank Adapters);
3. The latest version of GPT - GPT 4 [31] at the moment.

For arithmetic reasoning, was used GSM-8K [32], mathematical reasoning benchmarks and for Q&A reasoning was used Bamboogle [33], a dataset of compositional questions. With this tests were verified facts and observations regarding chain-of-thought (CoT) - as a form of L&R effect, the most complex and demanding to the size of the model. In general, the test results correlate well with the results from [17] and other facts from the literature overview [1,2].

Chain-of-thought was tested on the full line of available LLMs, in a variant with 9 few examples of question-answer, followed by a question whose answer is expected from the model, for example - 1 test out of 400 in Fig. 1.

In addition, the 9 examples demonstrated by each model were in 4 variants (in our tests in 4, in [17] - 9 variants), just one example from each variant to illustrate the type of test data:

STD (Standard prompting) Q: There were nine computers in the server room. Five more computers were installed each day, from monday to thursday. How many computers are now in the server room?

A: 29

CoT (Chain-of-Thought prompting) Q: There were nine computers in the server room. Five more computers were installed each day, from monday to thursday. How many computers are now in the server room?

A: There were originally 9 computers. For each day from monday to thursday, 5 more computers were installed. So 4 * 5 = 20 computers were added. Now 9 + 20 = 29 computers are now in the server room. The answer is 29.

Q: There are 15 trees in the grove. Grove workers will plant trees in the grove today. After they are done, there will be 21 trees. How many trees did the grove workers plant today?
A: There are 15 trees originally. Then there were 21 trees after the Grove workers planted some more. So there must have been 21 - 15 = 6 trees that were planted. The answer is 6.

Q: If there are 3 cars in the parking lot and 2 more cars arrive, how many cars are in the parking lot?
A: There are originally 3 cars. Then 2 more cars arrive. Now 3 + 2 = 5 cars are in the parking lot. The answer is 5.

Q: Leah had 32 chocolates and her sister had 42. If they ate 35, how many pieces do they have left in total?
A: Originally, Leah had 32 chocolates and her sister had 42. So in total they had 32 + 42 = 74. After eating 35, they had 74 - 35 = 39 pieces left in total. The answer is 39.

Q: Jason had 20 lollipops. He gave Denny some lollipops. Now Jason has 12 lollipops. How many lollipops did Jason give to Denny?
A: Jason had 20 lollipops originally. Then he had 12 after giving some to Denny. So he gave Denny 20 - 12 = 8 lollipops. The answer is 8.

Q: Shawn has five toys. For Christmas, he got two toys each from his mom and dad. How many toys does he have now?
A: Shawn started with 5 toys. He then got 2 toys each from his mom and dad. So he got 2 * 2 = 4 more toys. Now he has 5 + 4 = 9 toys. The answer is 9.

Q: There were nine computers in the server room. Five more computers were installed each day, from monday to thursday. How many computers are now in the server room?
A: There were originally 9 computers. For each day from monday to thursday, 5 more computers were installed. So 4 * 5 = 20 computers were added. Now 9 + 20 = 29 computers are now in the server room. The answer is 29.

Q: Michael had 58 golf balls. On tuesday, he lost 23 golf balls. On wednesday, he lost 2 more. How many golf balls did he have at the end of wednesday?
A: Michael started with 58 golf balls. He lost 23 on Tuesday, and lost 2 more on wednesday. So he had 58 - 23 = 35 at the end of Tuesday, and 35 - 2 = 33 at the end of wednesday. The answer is 33.

Q: Olivia has $23. She bought five bagels for $3 each. How much money does she have left?
A: Olivia had 23 dollars. She bought 5 bagels for 3 dollars each. So she spent 5 * 3 = 15 dollars. Now she has 23 - 15 = 8 dollars left. The answer is 8.

Q: Marilyn's first record sold 10 times as many copies as Harald's. If they sold 88,000 copies combined, how many copies did Harald sell?
A:

Fig. 1. This all is only 1 prompt (only 1 test from 400), in one form from 4 settings (standard, chain-of-thought, invalid, no relevant). Blue - 8 (few shot) demonstration (Q&A) how need reason and solve task in this setting. Green - question from GSM8K test. (Color figure online)

Invalid Reasoning Q: There were nine computers in the server room. Five more computers were installed each day, from monday to thursday. How many computers are now in the server room?

A: There were originally 9 computers. For each day from monday to thursday, 5 more computers were installed. Now 9 * 5 = 45 computers. Since 4 * 4 = 16, now 45 − 16 = 29 computers are now in the server room. The answer is 29.

No relevance Q: There were nine computers in the server room. Five more computers were installed each day, from monday to thursday. How many computers are now in the server room?

A: Haley is currently 23 in. tall. She grows at the rate of 10 in. every year for 4 years. So she will have grown by 10 * 4 = 40 in.. Her height after 4 years will be 23 + 40 = 63 in.. The answer is 63."

4   Experiment Results

First, the results of observing the L&R effect for each of the models:

1) bloomz-560M, bloomz-1 B1, bloomz-1.7B, bloomz-3B - the CoT effect was not observed. The answer of the model was just numbers, without chain-of-thought, and the answer was not correct.
2) bloomz-7B1 – the model's response was short fragments of reasoning and (or) incorrect answers. Probably, this allows us to state the first signs of the CoT effect, but on a model of this size - with a result of zero quality.
3) mt0-xxl 13B – the effect was not observed, but the model itself was from another line. The model's answer was only numbers or facts, without chain-of-thought, the answers were not correct.
4) bloom 176B – the CoT effect was observed in full. Reasoning reaching the answer - but often the answer was not correct, and the reasoning is sometimes inadequate.

5) Quanaco 33B - the CoT effect was observed in full. Model demonstrates very good results, despite the fact that all other models in the tables are superior in size. Definitely better than Quanaco 33B only GPT4 and comparable only text-davinci-003,

6) ChatGPT Plus with GPT 4 (the size of the model is not known, most likely not less than 175B as it was in the previous version) – the CoT effect was observed in full, correct reasoning and the correct result in almost all cases.

Second, here are the summary tables with tests – Table 1 and 2[1].

The omission of data in Tables 1, 2 in the row № 1 for the models text-davinci-2 and text-davinci-3 is explained by the fact that the data were taken for comparative analysis from [17], and the authors did not conduct such an experiment.

Table 1. Percentage of correct answers on the GSM-8K (arithmetic reasoning). Right reasoning (CoT) and Invalid reasoning give the same or almost the same result in all models. This means that models don't learn at the Inference but use prior reasoning patterns from the Train.

№	Test setting	Bloom 176B	Quanaco 33B LLAMA QLoRA	text davinci 002	text davinci 003	GPT4 ChatGPT Plus
1	Without any examples of resoning (without demo)	0%	40%	–	–	100%
2	STD (Standard prompting) without thoughts	0%	0%	0%	10%	80%
3	CoT: Chain-of-Thought right reasoning	10%	50%	40%	60%	90%
4	Invalid Reasoning	10%	50%	40%	80%	100%
5	No relevance	0%	0%	10%	10%	100%

And now the detailed results and conclusions from the experiments will be formulated:

In full agreement with the literature [1,2] the L&R effect is emergent and is not observed in models up to a certain model size.

The lower limit of the observation of L&R effect signs depends not only on the size of the model, but also on the model itself. Signs of the effect, although weak, observed in bloomz-7B1, do not appear in mt0-xxl 13B. As it was described in the studied sources [1,2].

As we assumed earlier, in the section on hypotheses, if there is no reasoning pattern already formed at Train stage in the model, the effect will emerge weakly (in case of falling into a similar area of reasoning within the language space of reasoning), or it will not be observed at all – which emerges in the fact that the

---

[1] Tests are available here.

Table 2. Percentage of correct answers on the Bamboogle Test (Q&A reasoning about facts). The same patterns Right reasoning (CoT) and Invalid reasoning are observed as in the previous table.

№	Test setting	Bloom 176B	Quanaco 33B LLAMA QLoRA	text davinci 002	text davinci 003	GPT4 ChatGPT Plus
1	Without any examples of resoning (without demo)	40%	90%	–	–	100%
2	STD (Standard prompting) without thoughts	30%	50%	30%	50%	80%
3	CoT: Chain-of-Thought right reasoning	40%	100%	50%	90%	100%
4	Invalid Reasoning	40%	80%	50%	90%	100%
5	No relevance	30%	20%	50%	60%	100%

answers and reasoning will be inadequate to the question (we see all this in the case of Bloom on GSM8K).

The most advanced model, ChatGPT4 – most often simply ignores the demonstrated examples, and the worse the example (that is, from the no relevance category), the better accuracy. Accordingly, the model explicitly uses a priori knowledge and reasoning methods that initially already reach the limit level (without the need for CoT on test examples).

The prohibition of explicit reasoning, which is "standard" prompting, significantly harms all the models.

Completely inappropriate examples of "no relevance" - damage all models except ChatGPT4.

ChatGPT4, although it ignores complete nonsense ("no relevance"), but it can worsen the result with correct CoT, which probably correspond less to the task than its own chains of reasoning. That is, the correct CoT from the examples – position ChatGPT4 (in 10% of cases) on less suitable patterns of reasoning than its own, from prior. As a result, the correct CoT examples sometimes interfere with ChatGPT4 on GSM8K.

As can be seen from lines 3 and 4 in each of Tables 1 and 2 and in full agreement with observations in [17] - Right reasoning and Invalid reasoning give the same or almost the same result in all models, but as tests have shown, often on a different questions That is, the demos are only a calibration points for pre-train patterns of reasoning. This means that models don't learn at the Inference but use prior reasoning patterns from the Train.

Based on this, we think that the our findings and other available observations sufficiently support our hypothesis that LLMs do not learn at the Inference stage, but use the reasoning patterns they extracted from the languages at the Train stage.

5   Conclusion

After analyzing the facts and observations in other works and conducting experiments, the original questions were answered:

Question I. L&R effect - What is it?

A hypothesis was formulated that LLMs form 'Inner Language spaces of reasoning' during the Training stage. The use of reasoning patterns from these Language spaces at the Inference stage is mistakenly perceived as learning and reasoning at the Inference stage, whereas, it is merely the application of pre-existing knowledge without acquiring anything new.

This hypothesis fits well with all the facts and observations found in the literature (referenced in the Part 2) and is supported by our experiments (discussed in the Part 3).

Question II. Do LLMs learn and reason at the Inference stage?

To date, we have not found any evidence in either the existing literature or our own experiments that confirms the actual realization of the theoretical scenarios regarding the learning of LLMs at the Inference stage. These scenarios are discussed in very interesting papers such as [34–37].

But following references [17, 19, 38] we state-perhaps somewhat categorically-that LLMs an the Inference stage do not learn in the classical sense of the word. Rather, their ability to learn and reason is constrained by the capabilities acquired during the Train stage.

6   Future Directions

It would be desirable to approach the questions considered in this work from the perspective of building a general theory of language spaces and patterns of reasoning in LLMs formed on languages. It want to note that it's necessary to distinguish between two aspects of this phenomenon:

1. The formation, which occurs during the Training stage,
2. The usage, which occurs during the Inference stage,
   of language spaces of reasoning in LLMs.

These aspects can either be part of a single, integral description or independent descriptions.

An interesting and important topic for further research would be to investigate the actual nature of the dependence of the ability to learn and reason on the number of model parameters. The emergent properties have been studied previously [1], but due to the lack of a consistent series of models, questions remain both about the presence of exact thresholds for emergence and about the nature of the dependence of emergent properties on the number of model parameters. The unresolved nature of these questions leaves too much room for assumptions, e.g. [39] - Are emergent abilities of large language models a mirage?

It seems important to understand the true dynamics of emergence in the ascending series of the same model, but with a different number of parameters: 1, 2, 3, ... , 174, 175, 176, ... ,999, 1000B

Unfortunately, such a series of models does not exist today. This is the only way to really understand how the emergent properties depend on the number of model parameters. Ideally, it would be good to do this on several independent model lines of LLMs and compare the results.

References

1. Wei, J., et al.: Emergent abilities of large language models. arXiv preprint arXiv:2206.07682 (2022)
2. Brown, T., et al.: Language models are few-shot learners. In: Advances in Neural Information Processing Systems, vol. 33, pp. 1877–1901 (2020)
3. Dong, Q., et al.: A survey for in-context learning. arXiv preprint arXiv:2301.00234 (2022)
4. Shulgin, M., Makarov, I.: Scalable zero-shot logo recognition. IEEE Access **11**, 142702–142710 (2023)
5. Wei, J., et al.: Chain-of-thought prompting elicits reasoning in large language models. In: Advances in Neural Information Processing Systems, vol. 35, pp. 24824–24837 (2022)
6. Garcia, X., et al.. The unreasonable effectiveness of few-shot learning for machine translation. In: International Conference on Machine Learning, pp. 10867–10878. PMLR (2023)
7. Savchenko, A.V.: Fast inference in convolutional neural networks based on sequential three-way decisions. Inf. Sci. **560**, 370–385 (2021)
8. Ermakov, M., Makarov, I.: Few-shot logo recognition in the wild. In: Proceedings of the 22nd International Symposium on Computational Intelligence and Informatics and 8th IEEE International Conference on Recent Achievements in Mechatronics, Automation, Computer Science and Robotics (CINTI-MACRo), pp. 000393–000398. IEEE (2022)
9. Savchenko, A.V., Savchenko, L.V.: Three-way classification for sequences of observations. Inf. Sci. 119540 (2023)
10. Chan, S.C.Y., Dasgupta, I., Kim, J., Kumaran, D., Lampinen, A.K., Hill, F.: Transformers generalize differently from information stored in context vs in weights. arXiv preprint arXiv:2210.05675 (2022)
11. Chan, S., et al.: Data distributional properties drive emergent in-context learning in transformers. In: Advances in Neural Information Processing Systems, vol. 35, pp. 18878–18891 (2022)
12. Grachev, A.M., Ignatov, D.I., Savchenko, A.V.: Neural networks compression for language modeling. In: Shankar, B.U., Ghosh, K., Mandal, D.P., Ray, S.S., Zhang, D., Pal, S.K. (eds.) PReMI 2017. LNCS, vol. 10597, pp. 351–357. Springer, Cham (2017). https://doi.org/10.1007/978-3-319-69900-4_44
13. Savchenko, A., Alekseev, A., Kwon, S., Tutubalina, E., Myasnikov, E., Nikolenko, S.: Ad lingua: text classification improves symbolism prediction in image advertisements. In: Proceedings of the 28th International Conference on Computational Linguistics, pp. 1886–1892 (2020)

14. Abulkhanov, D., Sorokin, N., Nikolenko, S., Malykh, V.: LAPCA: language-agnostic pretraining with cross-lingual alignment. In: Proceedings of the 46th International ACM SIGIR Conference on Research and Development in Information Retrieval, SIGIR '23, New York, NY, USA, pp. 2098–2102. Association for Computing Machinery (2023)

15. Gerasimova, O., Severin, N., Makarov, I.: Comparative analysis of logic reasoning and graph neural networks for ontology-mediated query answering with a covering axiom. IEEE Access **11**, 88074–88086 (2023)

16. Wang, X., et al.: Self-consistency improves chain of thought reasoning in language models. arXiv preprint arXiv:2203.11171 (2022)

17. Wang, B., et al.: Towards understanding chain-of-thought prompting: an empirical study of what matters. arXiv preprint arXiv:2212.10001 (2022)

18. Saparov, A., He, H.: Language models are greedy reasoners: a systematic formal analysis of chain-of-thought. arXiv preprint arXiv:2210.01240 (2022)

19. Min, S., et al.: Rethinking the role of demonstrations: what makes in-context learning work? arXiv preprint arXiv:2202.12837 (2022)

20. Webson, A., Pavlick, E.: Do prompt-based models really understand the meaning of their prompts? arXiv preprint arXiv:2109.01247 (2021)

21. Reynolds, L., McDonell, K.: Prompt programming for large language models: Beyond the few-shot paradigm. In: Extended Abstracts of the 2021 CHI Conference on Human Factors in Computing Systems, pp. 1–7 (2021)

22. Liu, X., et al.: GPT understands, too. arXiv preprint arXiv:2103.10385 (2021)

23. Zhou, Y., et al.: Large language models are human-level prompt engineers. arXiv preprint arXiv:2211.01910 (2022)

24. Khattab, O., et al.: Demonstrate-search-predict: composing retrieval and language models for knowledge-intensive NLP. arXiv preprint arXiv:2212.14024 (2022)

25. Yao, S., et al.: Tree of thoughts: Deliberate problem solving with large language models. arXiv preprint arXiv:2305.10601 (2023)

26. Ouyang, L., et al.: Training language models to follow instructions with human feedback. In: Advances in Neural Information Processing Systems, vol. 35, pp. 27730–27744 (2022)

27. Scao, T.L., et al.: Bloom: a 176b-parameter open-access multilingual language model. arXiv preprint arXiv:2211.05100 (2022)

28. Muennighoff, N., et al.: Crosslingual generalization through multitask finetuning. arXiv preprint arXiv:2211.01786 (2022)

29. Dettmers, T., Pagnoni, A., Holtzman, A., Zettlemoyer, L.: QLORA: efficient finetuning of quantized LLMS. arXiv preprint arXiv:2305.14314 (2023)

30. Touvron, H., et al.: LLAMA: open and efficient foundation language models. corr, abs/2302.13971 (2023). 10.48550. arXiv preprint arXiv:2302.13971

31. OpenAI. Gpt-4 technical report (2023)

32. Cobbe, K., et al.: Training verifiers to solve math word problems. arXiv preprint arXiv:2110.14168 (2021)

33. Press, O., Zhang, M., Min, S., Schmidt, L., Smith, N., Lewis, M.: Measuring and narrowing the compositionality gap in language models. arxiv 2022. arXiv preprint arXiv:2210.03350

34. Akyürek, E., Schuurmans, D., Andreas, J., Ma, T., Zhou, D.: What learning algorithm is in-context learning? investigations with linear models. arXiv preprint arXiv:2211.15661 (2022)

35. Von Oswald, J., et al.: Transformers learn in-context by gradient descent. In: International Conference on Machine Learning, pp. 35151–35174. PMLR (2023)

36. Garg, S., Tsipras, D., Liang, P.S., Valiant, G.: What can transformers learn in-context? A case study of simple function classes. In: Advances in Neural Information Processing Systems, vol. 35, pp. 30583–30598 (2022)
37. Dai, D., Sun, Y., Dong, L., Hao, Y., Sui, Z., Wei, F.: Why can GPT learn in-context? Language models secretly perform gradient descent as meta optimizers. arXiv preprint arXiv:2212.10559 (2022)
38. Razeghi, Y., Logan IV, R.L., Gardner, M., Singh, S.: Impact of pretraining term frequencies on few-shot reasoning. arXiv preprint arXiv:2202.07206 (2022)
39. Schaeffer, R., Miranda, B., Koyejo, S.: Are emergent abilities of large language models a mirage? arXiv preprint arXiv:2304.15004 (2023)

Machine Translation for Russian-Khakas Language Pair: Translation Results in Low-Resource Setting

Anna Lebedeva[1,2]([⊠]) [iD]

[1] OK.TECH, Saint Petersburg, Russia
annlebedeva.spb@gmail.com
[2] HSE University, Moscow, Russia

Abstract. The article is dedicated to applying transfer learning approach to the translation task of a low-resource Russian-Khakas language pair. This study shows that using the Russian-Chuvash language pair for pre-training can significantly improve the model's performance. The study describes the process of collecting and preprocessing the data, tokenization, training and evaluation of the results.

Keywords: neural machine translation · low-resource · transfer learning

1 Introduction

Low-resource languages are usually defined as less studied, resource scarce, less computerized, less privileged, less commonly taught, or low density, among other explanations [1, 17, 19]. Preserving and developing those languages though is very important, because they represent a unique repository of culture, knowledge, and history and also provide a powerful sense of belonging to a community for it's speakers.

Khakas language is one of the languages spoken in Russia that has a very limited amount of digitized data and this can be considered low-resource. It is a Turkic language spoken by Khakas people, mainly from Khakas Republic in Russia. The Khakas language is part of the South Siberian subgroup of Turkic languages, along with Shor, Chulym, Northern Altai and others [5]. According to 2010 census, Khakas language is spoken by about 42,000 people in Republic of Khakassia.[1] The writing system for the Khakas language is based on the Cyrillic script. Khakas language uses all of the letters from the Russian alphabet, adding four letters of its own: Ғ, I, Ң, Ö, Ӱ and Ч.

There are currently no web resources that would provide an online translation from Khakas to any other language, which we seek to change in the future,

---

[1] https://cyberleninka.ru/article/n/demograficheskaya-moschnost-hakasskogo-yazyka-i-yazykovaya-loyalnost-zhiteley-hakasii.

© The Author(s), under exclusive license to Springer Nature Switzerland AG 2024
D. I. Ignatov et al. (Eds.): AIST 2023, CCIS 1905, pp. 54–66, 2024.
https://doi.org/10.1007/978-3-031-67008-4_5

providing the efficient algorithm to train the translation model.[2] In this paper we describe the results of applying transfer learning approach to the task of translating the Russian-Khakas language pair.

The pipeline included training the baseline model, choosing the language pair for pre-training, preprocessing, adjusting the model's parameters, and training and evaluating the model. The model was trained in both translation directions. It was possible to achieve state-of-the-art results for the Russian-Khakas pair, and the ideas for future improvement will be presented at the end of the study.

2   Related Work

The effectiveness of NMT models declines significantly when the amount of training data decreases, as noted by Koehn and Knowles [8]. One approach to enhance translation accuracy is to transfer knowledge acquired from a language pair with richer resources. Zoph et al. [22] propose a transfer learning method, that consists of training a recurrent model on a large parallel corpus of a rich language pair, and then using the weights of this model as an initial configuration for a low-resource language pair, with embeddings of the side where language stays the same being frozen. This reduces the need for large amounts of task-specific labeled data, which can be costly and time-consuming to acquire [22]. Transfer learning provides a good initialization point for the model, as it has already learned meaningful features from pre-training. This initialization helps models converge faster during fine-tuning on task-specific data, reducing training time and computational resources. In this case, the pre-trained model is often referred to as the parent model, and fine-tuned one - as the child model. Valeev et al. [20] also used transfer learning. They use the Russian-Kazakh pair for training the parent model and then fine-tune it on the Russia-Tatar data. Shared vocabulary between all three languages is used.

Another common approach used in improving model's translation quality for low-resource languages is called back-translation. According to Sennrich et al. [15], the method works by combining monolingual training data and a synthetic source sentence to create an approximate context vector. This is done through back-translation, where the monolingual target text is automatically translated into the source language. The training process involves blending the synthetic parallel text with the original parallel text, without any distinction or fixed network parameters. Importantly, the additional training examples only have synthetic content on the source side, while the target side is obtained from the monolingual corpus. Other scientists, for example, Knowles et al. [6] suggest to mark back-translated sentences, for instance, with a <BT> tag in the beginning of it, so that the model can see this tag and learn to rely on this sentences less than on original parallel sentences.

Table 1 compares the amounts of data that different researchers used for parent and child models respectively when applying transfer learning.

---

[2] https://github.com/AnnaLebedeva/khakas-translator.

Table 1. Sizes of the corpora used by different authors for transfer learning and back-translation

Article	Language pairs		Parent parallel, pairs	Child parallel, pairs	Child source monoling	Child target monoling
	parent	child				
Valeev et al. [20]	ru-kk	ru-tt	5,000,000	324,000	9,000,000	8,800,000
Knowles et al. [6]	de-cs	de-hsb	22,000,000	60,000	12,000,000	680,000
Zoph et al. [22]	en-fr	en-es	53,000,000	2,500,000	-	-
Kocmi et Bojar [7]	en-fi	en-et	2,800,000	800,000	-	-
Kocmi et Bojar [7]	en-cs	en-et	40,100,000	800,000	-	-
Hujon et al. [4]	en-fr	en-kha	1,000,000	36,601	-	-

3   Data

3.1   Choosing the Language for Parent Model

The idea behind choosing the language for pre-training the model included the following factors: a) we wanted it to also be in Cyrillic script since we use shared vocabulary between parent and child models, b) preferably Turkic too so that it has similar morphology and syntax with Khakas, c) the dataset has to be relatively large so that the model can learn well.

Table 2 compares the amounts of data available on the Internet for the Turkic languages that currently use Cyrillic script in terms of parallel data with Russian language. We looked at five such languages for which the biggest amount of data is available: Kazakh, Kyrgyz, Bashkir, Tatar and Chuvash. None of these languages belong to the same branch as Khakas, which is usually put in the same group with Shor, Chulym, Northern Altai and others [2,5]. But for the above languages there is a very limited amount of parallel data or no parallel data at all [11].

Table 2. Sizes of available parallel corpora of Turkic languages that use Cyrillic script (parallel sentences with Russian language)

Language	Size of the parallel corpora, pairs of sentences	Source	Predominant collection method
Kazakh	4,403,385	TIL	Crawling
	~700,000	OPUS	Crawling
Kyrgyz	293,652	TIL	Crawling
	~200,000	OPUS	Crawling
Tatar	270,462	TIL	Crawling
	~200,000	OPUS	Crawling
	161,000	Institute of Applied Semiotics	Crawling
Bashkir	523,719	TIL	Crawling
	~600,000	OPUS	Crawling
Chuvash	794,654	TIL	Crawling
	~1,000,000	OPUS	Crawling
	1,002,013	Chuvash Bilingual Corpus	Manualy aligned literary works

In Table 2 we present the sizes of the corpora of Turkic languages with Cyrillic script that we managed to find online. Most of the data comes from TIL corpus [11] or OPUS [18]. Apart from OPUS and TIL, about 161,000 parallel lines for Tatar were provided by The Tatarstan Academy of Sciences Institute of Applied Semiotics.[3] As for Chuvash, there is an available corpus of 1,002,013 parallel lines at Chuvash Bilingual Corpus.[4]

As we see, the richest language pair in terms of parallel sentences available is Russian-Kazakh (4,403,385 pairs in TIL corpus), then goes Russian-Chuvash (1,002,013 pairs in Chuvash Bilingual Corpus) followed by Russian-Bashkir (523,719 pairs in TIL corpus). The corpora from different sources weren't compared for duplicates, so it is not known whether they can be combined to form a bigger corpus or they partly or fully intersect. It is known though that the OPUS corpus for Chuvash consists fully of the data of Chuvash Bilingual Corpus, as it is stated in the description of the data on TIL website.

Although Russian-Kazakh language pair is richer, we chose Russian-Chuvash language pair for training the parent model, because we hypothesized that the quality of the manually aligned corpora will play an important role in training. While Chuvash corpus, according to the description, contains mainly literary works that were manually aligned and checked, Kazakh corpus, according to the authors, contains mainly crawled data (Tatoeba Challenge, UDHR, Bible, Ted Talks, Mozilla) [11]. The decision was also based on Hujon et al. [4] presenting good results for the corpora of similar sizes (1,000,000 and 36,601 lines for parent and child models accordingly).

Chuvash is a Turkic language, native language of the Chuvash people and an official language of Chuvashia.[5,6] Being the only surviving member of the Oghur branch of Turkic languages,[7] it is spoken by approximately 1,043,000 people in Russia.[8] Chuvash language is based on the Cyrillic script, using all of the letters from the Russian alphabet and adding four letters of its own: Ӑ, Ӗ, Ҫ and Ӳ. Almost all of these added letters except Ӳ are different from those specific for Khakas. Interestingly, Chuvash mostly has specific vowels, while Khakas mostly has specific consonants. Despite this fact, we think that the small difference in alphabets will not be a big problem for training the translation model, because the vocabulary will be shared between these two languages and will include all letters from both.

As a preprocessing step we checked the data for duplicates, shuffled the sentences and ended up with 1,000,013 pairs in the train set and 2,000 lines in the validation set.

---

[3] https://www.antat.ru/en/ips/.
[4] https://ru.corpus.chv.su/content/about.html.
[5] http://www.cv-haval.org/ru/node/54.
[6] http://cvlat.blogspot.com/2010/11/blog-post.html.
[7] https://academic.oup.com/pages/op-migration-welcome?login=true.
[8] https://minlang.iling-ran.ru/lang/chuvashskiy-yazyk.

3.2   Data for Child Model

Data Source. Our dataset for child model was based on TIL Corpus of Russian-Khakas parallel sentences [11], which consists of 60,295 lines in the train set, and 1,000 lines in validation and test sets accordingly.

Preprocessing. Originally this dataset had spaces on both sides of punctuation marks and also trash symbols like '⁕', '& # 91' or random numbers. We removed extra spaces before the punctuation marks for the sake of more correct output when new sentences with correct punctuation are given to the model. We also cleaned the data from trash symbols and random numbers. Almost all of the Khakas sentences in the TIL corpus lack punctuation at all, but some sentences belong to literary works that were provided to us by the Electronic corpus of the Khakas language,[9] so we managed to restore the punctuation for those sentences, which made about half of the dataset.

4   Methods

Byte-Pair Encoding. Neural machine translation (NMT) models typically operate with a fixed vocabulary, but translation requires using unknown and rare words. This is especially relevant for agglutinative languages where words are formed through the combination of smaller morphemes, which is why translation models require mechanisms that operate below the word level. To handle this, previous approaches use dictionaries for unknown words. Byte-Pair Encoding approach (BPE) that we use in this work, introduced by Sennrich et al. [16], implements encoding words as sequences of subword units. This tool uses byte-pair encoding algorithm for the task of word segmentation. The desired number of merging operations can be set as an option.

Transfer Learning. In this study we use the approach to Transfer learning proposed by Kocmi and Bojar [7]: the training procedure involves initially training the parent language pair for a certain number of iterations, followed by a transition to the child language pair without resetting any (hyper)parameters. This approach shares similarities with the transfer learning method introduced by Zoph et al. [22], but uses the shared vocabulary as suggested by Nguyen and Chiang [12].

Evaluation Metrics. Despite being considered to have a low correlation with human evaluation [10], the Bilingual Evaluation Understudy (BLEU) is still a widely used automatic evaluation metric for machine translation. It measures the similarity between the machine translation output and one or more reference translations based on n-gram precision. The BLEU score ranges between 0 and 1, with a higher score indicating a better translation. We will use BLEU for evaluation alongside with ChrF (Character F-Score) [13]: an evaluation method

---

[9] https://khakas.altaica.ru/.

that measures character n-gram overlap instead of word n-grams, as seen in the BLEU metric. ChrF employs the F-score, which combines character n-gram precision (ChrP) and character n-gram recall (ChrR). It was not possible to use more up-to-date frameworks like COMET evaluation metric, because there is no pre-trained model for the Russian-Khakas language pair, and training COMET metric for evaluation of these languages requires human annotated data that is also absent.

5 Experiments

Training Setting. All the research was done using Sockeye toolkit [3]. Sockeye is an open-source sequence-to-sequence framework for Neural Machine Translation. The current version of Sockeye - Sockeye 3 - is based on Transformer, a typical encoder-decoder architecture with attention mechanisms [21]. It utilizes a classic 6-encoder and 6-decoder layer configuration with 8 attention heads on both encoder and decoder sides.

As for customizing hyperparameters, we only set the following: max-num-checkpoint-not-improved 30, bucket-width 10, keep-last-params 1, batch-size 2048, max-num-epochs 100, weight-tying none, optimized-metric bleu, cache-metric bleu, source-factors-combine concat. These parameters were the same for baseline, parent and child models.

Models were trained on one Nvidia Tesla V100 GPU with 32 GB of memory on The HSE University Supercomputer Modeling Unit [9].

For the experiments part we first cleaned the data as described in Subsect. 3.2. Apart from generally pre-training the model on Chuvash data, we experimented with the following parameters: changing the amount of Khakas data we build vocabularies on (column 'x times Khakas'), maximum sequence length (column 'maxl'), bpe-dropout customization (column 'dropout').

Baseline. Creating a baseline consisted of training Sockeye model with the above parameters on Russian-Khakas data only, without any pre-training. Scores for this model are in Table 3.

Vocabularies for Transfer Learning. Since the vocabulary is shared between languages that change from pre-training to fine-tuning and considering the sizes of the corpora, we duplicated Khakas data several times to make it of compatible size to Chuvash data, so that Khakas tokens are equally represented in the vocabulary, as suggested in Knowles et al. [6]. We first experimented with making amounts of Khakas and Chuvash data equal by duplicating Khakas data 16 times, but it happened that making the Khakas part even larger - in our case, duplicating it 26 times - improves the results of the final model (Models 2 and 3 in Table 3).

BPE-Dropout Customization. We applied BPE-dropout as described in Provi-lkov et al. [14]. The idea behind BPE-dropout is that during BPE-tokenization some merges are dropped, making the resulting sequences more robust. In our experiments, we applied dropout source-side as suggested in Knowles et al. [6] and for some described experiments we also performed dropout operation 5 times on the same data, concatenating the results to make our dataset larger and source sequences more diverse, as it was also suggested in Knowles et al. [6]. This allowed us to improve results in terms of both BLEU and ChrF metrics in both translation directions (Models 3 and 4 in Table 3)

Maximum Sequence Length. The reason behind experimenting with maximum sequence length is that while 99% of the tokenized sequences in Russian-Chuvash data have lengths under 100 tokens, 99% of Russian-Khakas sequences have lengths under 75 tokens. This gave us the idea that adjusting this parameter may improve the model's results. In fact, changing this parameter from 100 (typical for parent data) to 75 (typical for child data) improved ChrF score for Ru-Kh direction and both BLEU and ChrF scores for Kh-Ru direction (Models 4 and 5 in Table 3)

Table 3. Translation results of the presented models

Md	x times Khakas	maxl	dropout	ru-cv		cv-ru		ru-kh		kh-ru	
				BLEU	ChrF	BLEU	ChrF	BLEU	ChrF	BLEU	ChrF
1	-	75	src-side	-	-	-	-	19	41.1	11.6	37.3
2	16	100	src-side	12.8	53.2	18.4	49.1	25.5	55.5	12.7	43.0
3	26	100	src-side	13	53.4	16	46.4	25.9	56	16.6	45.8
4	26	100	src&x5	13	53.4	20.3	49.4	**29.6**	55.5	17	46.8
5	26	75	src&x5	12.4	53.8	17.1	49.5	28.5	**57.6**	**17.2**	**47.1**

6   Results

6.1   Metrics

The results of the applied Transfer Learning approach to the task of machine translation of Khakas language showed results comparable to the only Russian-Khakas translator available on the Internet[10] (Table 4). The author of the model also used back-translation method to improve the model's performance. We didn't use back-translation in our study, but this is a prospect for future improvements. However, even without back-translation, our model outperforms the existing model in Ru-Kh direction by 7.26 BLEU points. Unfortunately, the weights of the above model are not publicly available so there is no possibility to check the metrics on the same test data.

---

[10] https://github.com/adeshkin/khakas_russian_translator.

Table 4. Comparing the best results to the existing translator

Model	Ru-Kh	Kh-Ru
github/adeshkin TL	17.2	17.78
github/adeshkin TL+BT	20.24	18.73
Our best model	**28.5**	17.2

Table 5 shows the metrics different researchers from Table 1 achieved in a similar setting on their language pairs comparing to our research. Forward direction means translating from shared language to changing language: for example, from Russian to Chuvash for parent model and from Russian to Khakas for child model. Backward direction means the opposite. In some papers the results are only presented for one direction.

As we see from Table 5, it is not obvious that the transfer learning approach works better when the source language or the target language is shared: sometimes final results are better in forward direction, sometimes in backward direction. Mirzakhalov et al. [11] suggest that the morphology of a language can play an important role in translation quality, i.e. the model can show poorer results when translating from analytic languages into morphologically-rich languages.

Table 5. Comparison of results of applying the transfer learning approach to different language pairs with our results

Article	Language pairs		Parent parallel, pairs	Child parallel, pairs	BLEU scores	
	parent	child			forward	backward
Valeev et al. [20]	ru-kk	ru-tt	5,000,000	324,000	25.45	28.59
Knowles et al. [6]	de-cs	de-hsb	22,000,000	60,000	55.5	54.1
Zoph et al. [22]	en-fr	en-es	53,000,000	2,500,000	-	31.0
Kocmi et Bojar [7]	en-fi	en-et	2,800,000	800,000	19.74	24.18
Kocmi et Bojar [7]	en-cs	en-et	40,100,000	800,000	20.41	-
Hujon et al. [4]	en-fr	en-kha	1,000,000	36,601	51.11	-
Our research	ru-chv	ru-kh	1,000,013	60,295	**28.5**	**17.2**

6.2   Case and Error Analysis

Case Analysis. Since we are not speakers of Khakas, we could visually evaluate the translations in Khakas-Russian direction. The visual analysis of the translations showed that the model is sometimes capable of translating full sentences correctly, even though mostly not with exact words as in reference translation, but similar in meaning. Here are some examples (S = Source sentence, R = Reference sentence, T = Translated sentence). The translation to English is in brackets:

S: Хайзы харындас-пичелер, туған-чағыннар, чыылызып, колхозха кірер-кірбестеңер алай аалдаң пасха чирзер тизердеңер чӧптес турғаннар.

R: Некоторые братья и сёстры, родственники, собравшись, обсуждали, нужно ли входить в колхоз или необходимо уехать из села в другое место. (Some brothers and sisters and relatives gathered and discussed whether it was necessary to join the collective farm or whether it was necessary to leave the village to another place.)

T: Кто-то из них родственники и сестёр, собравшись, уговаривали о том, как или не входить в колхоз, или переправляться из деревни в другое место. (Some of them relatives and sisters, having gathered, persuaded them how to either not enter the collective farm, or to cross from the village to another place.)

S: Сах андох тигір чалынни тӱскен, кӱгӱрт кӱзӱреен паза табыстар істіле тӱскеннер. Чир уғаа тың тітіреп сыххан. Кізілер чир ӱстӱнде чуртааннаң сығара, чир пір дее іди тітіребеен полған. Уғаа тың тітіреп сыххан! Уғаа илбек тітірес!

R: Тогда засверкали молнии, зазвучали голоса, загремел гром, и произошло сильное землетрясение. Такого мощного землетрясения не было за все время существования людей на земле! Столь великим было это землетрясение! (Then lightning flashed, voices sounded, thunder roared, and a strong earthquake occurred. There has never been such a powerful earthquake in the entire existence of people on earth! How great was this earthquake!)

T: И произошли молнии, громы и голоса, и сделалось великое землетрясение, какого не бывало с тех пор, как люди на земле. Такое землетрясение! Так великое!(And there were lightnings, thunders, and voices, and there was a great earthquake, such as has not happened since people were on the earth. Such an earthquake! So great!)

S: Солдаттар пазох сых килділер

R: Солдаты снова вышли. (The soldiers came out again.)

T: Солдаты опять вышли. (The soldiers came out again [another word for "again" was used but with a very similar meaning].)

S: Худай Пабабысха хачан даа сабланыс ползын. Аминь.

R: Богу же и Отцу нашему слава во веки веков! Аминь. (Glory to our God and Father forever and ever! Amen.)

T: Хвала Богу и Отцу во веки веков. Аминь.(Praise be to God and the Father forever and ever. Amen.)

Error Analysis. We divided the resulting sentences in several categories: correctly translated; those where one word is missing from the reference; those where one word is added that is not in the reference; those where some words are translated wrong, but the general sense of the sentence remains; those where some words are translated wrong or cases are used incorrectly, so the general sense of the sentence is lost. We would like to note that even the sentences from the last category always have most of the words translated correctly. Table 6 presents the statistics of these evaluation results.

Table 6. Translation statistics of the resulting translation model

Translation result	Frequency of occurrence, percent
Correctly translated sentences	20
One word missed	16
One extra word added	6
Some words translated wrong but the sense remains	28
Some words translated wrong and the sense is lost	32

Here are the examples of the model missing some words:

S: - Тооза тартын полбаза чаалазарға килем
R: - Если не сможет до конца натянуть, воевать приду. (If it can't pull it all the way, I'll come to fight.)
T: - Если не натянет - воевать приду. (If it doesn't pull, I'll come to fight.)

S: Пір хатап ӱс харындас аңнап партырлар
R: Один раз три брата на охоту пошли, говорят. (Once three brothers went hunting, they say.)
T: Однажды три брата на охоту поехали. (One day three brothers went hunting.)

S: Ана іди мині ӱгредерге ыспааннар.
R: Вот так меня не отправили на учёбу. (That's why I wasn't sent to study.)
T: Вот так меня не послали. (That's how they didn't send me.)

About 30% of the times the model translates some of the words correctly, but the general sense of the sentence is lost:

S: Хайдағ андағ ниме мағаа анда таап алды полар, хайди ла хатығлирға итче полар?
R: Интересно, что она там для меня выискала, какое наказание готовит? (I wonder what she found out for me there, what punishment she is preparing?)
T: Что это за вещь там нашла меня, наверно, как только хочет наказать? (What kind of thing was there that found me, probably as if it wanted to punish me?)

S: Хачан ізік хыринда прайзы апсыр сыхханнарындох, Алиса кем килгенін сизін салған.
R: и, соответственно, Алиса догадалась, кого увидит, как только у дверей

все зачихали. (and, accordingly, Alice guessed whom she would see, as soon as everyone sneezed at the door.)

T: Когда у дверей все появились в молодости, Алиса заметила, кто пришел. (When everyone appeared at the door in their youth, Alice noticed who had come.)

7   Conclusion and Future Work

The application of the transfer learning approach showed good results in improving the translation model's performance on low-resource Russian-Khakas data. Future studies may include the following experiments:

1. Adjusting the number of BPE merging operations the way so that the tokenization is as morphological as possible, meaning that the words are divided into morphemes closest to their morphological structure. The question of whether the tokenization must repeat the morphology is not fully discussed in the literature and is one path of future studies.
2. Another way to broaden the study will be to explore the effect of increasing the size of the child corpus. Unfortunately, the parallel data for the Russian-Khakas pair is very limited. But apart from the TIL corpus there are around 7 literary pieces that are translated from Russian to Khakas also provided by the Electronic corpus of the Khakas language, but they are not aligned, so they can not make a parallel corpus yet. Using alignment tools like lingtrain aligner[11] may help match parallel lines of such books, and they showed promising results when we started using this tool. There is a possibility we will expand the parallel corpus this way and test how much it affects the model's performance.
3. Trying back-translation as suggested by Sennrich et al. [15]. This method can be used in our case since there are books in Khakas available at the Electronic corpus of the Khakas language, and also web scraping of newspapers is possible.
4. Another way of experiments may include trying a different Turkic language for the parent model, for example, Kazakh.
5. Using the ensemble of the described approaches may help achieve better quality of the resulting model.

Acknowledgements. We thank Ekaterina Enikeeva, Alexander Antonov, Vera Maltseva, Mikhail Lebedev, Daria Galimzianova and Igor Philatov for discussion and feedback.

---

[11] https://github.com/averkij/lingtrain-aligner.

References

1. Cieri, C., Maxwell, M., Strassel, S., Tracey, J.: Selection criteria for low resource language programs. In: Proceedings of the Tenth International Conference on Language Resources and Evaluation (LREC 2016), pp. 4543–4549. European Language Resources Association (ELRA), Portorož, Slovenia (2016). https://aclanthology.org/L16-1720
2. Gadzhieva, N.: On the issue of classification of Turkic languages and dialects. Theoretical Foundations of the Classification of World Languages.–M.: Science, pp. 100–126 (1980). (in Russian)
3. Hieber, F., et al.: Sockeye 3: Fast Neural Machine Translation with PyTorch. arXiv preprint arXiv:2207.05851 (2022)
4. Hujon, A.V., Singh, T.D., Amitab, K.: Transfer learning based neural machine translation of English-Khasi on low-resource settings. Procedia Comput. Sci. **218**, 1–8 (2023)
5. Johanson, L., Csató, É.I.: The Turkic Languages. Routledge, Milton Park (2015)
6. Knowles, R., Larkin, S., Stewart, D., Littell, P.: NRC systems for low resource German-Upper Sorbian machine translation 2020: transfer learning with lexical modifications. In: Proceedings of the Fifth Conference on Machine Translation, pp. 1112–1122 (2020)
7. Kocmi, T., Bojar, O.: Trivial transfer learning for low-resource neural machine translation. In: Proceedings of the Third Conference on Machine Translation: Research Papers, Brussels, Belgium, pp. 244–252. Association for Computational Linguistics (2018). https://doi.org/10.18653/v1/W18-6325. https://aclanthology.org/W18-6325
8. Koehn, P., Knowles, R.: Six Challenges for Neural Machine Translation (2017)
9. Kostenetskiy, P., Chulkevich, R., Kozyrev, V.: HPC resources of the higher school of economics. In: Journal of Physics: Conference Series, vol. 1740, p. 012050. IOP Publishing (2021)
10. Macháček, M., Bojar, O.: Results of the WMT14 metrics shared task. In: Proceedings of the Ninth Workshop on Statistical Machine Translation, pp. 293–301 (2014)
11. Mirzakhalov, J., et al.: A large-scale study of machine translation in Turkic languages. In: Proceedings of the 2021 Conference on Empirical Methods in Natural Language Processing, pp. 5876–5890 (2021)
12. Nguyen, T.Q., Chiang, D.: Transfer learning across low-resource, related languages for neural machine translation. arXiv preprint arXiv:1708.09803 (2017)
13. Popović, M.: chrF: character n-gram F-score for automatic MT evaluation. In: Proceedings of the Tenth Workshop on Statistical Machine Translation, pp. 392–395 (2015)
14. Provilkov, I., Emelianenko, D., Voita, E.: BPE-dropout: simple and effective subword regularization. arXiv preprint arXiv:1910.13267 (2019)
15. Sennrich, R., Haddow, B., Birch, A.: Improving Neural Machine Translation Models with Monolingual Data (2016)
16. Sennrich, R., Haddow, B., Birch, A.: Neural Machine Translation of Rare Words with Subword Units (2016)
17. Singh, A.K.: Natural language processing for less privileged languages: where do we come from? Where are we going? In: Proceedings of the IJCNLP-08 Workshop on NLP for Less Privileged Languages (2008). https://aclanthology.org/I08-3004

18. Tiedemann, J.: Parallel data, tools and interfaces in OPUS. In: Proceedings of the Eighth International Conference on Language Resources and Evaluation (LREC 2012), pp. 2214–2218. European Language Resources Association (ELRA), Istanbul, Turkey (2012)
19. Tsvetkov, Y.: Opportunities and challenges in working with low-resource languages. Slides Part-1 (2017)
20. Valeev, A., Gibadullin, I., Khusainova, A., Khan, A.: Application of Low-resource Machine Translation Techniques to Russian-Tatar Language Pair (2019)
21. Vaswani, A., et al.: Attention is all you need. In: Advances in Neural Information Processing Systems, vol. 30 (2017)
22. Zoph, B., Yuret, D., May, J., Knight, K.: Transfer learning for low-resource neural machine translation. In: Proceedings of the 2016 Conference on Empirical Methods in Natural Language Processing, Austin, Texas, pp. 1568–1575. Association for Computational Linguistics (2016). https://doi.org/10.18653/v1/D16-1163. https://aclanthology.org/D16-1163

Automatic Aspect Extraction
from Scientific Texts

Anna Marshalova[1(✉)], Elena Bruches[1,2], and Tatiana Batura[2]

[1] Novosibirsk State University, Novosibirsk, Russia
a.marshalova@alumni.nsu.ru
[2] A. P. Ershov Institute of Informatics Systems, Novosibirsk, Russia

Abstract. Being able to extract from scientific papers their main points, key insights, and other important information, referred to here as aspects, might facilitate the process of conducting a scientific literature review. Therefore, the aim of our research is to create a tool for automatic aspect extraction from Russian-language scientific texts of any domain. In this paper, we present a cross-domain dataset of scientific texts in Russian, annotated with such aspects as Task, Contribution, Method, and Conclusion, as well as a baseline algorithm for aspect extraction, based on the multilingual BERT model fine-tuned on our data. We show that there are some differences in aspect representation in different domains, but even though our model was trained on a limited number of scientific domains, it is still able to generalize to new domains, as was proved by cross-domain experiments. The code and the dataset are available at https://github.com/anna-marshalova/automatic-aspect-extraction-from-scientific-texts.

Keywords: Aspect extraction · Scientific information extraction · Dataset annotation · Sequence labelling · BERT fine-tuning

1 Introduction

As the number of published research papers increases, it becomes more and more challenging to keep abreast of them. Therefore, there is a growing need for tools that can automatically extract relevant information from scientific texts.

Unfortunately, although Russian is among the languages most commonly used in science [33], there is only a sparse amount of tools for automatic information extraction from Russian scientific texts. To make matters worse, most of them focus on certain domains, e.g. medicine [24,31] or information technologies [2,7].

To address this, our study aims to create a tool capable of extracting the main points of the research, which we refer to as aspects, from Russian scientific texts of any domain. It could be used to select, summarize, and systematize papers and is likely to make scientific literature reviewing more efficient.

The main contributions of this paper are summarized as follows:

© The Author(s), under exclusive license to Springer Nature Switzerland AG 2024
D. I. Ignatov et al. (Eds.): AIST 2023, CCIS 1905, pp. 67–80, 2024.
https://doi.org/10.1007/978-3-031-67008-4_6

1. we present a dataset containing abstracts for Russian-language scientific papers on 10 scientific domains annotated with 4 types of aspects: Task, Contribution, Method, and Conclusion;
2. we provide an algorithm for automatic aspect extraction.

The dataset as well as the algorithm implementation are publicly available and may be of some use to other researchers.

The rest of the paper is organized as follows. Section 2 gives some background information about the task of aspect extraction, points out the factors influencing the selection of aspects to identify, and reviews datasets and methods for both Russian and English that could be used for this task or similar tasks. Section 3 describes our dataset and explains why the above-mentioned 4 aspects were chosen in our work. Section 4 outlines the details of our approach, based on using BERT, fine-tuned on our data. Section 5 describes our experiments and presents their results. Finally, Sect. 6 discusses the principal findings and limitations of the study and suggests the broader impact of the work.

2   Related Work

2.1   Background

Pieces of information extracted from scientific papers do not have a conventional name. They are referred to as key-insights [26], core scientific concepts, scientific artifacts [15], or scientific discourse categories [29]. Like [14] and [18] we will call them *aspects* of a paper.

Moreover, there is currently no consensus on which information should be considered important and extracted from papers. We have discovered that the sets of aspects chosen by different researchers largely depend on the task and the data considered in their work.

Firstly, in most related works, aspect extraction is solved as a sentence classification or named-entity recognition (NER) task. This largely affects the aspects considered in the work. For example, studies devoted to NER tend to consider aspects expressed in short phrases, e.g. Method and Tool. In studies on sentence classification, higher-level aspects, such as Background, Goal and Conclusion, are usually considered. In order to take into account both types of aspects, we propose to identify as aspects phrases of any length within a sentence. A similar approach to text unit identification is used in the task of argumentative zoning [34], which, however, focuses on the argumentative structure of texts, which also influences the list of extracted types of information.

Secondly, the set of aspects depends on the size of the texts used. The selected texts can be both full papers and their abstracts. In our work, the latter are used, but even though abstracts reflect main aspects of papers, some information might still be missed, e.g. information about related work [29] and further work [37]. As a result, sets of aspects extracted from full paper texts usually include a larger number of aspects.

Finally, the aspects identified in papers depend on their scientific domain, as there are a number of domain-specific aspects. For example, papers on evidence-based medicine are traditionally structured according to the PICO criterion[1], which strongly influences the sentence categories identified in such texts [6,20]. [19] studies papers on machine learning, so such entities as Dataset and Metric are proposed to be extracted. In this regard, the development of a set of aspects that can be used for different domains is rather challenging.

2.2   Datasets

The dataset presented in this paper has no direct analogues, as currently there are no datasets designed specifically for aspect extraction from Russian scientific texts of different domains. Let us give a view of its closest analogues.

For Russian, there are datasets for NER and relation extraction from texts on various topics. These may include scientific texts, such as abstracts of papers [7,24], as well as texts of other genres, for example, medical records [10,11,21,27] or publications and forum entries from domain-specific social media [5,25,32].

Datasets for information extraction from English-language scientific texts include the ones for sentence classification in medical abstracts [6,8,18,20], computer science abstracts [13], papers on computational linguistics [14,37], and computer graphics [29]. There are also datasets for NER and relation extraction from English scientific texts on computer science, materials science, physics [1], and machine learning [19]. Finally, there are datasets for argumentative zoning, containing full papers [34].

2.3   Methods

Rule-based methods of information extraction from scientific texts usually involve extracting text segments containing certain keywords and n-grams or lexico-grammatical patterns [2,14]. The main disadvantage of rule-based approaches is the effort required for the creation of rules and templates, which are also usually domain-specific.

The most popular classical machine learning methods used for this task include Conditional Random Fields (CRF) [15,20,32] and Support Vector Machines (SVM) [6,29,31], trained on manually constructed features.

As for deep-learning methods, some researchers use convolutional neural networks and recurrent neural networks with pretrained static embeddings such as GloVe and word2vec [8,10,13,21].

However, the state-of-the-art method for information extraction from scientific texts is based on pretrained masked-language models such as BERT [30,36] and SciBERT [18,19,37]. Both BERT [9] and SciBERT [3] have Russian analogues: ruBERT [23] and ruSciBERT [12], the former being used by some researchers for NER in biomedical abstracts [24] and texts on cybersecurity [35].

---

[1] Population/Problem (P), Intervention (I), Comparison (C) and Outcome (O).

3   Dataset Creation

The created dataset contains 200 abstracts to Russian-language scientific papers of 10 domains: Medicine and Biology, History and Philology, Journalism, Law, Linguistics, Mathematics, Pedagogy, Physics, Psychology, and Computer Science.

In these texts, we chose to identify 4 aspects: the task solved in the research, the authors' contribution, the employed methods, and the conclusion of the study. Examples[2] for each of the aspects are shown in Table 1.

Table 1. The list of aspects identified in our dataset with examples.

Aspect	Example
Task	*The paper is devoted to the task of <Task> improving the performance of small information systems </Task>.*
Contribution	*<Contrib> ancient settlements, fortifications and protective walls of the Margaritovka basin have been studied </Contrib>.*
Method	*Measurement of electrical and viscoelastic characteristics of erythrocytes was carried out by the method of <Method> dielectrophoresis </Method>.*
Conclusion	*The author comes to the conclusion that <Conc> bilateral unequal political alliances reflect local specifics </Conc>.*

As shown in the example of annotated text in Fig. 3a, aspects can be nested, i.e. an aspect can be a part of another aspect.

The annotation was performed by two of the authors of this paper: a bachelor student of computational linguistics and a senior researcher, guided by the annotation instructions, which were compiled according to the semantic and syntactic features of the texts. All disagreements were discussed and resolved jointly. The inter-annotator agreement was evaluated with a strict F1 score [17]:

$$F1 = \frac{2a}{2a + b + c}. \tag{1}$$

In this equation, a stands for aspects that were identified by both annotators, while b and c are aspects identified by only one of the annotators respectively. Macro-averaged over all 4 aspects, the score measured 0.92.

As a result, 836 aspects were identified, with half of them being the Contribution aspect (see Fig. 1). This is probably due to the fact that our dataset consists of abstracts, which are written to describe the authors' contribution.

---

[2] The texts are originally in Russian and were translated into English only to provide examples in the paper.

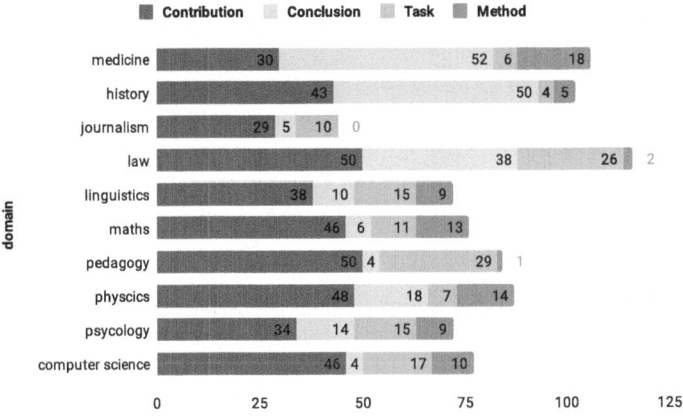

Fig. 1. Distribution of aspect types over different domains in our dataset.

While length of one text is 115 tokens on average, it fluctuates across different domains from 50 to 177 tokens. Table 2 shows that average text length positively correlates with the number of aspects identified in each domain.

Table 2. Average text length (tokens) in each domain compared to the number of aspects identified in these domains.

Domain	Average Text Length	Aspects Identified
Journalism	50	44
Psychology	86	72
Linguistics	90	72
Pedagogy	93	84
Mathematics	95	76
Computer Science	106	77
Physics	144	87
Medicine	146	106
Law	159	116
History	177	102

The number of aspects in one text varies from 0 to 13, and an average text contains 4 aspects. On average, one aspect consists of 12 tokens, but the length of an aspect largely depends on its type: Method and Task are similar to entities and are expressed in rather short phrases (3–5 tokens), whereas Contribution and Conclusion are expressed in whole sentences or clauses (12–16 tokens).

Contribution is sufficiently represented in all domains, while Conclusion prevails in medical texts since most of them describe the results of clinical studies.

In addition, this aspect is often found in abstracts of papers on History, which often contain analysis of archaeological discoveries. Papers on Pedagogy and Mathematics are more practically oriented (they tend to propose new methods rather than describe experiments and observations); therefore, their abstracts contain fewer conclusions.

Methods are most often mentioned in texts on the natural and exact sciences: Medicine, Mathematics, Physics, and Computer science; these might also include Linguistics and Psychology.

As for the Task aspect, we discovered that in some domains, especially the humanities, we are not talking about tasks, but rather about problematic issues or objects of research (e.g. *problems of dialogue, self-identification issues* or even *prepositions that function in journalistic discourse*). It was proposed to attribute them to the Task aspect as well.

4 Automatic Aspect Extraction

The pipeline of automatic aspect extraction performed by the presented tool is the following:

1. **Text pre-processing.** At this step, we tokenize the texts using the Natural Language Toolkit (NLTK) library [4] (v3.6.2).
2. **Aspect extraction processing.** At this step, we assign each token to corresponding aspects using the model described in Sect. 4.1.
3. **Aspect post-processing.** At this step, we use some heuristics to improve aspect boundary detection. These involve, for example, removing one-word aspects and excluding punctuation and conjunctions from the end and the beginning of an aspect. After that, tokens assigned to one aspect mention are united into spans. Aspects expressed by nominal phrases are put in the nominative case using spaCy [16] (v3.5) and pymorphy2 [22] (v0.9.1). Finally, the spans are detokenized, and outputted in the format shown in Fig. 3c.

4.1 Model

We solve aspect extraction as a sequence labeling task. We do not employ BIO encoding, which is commonly used for sequence labelling, as in our annotation, aspects of the same type never follow each other in a row. They are separated by at least a comma or conjunction, so there is no need to distinguish between B-tags and I-tags, although we use O-tags for tokens that do not belong to any aspect. To take nested aspects into account, we use multilabel classification: a token can be labeled with one aspect or two aspects, if one of them is nested. Hence, the model is trained to label each token in a text with one or two of the four aspects (Task, Contribution, Method, Conclusion) or with the tag 'O'.

We use the model architecture shown in Fig. 2. Namely, we use a pretrained BERT-like model fine-tuned on our dataset as an embedding layer and a fully-connected linear layer with a sigmoid activation function as a classifier. For each token, the model outputs an array of 4 values between 0 and 1, which can be interpreted as probabilities of assigning the token to each of the aspects. The token is labeled with up to two most probable aspects, whose probabilities are greater than the threshold of 0.5. If all of the values are less than the threshold, the token is labeled with the 'O' tag.

The model is trained with a batch size of 16, using binary cross-entropy as the loss function and Adam as the optimizer. The learning rate is set to 10^{-5}.

We use multilingual BERT as the pretrained BERT-like model to fine-tune it on our data. This choice, as well as the choice of the model's architecture, is based on the results of our experiments, described in Sect. 5.2.

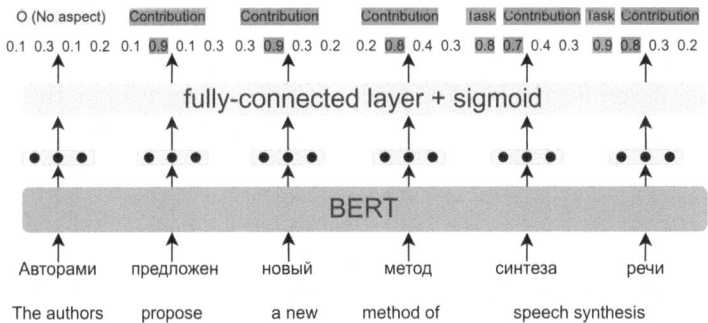

Fig. 2. The scheme of the proposed model.

5 Experiments and Results

5.1 Evaluation Method

For evaluation, we use precision, recall, and F1 for individual tokens, as well as Exact Match Ratio (EMR) of aspects—the ratio of the number of correctly extracted aspects to the total number of aspects, which demonstrates the quality of aspect boundary detection. We calculate these metrics for each aspect and use macro-averaging to get the final scores.

To make evaluation more consistent, we employ 5-fold cross-validation. For each experiment, we provide mean and standard deviation (STD) of metrics across the folds. Finally, to compare different models' performance, randomization tests [28] are conducted.

5.2 Experiments with Models

To find which model would be the most suitable for our task, we conducted a number of experiments, using BERT-like models compatible with Russian:

multilingual BERT (mBERT) [9], ruBERT [23], ruBERT-tiny2[3], and ruSciBERT [12]. The multilingual model performed best on our task (Table 3a). However, the difference between the performance of mBERT and ruBERT turned out to be statistically insignificant (p = 0.3). Therefore, although we chose mBERT as the baseline model for further experiments, ruBERT could have also been chosen.

A series of experiments were also conducted with the multilingual BERT. These involve using CRF as a classifier, adding Bidirectional LSTM (BiLSTM) layer between the embedding and classification layers, and freezing the weights of the embedding layer to train only the classifier. However, these modifications did not outperform the baseline model architecture with a fully-connected linear layer as a classifier (Table 3b). According to the randomization test, there is no statistically significant difference between the performance of the baseline model and the model with the BiLSTM layer (p = 0.28). However, training and inference are faster for the baseline model, so we consider it to be the best one.

Table 3. Comparison of Macro F1 scores obtained with different models.

(a) Macro F1 scores obtained with different pretrained models.

Model	Macro F1	
	Mean	STD
mBERT	**0.566**	0.035
ruBERT	0.546	0.071
rubert-tiny2	0.449	0.05
ruSciBERT	0.287	0.066

(b) Macro F1 scores obtained with different model architectures.

Model	Macro F1	
	Mean	STD
mBERT (baseline)	**0.566**	0.035
mBERT + CRF	0.278	0.027
mBERT + BiLSTM	0.54	0.023
mBERT (frozen weights)	0.381	0.04

Having compared the predictions of the models, we noticed that:

1. RuSciBERT extracts spans of aspects that are too long, which leads to low precision.
2. Most of the models extract nested aspects more often than needed and more often than the best model, which might mean that these models are less sure about their predictions, so they tend to assign tokens to two aspects. This also lowers the precision.

These observations partially explain the obtained results. However, for a more detailed analysis of how the models work and to find out the reasons why mBERT outperforms specialised monolingual models such as ruSciBERT, additional experiments are needed, which we plan to perform in the future.

Table 4 shows the metrics for the best model, which has the architecture shown in Fig. 2 and uses multilingual BERT as an embedding layer.

The best extracted aspect is Contribution, which can be explained by its frequency in the dataset. Exact match ratio measures lower than other metrics, indicating problems with aspect boundary detection.

---

[3] https://huggingface.co/cointegrated/rubert-tiny2.

Table 4. Metrics for the best model after post-processing with heuristics.

Tag	Precision		Recall		F1		EMR	
	Mean	STD	Mean	STD	Mean	STD	Mean	STD
O (no aspect)	**0.735**	0.019	0.602	0.101	0.658	0.055	–	–
Task	0.395	0.15	0.439	0.172	0.407	0.143	0.222	0.125
Contrib	0.683	0.104	**0.827**	0.11	**0.737**	0.047	**0.53**	0.096
Method	0.427	0.105	0.464	0.188	0.436	0.127	0.225	0.155
Conc	0.58	0.132	0.617	0.066	0.593	0.094	0.225	0.09
Macro	0.564	0.043	0.59	0.035	0.566	0.035	0.317	0.049

5.3   Cross Domain Experiments

Our corpus contains texts on only a limited set of topics, while compiling a corpus that would cover all domains is a rather complicated task. We conducted 10 experiments in order to find out how our model performs on unseen domains. Within each of them, the texts of one of the domains were used to test a model trained on the texts of the rest 9 domains. For these experiments, 5-fold cross-validation was used to split data into train and validation.

The results of the experiments, shown in Table 5, are comparable with the best model's performance and confirm the ability of our model to generalize to various subject areas.

Table 5. F1 scores for each of the cross-domain experiments. For each experiment, we used texts from one of the domains as a test set for a model trained on 9 other domains.

Domain	Macro F1	
	Mean	STD
Biology and Medicine	0.471	0.038
Computer Science	0.543	0.053
History and Philology	0.45	0.015
Journalism	0.526	0.042
Law	0.473	0.02
Linguistics	0.472	0.013
Math	0.513	0.028
Pedagogy	0.529	0.066
Physics	0.536	0.038
Psychology	0.51	0.068

5.4 Analysis of Automatic Annotation

Task Contrib Method Conc

В работе рассматривается задача стабилизации робота типа акробот « Acrobot » методом обратной связи линеаризацией. Рассматриваемая модель акробота представляет собой роботизированную конструкцию двухзвенного робота с единственный приводом в вертикальной плоскости (локте). Рассматриваемая модель представляет интерес, определяемый возможностью аналитического исследования динамических уравнений с незначительным числом параметров. Получено решение задачи стабилизации конструкции акробота линейно-квадратичными регуляторами. Однако найденные решения не обеспечивают робастность параметров, что затрудняет их практическое использование. Приведены результаты вычислительных экспериментов.

(a) A manually annotated text.

Task Contrib Method Conc

The paper deals with the problem of stabilization of an Acrobot-type robot using the feedback method with partial linearization. The considered Acrobot model is a robotic two-link robot with a single drive in the vertical plane (elbow). The considered model is of interest, determined by the possibility of an analytical study of dynamic equations with a small number of parameters. The solution was provided to the problem of the stabilization of the structure with linear-quadratic regulators. However, the proposed solutions did not allow obtaining robust values of the ranges of parameters that provide the specified quality control. The results of numerical experiments are presented.

(b) A manually annotated text, translated into English.

ЗАДАЧИ
1. Задача стабилизации робота типа акробот «Acrobot»
2. Стабилизация конструкции акробота
МЕТОДЫ
1. Метод обратной связи линеаризацией
2. Линейно-квадратичные регуляторы
ВКЛАД
1. Получено решение задачи стабилизации конструкции акробота линейно-квадратичными регуляторами
2. Приведены результаты вычислительных экспериментов
ВЫВОД
1. Не обеспечивают робастность

(c) Aspects extracted from the text.

TASKS
1. The problem of stabilization of an Acrobot-type robot
2. Stabilization of the structure
METHODS
1. The feedback method with partial linearization
2. Linear-quadratic regulators
CONTRIBUTION
1. The solution was provided to the problem of the stabilization of the structure with linear-quadratic regulators
2. The results of numerical experiments are presented
CONCLUSION
1. Did not allow obtaining robust values

(d) Aspects extracted from the text, translated into English.

Fig. 3. Example of automatic aspect extraction compared with manual annotation.

The metrics represent how our model's predictions coincide with manual annotation. However, although the annotation was performed according to the instructions, it still has some controversial points, which means some other variants of annotation might also be possible. Having analyzed examples of annotation produced by the model, we have noticed that some of them are still viable even if they differ from those in the dataset. For example, as shown in Fig. 3c, the model sometimes includes the word 'задача' (the problem of) into the Task aspect. We tend to exclude such phrases in the manual annotation, but the result produced by the model can hardly be considered wrong.

Another example is the extracted conclusion shown in Fig. 3c. It is shorter than expected and does not fully disclose the conclusion of the original text, but still represents its main point. This can probably be considered a minor mistake.

We believe that in our case, general plausibility of results is more important than total compliance with the annotation, but at the same time, it is more difficult to measure due to its being rather subjective. In the future, we plan to explore ways to estimate it more formally.

6 Discussion

In this paper, we propose a tool for automatic aspect extraction from scientific texts in Russian and a cross-domain dataset for this task. The findings of this study have to be seen in light of the following limitations:

1. **Small size of the dataset.** The created dataset contains only 200 texts due to the rather small number of annotators and the labor intensiveness of the annotation process. However, even such an amount of data was sufficient to obtain primary results. Moreover, in the future, we plan to extend the dataset with more annotated texts, probably by using semi-supervised methods.
2. **Aspect boundary detection.** Exact match ratio is lower than the metrics for individual tokens, which means that we have to continue working on improving aspect boundary detection, e.g. by developing new heuristics.

Nevertheless, we hope that the conducted research contributed to the existing resources and methods of information extraction from scientific texts, as the proposed algorithm and dataset are the first to be cross-domain and Russian-language-oriented at the same time.

The developed tool can potentially be used to select, summarize, and systematize scientific publications, making academic literature reviewing more convenient and efficient, while the dataset can be used for further experiments in the task of aspect extraction. Both are publicly available and, we hope, will be useful for other researchers.

7 Conclusion

This work presents a cross-domain dataset of Russian-language scientific texts with manual aspect annotation as well as a tool for automatic aspect extraction from Russian scientific texts of any domain.

The created dataset contains 200 texts annotated with 4 aspects: Task, Contribution, Method, and Conclusion, which were chosen according to certain features of the texts in the dataset.

To implement the algorithm of automatic aspect extraction, we fine-tuned a pretrained masked language model on our data. The best results (Macro F1 = 0.57) were obtained with multilingual BERT combined with rule-based heuristics for post-processing. The model's robustness to different scientific domains was proven by cross-domain experiments.

References

1. Augenstein, I., Das, M., Riedel, S., Vikraman, L., McCallum, A.: SemEval 2017 task 10: ScienceIE - extracting keyphrases and relations from scientific publications. In: Proceedings of the 11th International Workshop on Semantic Evaluation (SemEval-2017), Vancouver, Canada, pp. 546–555. Association for Computational Linguistics (2017)
2. Batura, T., Bakiyeva, A., Charintseva, M.: A method for automatic text summarization based on rhetorical analysis and topic modeling. Int. J. Comput. **19**(1), 118–127 (2020)
3. Beltagy, I., Lo, K., Cohan, A.: SciBERT: a pretrained language model for scientific text. In: Proceedings of the 2019 Conference on Empirical Methods in Natural Language Processing and the 9th International Joint Conference on Natural Language Processing (EMNLP-IJCNLP), Hong Kong, China, pp. 3615–3620. Association for Computational Linguistics (2019)

4. Bird, S., Klein, E., Loper, E.: Natural Language Processing with Python: Analyzing Text with the Natural Language Toolkit. O'Reilly Media, Inc. (2009)
5. Blinov, P., Reshetnikova, A., Nesterov, A., Zubkova, G., Kokh, V.: Rumedbench: a Russian medical language understanding benchmark. In: Michalowski, M., Abidi, S.S.R., Abidi, S. (eds) AIME 2022. LNCS, vol. 13263, pp. 383–392. Springer, Cham (2022). https://doi.org/10.1007/978-3-031-09342-5_38
6. Boudin, F., Nie, J.Y., Bartlett, J.C., Grad, R., Pluye, P., Dawes, M.: Combining classifiers for robust PICO element detection. BMC Med. Inform. Decis. Mak. **10**(1), 1–6 (2010)
7. Bruches, E., Pauls, A., Batura, T., Isachenko, V.: Entity recognition and relation extraction from scientific and technical texts in Russian. In: 2020 Science and Artificial Intelligence Conference (SAI ence), pp. 41–45. IEEE (2020)
8. Dernoncourt, F., Lee, J.Y.: PubMed 200k RCT: a dataset for sequential sentence classification in medical abstracts. In: Proceedings of the Eighth International Joint Conference on Natural Language Processing (Volume 2: Short Papers). pp. 308–313, Taipei, Taiwan. Asian Federation of Natural Language Processing (2017)
9. Devlin, J., Chang, M.W., Lee, K., Toutanova, K.: BERT: pre-training of deep bidirectional transformers for language understanding. In: Proceedings of the 2019 Conference of the North American Chapter of the Association for Computational Linguistics: Human Language Technologies, Volume 1 (Long and Short Papers), Minneapolis, Minnesota, pp. 4171–4186. Association for Computational Linguistics (2019)
10. Dudchenko, A., Dudchenko, P., Ganzinger, M., Kopanitsa, G.D.: Extraction from medical records. In: pHealth, pp. 62–67 (2019)
11. Gavrilov, D., Gusev, A., Korsakov, I., Novitsky, R., Serova, L.: Feature extraction method from electronic health records in Russia. In: Conference of Open Innovations Association, FRUCT, pp. 497–500. FRUCT Oy (2020)
12. Gerasimenko, N., Chernyavsky, A., Nikiforova, M.: ruSciBERT: a transformer language model for obtaining semantic embeddings of scientific texts in Russian. Doklady Mathematics **106**, S95–S96 (2022)
13. Gonçalves, S., Cortez, P., Moro, S.: A deep learning classifier for sentence classification in biomedical and computer science abstracts. Neural Comput. Appl. **32**, 6793–6807 (2020)
14. Gupta, S., Manning, C.: Analyzing the dynamics of research by extracting key aspects of scientific papers. In: Proceedings of 5th International Joint Conference on Natural Language Processing. Asian Federation of Natural Language Processing, Chiang Mai, Thailand, pp. 1–9 (2011)
15. Hassanzadeh, H., Groza, T., Hunter, J.: Identifying scientific artefacts in biomedical literature: the evidence based medicine use case. J. Biomed. Inform. **49**, 159–170 (2014)
16. Honnibal, M., Montani, I.: spaCy 2: natural language understanding with Bloom embeddings, convolutional neural networks and incremental parsing (2017), to appear
17. Hripcsak, G., Rothschild, A.S.: Agreement, the f-measure, and reliability in information retrieval. J. Am. Med. Inform. Assoc. **12**(3), 296–298 (2005)
18. Huang, T.H.K., Huang, C.Y., Ding, C.K.C., Hsu, Y.C., Giles, C.L.: CODA-19: using a non-expert crowd to annotate research aspects on 10,000+ abstracts in the COVID-19 open research dataset. In: Proceedings of the 1st Workshop on NLP for COVID-19 at ACL 2020. Association for Computational Linguistics (2020)

19. Jain, S., van Zuylen, M., Hajishirzi, H., Beltagy, I.: SciREX: a challenge dataset for document-level information extraction. In: Proceedings of the 58th Annual Meeting of the Association for Computational Linguistics, pp. 7506–7516. Association for Computational Linguistics (2020)

20. Kim, S.N., Martinez, D., Cavedon, L., Yencken, L.: Automatic classification of sentences to support evidence based medicine. BMC Bioinformatics, vol. 12, pp. 1–10. BioMed Central (2011)

21. Kivotova, E., Maksudov, B., Kuleev, R., Ibragimov, B.: Extracting clinical information from chest x-ray reports: a case study for Russian language. In: 2020 International Conference Nonlinearity, Information and Robotics (NIR), pp. 1–6. IEEE (2020)

22. Korobov, M.: Morphological analyzer and generator for Russian and Ukrainian languages. In: Khachay, M.Y., Konstantinova, N., Panchenko, A., Ignatov, D.I., Labunets, V.G. (eds.) AIST 2015. CCIS, vol. 542, pp. 320–332. Springer, Cham (2015). https://doi.org/10.1007/978-3-319-26123-2_31

23. Kuratov, Y., Arkhipov, M.: Adaptation of deep bidirectional multilingual transformers for Russian language. In: Computational Linguistics and Intellectual Technologies Papers from the Annual International Conference «Dialog», Moscow, May 29 – June 1, 2019, Proceedings, pp. 333–339 (2019)

24. Loukachevitch, N., et al.: Nerel-bio: a dataset of biomedical abstracts annotated with nested named entities. Bioinformatics **39**(4), btad161 (2023)

25. Miftahutdinov, Z., Alimova, I., Tutubalina, E.: On biomedical named entity recognition: experiments in interlingual transfer for clinical and social media texts. In: Jose, J.M., et al. (eds.) ECIR 2020. LNCS, vol. 12036, pp. 281–288. Springer, Cham (2020). https://doi.org/10.1007/978-3-030-45442-5_35

26. Nasar, Z., Jaffry, S.W., Malik, M.K.: Information extraction from scientific articles: a survey. Scientometrics **117**, 1931–1990 (2018)

27. Nesterov, A., et al.: RuCCoN: clinical concept normalization in Russian. In: Findings of the Association for Computational Linguistics: ACL 2022, Dublin, Ireland, pp. 239–245. Association for Computational Linguistics (2022)

28. Noreen, E.W.: Computer-Intensive Methods for Testing Hypotheses. Wiley, New York (1989)

29. Ronzano, F., Saggion, H.: Dr. inventor framework: extracting structured information from scientific publications. In: Japkowicz, N., Matwin, S. (eds.) DS 2015. LNCS (LNAI), vol. 9356, pp. 209–220. Springer, Cham (2015). https://doi.org/10.1007/978-3-319-24282-8_18

30. Shang, X., Ma, Q., Lin, Z., Yan, J., Chen, Z.: A span-based dynamic local attention model for sequential sentence classification. In: Proceedings of the 59th Annual Meeting of the Association for Computational Linguistics and the 11th International Joint Conference on Natural Language Processing (Volume 2: Short Papers), pp. 198–203. Association for Computational Linguistics (2021)

31. Shelmanov, A., Smirnov, I., Vishneva, E.: Information extraction from clinical texts in Russian. In: Computational Linguistics and Intellectual Technologies Papers from the Annual International Conference «Dialog», Moscow, May 27–30, 2015, Proceedings, pp. 560–572 (2015)

32. Sirotina, A., Loukachevitch, N.: Named entity recognition in information security domain for Russian. In: Proceedings of the International Conference on Recent Advances in Natural Language Processing (RANLP 2019), pp. 1114–1120 (2019)

33. Skvortsova, I.A.: Russian language among the world languages. In: VIII Vinogradov Conference, pp. 171–173 (2022)

34. Teufel, S., et al.: Argumentative zoning: information extraction from scientific text. Ph.D. thesis, Citeseer (1999)
35. Tikhomirov, M., Loukachevitch, N., Sirotina, A., Dobrov, B.: Using BERT and augmentation in named entity recognition for cybersecurity domain. In: Métais, E., Meziane, F., Horacek, H., Cimiano, P. (eds.) NLDB 2020. LNCS, vol. 12089, pp. 16–24. Springer, Cham (2020). https://doi.org/10.1007/978-3-030-51310-8_2
36. Yamada, K., Hirao, T., Sasano, R., Takeda, K., Nagata, M.: Sequential span classification with neural semi-Markov CRFs for biomedical abstracts. In: Findings of the Association for Computational Linguistics: EMNLP 2020, pp. 871–877. Association for Computational Linguistics, Online (2020)
37. Zhang, C., Xiang, Y., Hao, W., Li, Z., Qian, Y., Wang, Y.: Automatic recognition and classification of future work sentences from academic articles in a specific domain. J. Informet. **17**(1), 101373 (2023)

User Review Summarization in Russian

Artem Prisiazhniuk[1]([✉]) [iD] and Valentin Malykh[2]

[1] National Research University Higher School of Economics, Moscow, Russia
artemprisiazhniuk@gmail.com
[2] ITMO University, Moscow, Russia
valentin.malykh@phystech.edu

Abstract. The prevalence of online shopping has made it the foremost method for purchasing various goods. The online customer reviews play a crucial role in providing valuable insights into customers' interests and knowledge of the product.

Recent articles have focused on extracting sentiment and aspect information from reviews and incorporating it into the review summarization process. Unfortunately, the majority of these methods have primarily worked with the English language and have been evaluated exclusively on English language datasets.

Based on recent advances on the topic, this work researches review summarization methods in application to the Russian language. We collect a corpus of Russian reviews and evaluate the models on manually created summaries and summaries from an existing English language dataset in the same domain. Specifically, the best ROUGE-1/2/L scores of 33.87/2.87/12.20 on the collected data are achieved by fine-tuned Ace-Sum and the best ROUGE-1/2/L scores of 30.67/6.88/18.55 on Space dataset are achieved by PlanSum. Additionally, we investigate the impact of fine-tuning these models on a small subset of data entries from different dataset.

Keywords: Deep Learning · Natural Language Processing · Opinion Summarization · Review Summarization · Unsupervised Learning

1 Introduction

Opinion summarization refers to the task of extracting and representing salient opinions expressed within human-written textual data consisting of multiple statements (e.g. product reviews). Its main goal is to transform a collection of reviews of an entity into a concise summary which consists of aggregated points of view of the reviewers and emphasize general specifics of the item.

The researchers in the field explore both supervised and unsupervised settings. While supervised setting is widely used due to its effectiveness, it requires golden summaries for training which can be extremely difficult and resource intensive to produce.

Moreover, the majority of existing studies focuses solely on the English data, neglecting non-English languages that lack publicly accessible resources. This

© The Author(s), under exclusive license to Springer Nature Switzerland AG 2024
D. I. Ignatov et al. (Eds.): AIST 2023, CCIS 1905, pp. 81–95, 2024.
https://doi.org/10.1007/978-3-031-67008-4_7

hampers the opportunities for broadening our understanding of cross-linguistic properties of opinion summarization. Therefore, exploring ways to extend experiments beyond the English language remains a crucial task for the research.

This work explores best-performing weakly-supervised and unsupervised methods of opinion summarization. We collect the dataset of the Russian language reviews and manually create golden summaries for every entity. We research the possible options of preprocessing and seed words extraction for weakly-supervised setting. Then we adapt the methods for work with the Russian language reviews, train them on the collected corpus and compare them on both manually labeled examples and translated test splits from the Space dataset [3].

We make our code and data publicly available.[1]

2   Related Works

In recent years the researchers suggested a lot of methods for opinion summarization. Some of them suggest using automatic data aggregation while others use weak supervision in the form of seed words for further major topics identification. Some articles suggest using information about such topics or aspects to model summary generation.

Using word embeddings and attention mechanism to find coherent and meaningful aspects was first proposed in [9]. In contrast to previous approaches the majority of which is based on LDA algorithm, the authors present their deep learning model - Attention-based Aspect Extraction. Despite focusing on aspect extraction rather than on summarization, this article inspired the research of the usage of aspect information for opinion summarization.

In an extension of the previous article [3] the authors use weak supervision in the form of manually collected aspect seed words and replace one seed matrix with multiple matrices for all the aspects to improve the generalization of the approach.

Another article [1] experiments with aspect extraction method from the previous article and propose Controller Induction Model which helps to control the coverage of aspects in generated summaries. The researchers propose a transformer-based architecture which learns the token-, sentence- and document-level aspect embeddings, which are further used for aspect-controllable summary creation.

The authors then fine-tune a pretrained sequence-to-sequence T5 model [16] to generate opinion summaries using tokens and sentences based on their embeddings.

One of the well performing extractive approaches is proposed in [4]. Based on the VQ-VAE [15], Quantized Transformer (QT) uses transformer encoder, vector quantizer and transformer decoder to create a summary from the set of reviews of an entity. The encoder creates multi-head representations for the

---

[1] https://github.com/artemprisiazhniuk/aspects_summarization.

sentences in review corpus, the quantizer head-wise quantizes the representations and the decoder reconstructs the original sentences. Summarization algorithm uses quantization properties to aggregate the opinions and extract the most popular ones to create the summary.

A novel unsupervised approach to opinion summarization [2] incorporates content planning to generate coherent and informative summaries. It consists of content planning, where a set of important aspects is identified from the input reviews, and summary generation, where a summary is generated based on the identified aspects. The content planning stage is designed to improve the relevance and coherence of the generated summaries by ensuring that they cover the most salient aspects of the input reviews.

QT research is followed in [5]. The article further explore the heuristics to use with the transformer architecture. The method works in unsupervised setting and does not require keywords extraction. It produces a multi-head representation of review sentences which is further used for sentence reconstruction as well as in QT approach but proposes a performance-improving heuristic.

3   Data

The standard datasets for opinion summarization are Rotten Tomatoes or RT [19], Yelp [7], Amazon [6], Oposum [3] and Space [4], each presenting data in movies, businesses, products, fashion and hotels domains respectively.

Unfortunately, there are no publicly available datasets for review summarization in Russian and the majority of Russian online services do not allow to use their data. Therefore, the data was collected from the open internet source (tripadvisor.ru) with the help of web-scraping.

We collected around 1 million of reviews of hotels from 11 cities, namely, Amsterdam, Barcelona, Berlin, Istanbul, Lisbon, London, Madrid, Paris, Rome, Tbilisi and Yerevan and structured the data in a convenient way.

The labeling process for the validation and test splits of our dataset involved manual annotation. Similar to Space dataset (50 hotels; 25 for development, 25 for testing), we created summaries for 56 hotels in validation and test splits; 28 for development, 28 for testing. We wanted to compare the performance of the models trained on the collected Russian data and the existing English data, which is why we approximated the labeling of Space dataset [4].

The only instruction was to write a summary based on the given set of aspects. The set of aspects was taken from [1], translated and fixed for manual summary construction. Every hotel was labeled based on 50 reviews stratified by rating to ensure that both low- and high-rating reviews are presented in the data. We manually filtered the concepts and phrases corresponding to specific aspects and chose the most emphasized and repetitive of them as a summary basis. We then wrote simple summaries from the collected phrases.

We did not evaluate the quality of produced reviews as they are human-written. We admit that the human-written summaries might provide some restrictions on the operation of models and leave this for further research.

Table 1 shows the overall information about opinion summarization datasets, including review corpus size, number of domains, number of aspects (used in some methods), number of test examples. Here "domains" refer to the entries' categories reflected in the dataset, e.g. for Oposum dataset the domains are Laptop Bags, Bluetooth Headsets, Boots, Keyboards, Televisions and Vacuums. "Aspects" refer to the pre-selected properties of the entries, e.g. for Space dataset the aspects are building, cleanliness, food, location, rooms and service.

Table 1. Information about the opinion summarization datasets

Dataset	RT	Yelp	Amazon	Oposum	Space	Collected data
#reviews	250k	1M	4.5M	4.1M	1.1M	1.1M
#domains	1	1	6	6	1	1
#aspects	-	-	-	18	6	6
#test examples	1250	200	60	60	50	56

Table 2 provides basic statistics about the collected data. "Char" stands for character, "Rev" stands for review, "Sum" stands for summary, "Sent" stands for sentence.

Table 2. Basic statistics about the collected data (average)

Split	Char/Rev	Char/Sum	Sent/Rev	Sent/Sum	Rev/Hotel
Train	516	-	2	-	153
Dev	523	162	2	4.9	195
Test	520	154	2	5.3	175

The expected input is the set of reviews of one entity. The expected output is the summary which includes the most representative opinions from the set of reviews. Sample reviews from the collected data and summaries generated by models could be found in Reviews and Summaries sections in Appendix.

4   Methodology

We chose both abstractive and extractive models with highest metrics values on Space and Amazon datasets which utilize aspect information for the correct comparison on the Russian language dataset.

4.1 Abstractive Methods

AceSum [1] is one of the best performing abstractive models. The method utilizes weak supervision in the form of aspect seed words which are used to identify aspects in tokens, sentences and documents and to create supervised synthetic dataset for summarization training in accordance with identified aspects.

In the original AceSum paper a sequence-to-sequence transformer T5 (small) serves as a summarization model. In order to use this approach in a multi-lingual setting we employed mT5 (small) version which can analyze and generate non-ASCII characters.

Another well performing abstractive method is PlanSum [2]. Unlike AceSum it works in unsupervised setting and does not require any additional markup. The authors calculate aspect and sentiment distributions and use them to sample reviews for the summary.

To modify the abstractive models for the Russian language data processing, we utilized `DeepPavlov/rubert-base-cased` tokenizer rather than the distilroberta-base used in the initial design. Additionally, we adapted all input-output functions to accommodate Cyrillic characters. The models were trained on the collected Russian language corpus with the hyper-parameters from original articles. However, the batch size had to be reduced because of computational limits.

We were concerned that the style of manual summary creation may affect abstractive models' performance on the test set with summaries created via aggregation of crowd-sourcing answers. This is why we explored the impact of fine-tuning the models on the small part of Space dataset. The results can be seen in Table 3.

4.2 Extractive Methods

A variant of VQ-VAE for aspect-based summarization called Quantized Transformer (QT) is proposed in [4]. The transformer encoder forms a multi-head representation of the sentence, the quantizer maps each sentence head to its nearest head vector and the transformer decoder is trained for sentence reconstruction. These parts are then used to create a summary.

Semantic Autoencoder (SemAE) [5] is an unsupervised approach to opinion summarization that uses sparse coding to extract the most informative sentences from the input reviews. The proposed method involves representing the input reviews as a bag-of-words and using sparse coding to extract the most informative sentences from the input reviews based on their relevance to the overall sentiment expressed in the reviews. The extracted sentences are then combined to form a summary.

TransSum [18] and COOP [10] methods which outperform PlanSum on the Amazon dataset were not taken into consideration because TransSum utilizes self-supervised learning for model training and COOP experimentally changes the way of aggregating reviews and do not use benefits of the recent articles.

To make the extractive methods suitable for handling the Russian language data, we followed the suggestion from the authors and utilized a trainable tokenizer called SentencePiece [8]. The tokenizer was trained on the entirety of our collected corpus without any preprocessing steps.

Our experiments showed that optimal results for both models were achieved by training for 10 epochs. The hyperparameter tuning showed that the hyperparameters proposed in the original paper work best with our data. The results can be seen in Table 3.

These extractive approaches are based on the transformer architecture and utilize its properties to accomplish the task. Their technical implementations impose restrictions on fine-tuning on a small portion of data, that is why we do not experiment with fine-tuning the methods on Space dataset.

5   Experiments

5.1   Metrics

To thoroughly examine the outcome of summary generation, we employ the widely-used ROUGE metric [12], which has been employed by numerous researchers in the field. By utilizing this metric, we aim to provide an accurate evaluation of the generated summaries and compare them with other approaches.

5.2   Data Preprocessing

Most opinion summarization papers focus on the development of frameworks and employ only basic preprocessing in the form of review selection. It includes removing excessively brief or verbose reviews and reviews containing special symbols. We examined preprocessing strategies commonly adopted in opinion summarization and executed a series of experiments to evaluate the impact of varying preprocessing configurations.

The methods achieve the best quality with no additional preprocessing, which is consistent with the original papers' methodology. The detailed experiments on data preprocessing can be found in Data Preprocessing section in Appendix.

5.3   Seed Words Extraction

AceSum [1] is a method of opinion summarization in a weakly-supervised setting where the aspect seed words are required to map documents, sentences and tokens to a specific aspect to further create their aspect embedding and use it to reconstruct documents, sentences and tokens or to compare with other parts of documents. The authors propose manually selecting seed words, because it leads to better results.

We experimented with different automatic seed words extraction techniques which could eliminate the need for the manual seed words extraction. Unfortunately, none of the tested solutions outperformed manual seed words selection or produced reasonable results, therefore, in our main experiments we use manually selected seed words. The detailed experiments on seed words extraction can be found in Seed Words Extraction section in Appendix.

5.4   Review Summarization

In order to find the best approach for the use on the collected data we train PlanSum, AceSum, QT and SemAE on the data and evaluate them on both the collected data and Space dataset as it represents the same domain of data.

Table 3. Evaluation on the datasets

Dataset	Model	ROUGE-1	ROUGE-2	ROUGE-L
Russian	AceSum	23.24	1.65	13.42
	PlanSum	20.57	2.22	12.69
	AceSum-FT	**33.87**	**2.87**	**18.20**
	PlanSum-FT	24.86	1.91	16.57
	QT	13.06	0.41	8.72
	SemAE	17.90	1.66	12.08
Space	AceSum	14.28	3.52	10.09
	PlanSum	**30.67**	6.88	18.55
	AceSum-FT	21.64	**7.07**	14.39
	PlanSum-FT	28.36	5.82	**19.44**
	QT	17.69	2.45	11.30
	SemAE	24.95	4.53	14.55

The models were evaluated on two distinct datasets - the collected data and the Space dataset. Table 3 shows the evaluation results, where "-FT" means fine-tuned on the part of Space dataset. AceSum and PlanSum methods exhibited similar levels of competency on the collected data, while PlanSum demonstrated superiority over AceSum on the Space dataset. Interestingly, fine-tuning of the models using the part of Space dataset led to significant improvements in the overall quality. AceSum strongly outperformed PlanSum on the collected data, but showed the inferior performance on the Space dataset. SemAE surpassed the performance of QT on both the Russian language data and Space dataset. However, it did not manage to outperform abstractive methods.

Table 4 shows summaries generated by the methods. More examples can be found in Summaries section in Appendix. As we can see, AceSum and PlanSum try to cover aspects from the fixed set, while QT and SemAE extract extra specific details from the reviews.

Table 4. Summaries generated by different methods

Model	Summary
Reference	Красивое здание и интерьер, удачное расположение. Дружелюбный персонал, приятное обслуживание. Чистые просторные номера с хорошим видом.
AceSum	Отель был очень дружелюбным и очень удобным. Завтрак был очень вкусным, и персонал был дружественным и внимательным.
PlanSum	Обслуживание было превосходным. Номера очень большой и очень чистые, с прекрасной ванной комнатой. Это не большой, как в нескольких минутах ходьбы от центра города.
QT	Детям очень понравились 2 бассейна крыша и внутренний двор. Номер очень элегантный и имеет все, в чем вы нуждаетесь. Это около такси, автобусной остановки, остановки автобусного тура и всех магазинов высокого класса, о которых вы могли думать. Замечательный отдых. Главный туалет, прикрепленный к главной спальне, хорош, но не практичен. Единственное чего не хватает это халатов.
SemAE	Есть много достопримечательностей в нескольких минутах ходьбы. Цены в сравнении с тем, что вы получаете великолепные! настоятельно рекомендуется! идеальный. Номер довольно простой, как и следовало ожидать. В отеле нет стойки регистрации или персонала, и вы получите доступ к своей комнате, используя код доступа, который вы получите по электронной почте. Вид на улицу был в порядке. Расположение удобно

5.5   Manual Analysis

We conducted a manual analysis of the generated summaries. As expected abstractive summaries contain more specific information about the given set of aspects, while extractive summaries contain more general information as they are compose using phrases extracted from reviews. QT and SemAE tend to include a lot of excess and sometimes personal information like names and number of kids of a reviewer. PlanSum tends to include additional information that does not straightforwardly relate to the aspects, like the size of bathroom. AceSum sticks to the aspects discovered in reviews, the aspects follow a particular order and are described using one of a few adjectives for every aspect. Some of the summaries produced by AceSum and PlanSum cover aspects which are not presented in the reference summary. It may be the cause of either the abundance of reviews or the quality of references as they are human-written. Generally, PlanSum produces more human-readable texts, while AceSum covers aspects more precisely.

6   Conclusion

In this work we explored the application of modern solutions to the Russian language opinion summarization. We managed to collect around one million reviews of hotels and annotate part of the dataset manually for evaluation purposes. We researched preprocessing and seed word extraction techniques and compared models from unsupervised and weakly-supervised settings. The best performing models among both abstractive and extractive approaches were adapted for the Russian language analysis and trained to summarize the opinions in the hotel domain.

Abstractive approaches outperform extractive approaches on the collected Russian language dataset (in contrast to findings for English data presented in [5]) and sometimes benefit from fine-tuning on a small part of the translated Space dataset. Specifically, the best performing model on the manually annotated data is AceSum-FT [1] with ROUGE-1/2/L scores of 33.87/2.87/12.20 and the best performing model on the translated Space dataset is PlanSum [2] with ROUGE-1/2/L scores of 30.67/6.88/18.55.

Further studies might provide the explanations of the prevalence of abstractive methods over extractive ones on the Russian language data and stylistic limitations of human-written summaries.

Appendix

Data Preprocessing

We explored several data preprocessing configurations found in different articles: "None" - only review selection; (1) - remove numbers, punctuation (except ".", "!" and ","), stopwords and emoticons; (2) - remove numbers, punctuation (except ".", "!" and ",") and emoticons; (3) use only reviews without latin letters (as we evaluate a cyrillic language), this removes geographical titles and also translations from other languages with misprints. We trained PlanSum [2] and SemAE [5] as representatives of abstractive and extractive approaches for several epochs on the collected data to compare the means of preprocessing. The authors of the methods did not apply any additional preprocessing in the original papers, they only propose selection of reviews based on their length and presence of special symbols. The results of comparison can be found in Table 5.

The methods achieve the best quality with no additional preprocessing, which is consistent with the original papers methodology. It appears that more preprocessing stages tend to remove information valuable for the task.

Table 5. Evaluation results of different preprocessing methods

Model	Configuration	ROUGE-1	ROUGE-2	ROUGE-L
PlanSum	None	16	1.6	10.4
	(1)	11.2	1.1	7.2
	(2)	11.8	1.2	7.7
	(3)	12.5	1.2	8.1
SemAE	None	15.1	2	10
	(1)	12.5	0.6	8.8
	(2)	13.2	0.4	8.7
	(3)	14	1.1	9.2

Seed Words Extraction

In order to find the most suitable aspect terms extraction method for the Russian language we experimented with different kinds of word embeddings. The usage of general purpose word embeddings leads to seed words not suitable for specific domain, various solutions were tested to fix this problem.

[20] suggested using only part of the words in reviews according to their frequencies and importance to extract more reasonable keywords. We used YAKE library [11] to explore this solution.

This problem can also be solved using supervised machine learning. [14] trained a machine learning model on a labeled SemEval 2014 dataset of the Russian language reviews to identify aspect terms based on the context in which they appear.

Another approaches suggested using syntactic parsing to choose aspect terms based on syntax and grammatical structures. As suggested by [13,17] and others we used rule-based approach with the help of Spacy library.

Unfortunately, none of the mentioned solutions produced a reliable result, therefore, we appraised part of the reviews manually and found the seed words most representative of the fixed hotel domain aspects "здание", "чистота", "еда", "расположение", "номер", "обслуживание". This does not allow for unexpected aspect discoveries but provides reliable seed words for weakly-supervised setting for the hotel domain.

Reviews

Table 6 shows sample reviews.

Summaries

Tables 7 and 8 show summaries of one hotel generated by different models.

Table 6. Sample reviews from the collected dataset

Review	Rating
Очень крутой и модный отель, наверно самый лучший в берлине на текущий момент. Дизайн отеля и качество предоставляемых услуг на очень высоком уровне. Очень хороший полноценный спортзал, вкусные завтраки. Один из лучших отелей в которых я останавливался.	5
Отель был потрясающим , как и персонал. Это был наш второй визит , и мы обязательно остановимся здесь снова! Утром завтрак был вкусным. Рядом со всем , что может предложить Амстердам. Номера очень уникальные , а кровати очень удобные!	5
Удивительное обслуживание , персонал был фантастическим , чтобы помочь нам с нашими потерянными чемоданами. Расположение очень хорошее , и уборка сделала все возможное , чтобы помочь мне с розеткой для моего телефона. Очень чистый и современный. Завтрак был отличным , и мы чувствовали себя там очень комфортно. Я определенно рекомендую этот отель.	5
Действительно хороший современный отель. В хорошем районе с трамвайной остановкой прямо перед ним. В самом отеле было тихо и был общий холл и балкон. Персонал был очень дружелюбным и украсил мою комнату на мой день рождения. Рекомендую остановиться здесь. Отличное соотношение цены и качества!	4
Очень хороший номер с отличными удобствами , такими как хромкаст и динамик Bluetooth. Хорошая кровать и душ. Свет очень уютный , но иногда в комнате слишком темно. Завтрак вкусный даже с личным печеным яйцом. Также доступен безглютеновый хлеб и биойогурт.	4
Остановился там в апреле 2016 года , отель классический для командировок , цена качество. Номер хороший , чистый , убирались каждый день. Единственное вид из окна на стройку , но это не беда. Рядом есть трамвайная остановка чтобы добраться до центра , ехать придется минут 20 , это отнимает много времени на пердвижения и утомляет. Рядом также есть два супермаркета и рынок выходного дня. В целом понравилось , но если приехали в первый раз и погулять , то лучше конечно останавливаться в центре.	4
Отель расположен в 10 мин пешком от метро и в 20 25 мин от моря. Рядом есть супермаркет и променад с ресторанчиками. В 15 мин находится улица диагональ и большой торговый центр . Завтраки очень скудные и однообразные.	3
Этот отель отличается от фотографий , он довольно старый. Номер был совсем рядом с улицей , так что там было довольно шумно. Соседний номер был занят , и нам все было слышно смех , телевизор , душ ... Не рекомендую.	3
Расположение отеля хорошее. но немного трудно добираться. Очень близко кинотеатр всего в нескольких шагах от отеля. Персонал услужливый. Еда средняя. Цена относительно средняя.	3

Table 7. Summaries generated by different methods

Model	Summary
Reference	Уютный отель с хорошим расположением и красивой террасой. Персонал дружелюбный и внимательный, хорошее обслуживание. Номера простые, но большие и чистые. Вкусный завтрак.
AceSum	Отель был очень чистым и чистым. Завтрак очень удобный и удобный. номер очень дружелюбный. Номер был чистым, и хорошим.
PlanSum	Расположение центральное, и вы находитесь в нескольких минутах ходьбы от фантастических баров и ресторанов, пляжа и многочисленных ресторанов на улице квартала Марэ. Сами номера были очень тихими и очень чистые, хорошо оборудованной и безупречно чистыми. Однако были некоторые проблемы с сантехникой и постельным бельем, стиральную кровать и отдельная спальня. В целом, номер был очень чистым и опрятным люксом.
QT	Кровати не очень удобные , но вы не будете тратить очень долго в номере . Отличный хостел , отличная атмосфера , отличная кухня , и вы сэкономите много денег , готовя себе еду , есть замечательная терраса , когда солнечно , и все это в центре Мадрида по очень низкой цене . Кухня казалась хорошей , хотя я ею особо не пользовался , но терраса , о , терраса !
SemAE	Завтрак включен и очень хорошие места общего пользования , а также огромный Terraza ! Ели , персонал великолепный , чистые номера , Wi Fi отличный , ванные комнаты , завтрак отличный ! Два владельца были такими милыми и дружелюбными . Отель и номера очень чистые , а также кухня . Полностью оборудованная кухня , удобные кровати , быстрый Wi Fi и великолепная открытая терраса . Хорошая кухня и места общего пользования , включая хорошую террасу

Table 8. Summaries generated by different methods

Model	Summary
Reference	Уютный чистый отель с хорошим расположением. Услужливый и дружелюбный персонал. Чистый и удобный номер, вкусный завтрак.
AceSum	Отель очень чистый и дружелюбный, и персонал был очень вежливым и услужливым. Номера очень удобные и удобные, очень просторные, а номера были чистыми и просторными.
PlanSum	Персонал был дружелюбным, и услужливым. Отель находится в нескольких минутах ходьбы от фантастических баров и ресторанов на улице. Номера были очень чистыми, хорошо оформлены и очень чистые, с прекрасными удобствами и постельным бельем, микроволми хорошо оборудованным и удобной ванной комнатой. Завтрак тоже был очень хорош с большим выбором свежих продуктов.
QT	Если Вам нужно быть в Западном Берлине и иметь условия для работы быстрый wi fi , приспособленный стол , кресло и розетки , а также террасы , чтобы переключиться , то рекомендую . Не были в Германии , прежде чем поэтому , если вы прибыли на ресепшен , теплый , дружелюбный персонал , заставил нас сразу же расслабиться . Хороший и хорошо расположенный отель близко к двум станциям Zoologisher Garten и Uhlandstrasse
SemAE	Отель удивительный , очень красивые номера и сотрудник ресепшен все очень приветливые и дружелюбные ! У нас был номер , выходящий во двор . Очень приятные номера ! И холл , и номер , и отношение , и расположение . Отличный отель , хорошее расположение и персонал и номера очень хорошие . Чистые номера , со свежим постельным бельем каждый день . Очень понравилось расположение , обслуживание и сам номер . Отличные номера , персонал

References

1. Amplayo, R.K., Angelidis, S., Lapata, M.: Aspect-controllable opinion summarization. In: Proceedings of the 2021 Conference on Empirical Methods in Natural Language Processing. Online and Punta Cana, Dominican Republic: Association for Computational Linguistics, pp. 6578–6593 (2021). https://doi.org/10.18653/v1/2021.emnlp-main.528. https://aclanthology.org/2021.emnlp-main.528
2. Amplayo, R.K., Angelidis, S., Lapata, M.: Unsupervised Opinion Summarization with Content Planning. CoRR abs/2012.07808 (2020). arXiv: 2012.07808. https://arxiv.org/abs/2012.07808
3. Angelidis, S., Lapata, M.: Summarizing opinions: aspect extraction meets sentiment prediction and they are both weakly supervised. In: Proceedings of the 2018 Conference on Empirical Methods in Natural Language Processing. Brussels, Belgium: Association for Computational Linguistics, pp. 3675–3686 (2018). https://doi.org/10.18653/v1/D18-1403. https://aclanthology.org/D18-1403
4. Angelidis, S., et al.: Extractive Opinion Summarization in Quantized Transformer Spaces (2020). https://doi.org/10.48550/arXiv.2012.04443
5. Chowdhury, S.B.R., Zhao, C., Chaturvedi, S.: Unsupervised extractive opinion summarization using sparse coding. In: Proceedings of the 60th Annual Meeting of the Association for Computational Linguistics (Volume 1: Long Papers). Dublin, Ireland: Association for Computational Linguistics, pp. 1209–1225 (2022). https://doi.org/10.18653/v1/2022.acllong.86. https://aclanthology.org/2022.acl-long.86
6. Bražinskas, A., Lapata, M., Titov, I.: Unsupervised opinion summarization as copycat-review generation. In: Proceedings of the 58th Annual Meeting of the Association for Computational Linguistics. Online: Association for Computational Linguistics, pp. 5151–5169 (2020). https://doi.org/10.18653/v1/2020.acl-main.461. https://aclanthology.org/2020.aclmain.461
7. Chu, E., Liu, P.: MeanSum: a neural model for unsupervised multi-document abstractive summarization. In: Chaudhuri, K., Salakhutdinov, R. (eds.) Proceedings of the 36th International Conference on Machine Learning, vol. 97. Proceedings of Machine Learning Research. PMLR, pp. 1223–1232 (2019). https://proceedings.mlr.press/v97/chu19b.html
8. Google. SentencePiece. In: GitHub repository (2018). https://github.com/google/sentencepiece
9. He, R., et al.: An unsupervised neural attention model for aspect extraction. In: Proceedings of the 55th Annual Meeting of the Association for Computational Linguistics (Volume 1: Long Papers). Vancouver, Canada: Association for Computational Linguistics, pp. 388–397 (2017). https://doi.org/10.18653/v1/P17-1036. https://aclanthology.org/P17-1036
10. Iso, H., et al.: Convex aggregation for opinion summarization. In: Findings of the Association for Computational Linguistics: EMNLP 2021. Punta Cana, Dominican Republic: Association for Computational Linguistics, pp. 3885–3903 (2021). https://doi.org/10.18653/v1/2021.findings-emnlp.328. https://aclanthology.org/2021.findings-emnlp.328
11. LIAAD. Yet Another Keyword Extractor (Yake). In: GitHub repository (2018). https://github.com/LIAAD/yake
12. Lin, C.-Y.: ROUGE: a package for automatic evaluation of summaries. In: Text Summarization Branches Out. Barcelona, Spain: Association for Computational Linguistics, pp. 74–81 (2004). https://aclanthology.org/W04-1013

13. Gupta, D., Venugopalan, M.: An unsupervised hierarchical rule based model for aspect term extraction augmented with pruning strategies. Procedia Comput. Sci. **171**, 22–31 (2020). https://doi.org/10.1016/j.procs.2020.04.303

14. Mayorov, V., et al.: A high precision method for aspect extraction in Russian. In: Dialogue (2014). https://www.dialog-21.ru/media/1114/mayorovvdetal.pdf

15. van den Oord, A., Vinyals, O., Kavukcuoglu, K.: Neural Discrete Representation Learning. CoRR abs/1711.00937 (2017). arXiv: 1711.00937. http://arxiv.org/abs/1711.00937

16. Raffel, C., et al.: Exploring the Limits of Transfer Learning with a Unified Text-to-Text Transformer (2019). https://doi.org/10.48550/ARXIV.1910.10683. https://arxiv.org/abs/1910.10683

17. Salwa, A., et al.: Aspect Extraction Performance with POS Tag Pattern of Dependency Relation in Aspect-based Sentiment Analysis, pp. 1–6 (2018). https://doi.org/10.1109/INFRKM.2018.8464692

18. Wang, K., Wan, X.: TransSum: translating aspect and sentiment embeddings for self-supervised opinion summarization. In: Findings of the Association for Computational Linguistics: ACL-IJCNLP 2021. Online: Association for Computational Linguistics, pp. 729–742 (2021). https://doi.org/10.18653/v1/2021.findings-acl.65. https://aclanthology.org/2021.findings-acl.65

19. Wang, L., Ling, W.: Neural network-based abstract generation for opinions and arguments. In: Proceedings of the 2016 Conference of the North American Chapter of the Association for Computational Linguistics: Human Language Technologies, San Diego, California: Association for Computational Linguistics, pp. 47–57 (2016). https://doi.org/10.18653/v1/N16-1007. https://aclanthology.org/N16-1007

20. Xu, Z., Zhang, J.: Extracting keywords from texts based on word frequency and association features. Procedia Comput. Sci. **187** (2021). 2020 International Conference on Identification, Information and Knowledge in the Internet of Things, IIKI 2020, pp. 77–82. ISSN: 1877-0509. https://doi.org/10.1016/j.procs.2021.04.035. https://www.sciencedirect.com/science/article/pii/S1877050921008139

Tuning-Free Discriminative Nearest Neighbor Few-Shot Intent Detection via Consecutive Knowledge Transfer

Maksim Savkin[(✉)] [iD] and Vasily Konovalov [iD]

Moscow Institute of Physics and Technology, Dolgoprudny, Russia
{savkin.mk,vasily.konovalov}@phystech.edu

Abstract. Few-shot intent classification and out-of-scope (OOS) detection are core components of task-oriented dialogue systems. Solving both tasks can be challenging because of limited data availability. In this study, we aim to develop a few-shot intent classification model capable of OOS detection that does not require fine-tuning on target data. We adopt the discriminative nearest neighbor classification architecture and replace the fine-tuning phase with a consecutive pre-training approach involving natural language inference and paraphrasing tasks. Our approach leverages the training set for predictions, offering a quick and convenient way to adjust the model's behavior by modifying a set of few labeled examples. When compared to methods that do not require fine-tuning, the developed model exhibits higher scores on various few-shot intent classification datasets.

Keywords: Intent Classification · Out-of-scope detection · Few-shot Learning · Knowledge Transfer · Natural Language Inference · Paraphrasing · Nearest Neighbor · Transformers · Natural Language Processing

1 Introduction

Intent detection is the task of identifying user intent given an utterance. Naturally it appears in goal-oriented dialogue systems, where it is important to also detect out-of-scope (OOS) [15] utterances that do not belong to any predefined intent classes. Addressing both intent classification and OOS detection is crucial for generating an appropriate response to user's request. Dealing with these tasks simultaneously can be challenging, especially in data-scarce environments, but still in this paper we consider using a few-shot setting, since manually annotating a large intent classification dataset can be costly and time-consuming.

A common approach to text classification involves utilizing a transformer-encoder model as a feature generator and a classification head with fully-connected and softmax layers, which output a vector of probabilities. In such architectures pre-training is critical as it improves their generalization abilities and facilitates adaptation to specific domains [9]. However, despite the

© The Author(s), under exclusive license to Springer Nature Switzerland AG 2024
D. I. Ignatov et al. (Eds.): AIST 2023, CCIS 1905, pp. 96–110, 2024.
https://doi.org/10.1007/978-3-031-67008-4_8

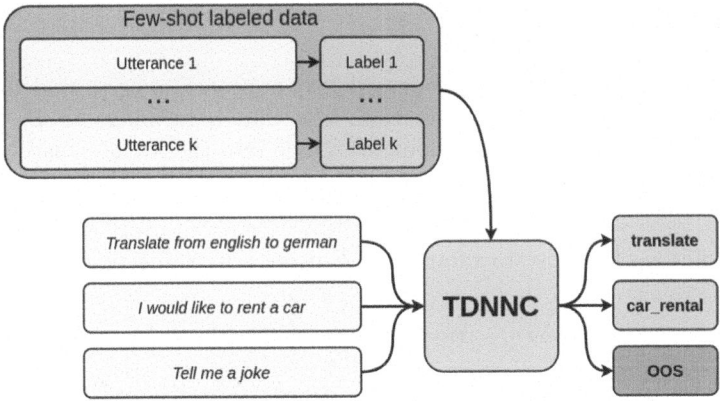

Fig. 1. A diagram of intent prediction in TDNNC.

strong performance of pre-trained transformers [15] such as BERT [5] and RoBERTa [16] on in-scope examples, recent research [25] has revealed their struggle with out-of-scope utterances. This issue arises because these models primarily learn to discriminate among in-scope intents.

In discriminative nearest neighbor classification (DNNC) [24] approach the distance metric between embeddings is replaced with a deep cross-encoder model, which takes a pair of utterances as input and predicts the probability of them belonging to the same intent class, Fig. 2. By eliminating the embedding bottleneck and comparing input utterances with training examples during both training and inference, DNNC achieves state-of-the-art scores on multiple intent classification benchmarks. It demonstrates excellent accuracy in both in-scope classification and OOS detection. However, this approach has a drawback of learning on a squared-size dataset, resulting in high demands on GPU memory during training. Existing tuning-free methods, on the other hand, still struggle from poor embeddings and have low out-of-scope detection quality.

In this paper, we aim to address the issue of high training requirements by proposing a tuning-free modification of the DNNC architecture (TDNNC). TDNNC inputs a few-shot set of labeled data and utterances to classify and predicts labels without any additional fune-tuning, see Fig. 1. Inspired by a success in transferring knowledge from natural language inference (NLI) task [4,7,11,13,20], we have compared the effect of pre-training on different binary classification tasks. We have also experimented with augmentation on a small pre-training dataset in order to enhance the OOS detection quality. We conducted experiments on multiple few-shot intent classification datasets that include OOS examples. Our results demonstrate an improvement in prediction quality over the existing methods. Additionally, we compare our model with several strong fine-tuned baselines, including RoBERTa and the original DNNC model. TDNNC model was implemented in the DeepPavlov [2] library.

2 Background

The conventional k-nearest neighbours (kNN) approach aims to find the most suitable match for the input utterance u within a training set E. This is accomplished by projecting the input utterance into the feature space of some transformer-encoder model

$$h_u = BERT(u) \tag{1}$$

and then finding the closest example $e \in E$ using a fixed metric S:

$$kNN(u) = class\left(\underset{e_i \in E}{argmax}\left(S(h_u, h_{e_i})\right)\right) \tag{2}$$

, where $S \in \mathbb{R}^d \times \mathbb{R}^d \to \mathbb{R}$ – is some similarity measure in high-dimensional space such as cosine similarity; $class(e_i)$ takes a training example as input and outputs it's label.

As mention before, this method suffers from poor embeddings, which are unable to effectively discriminate out-of-scope examples from in-scope classes. To address this issue, the authors of the DNNC paper [24] proposed a solution that combines the steps of generating embeddings and calculating distances into a single cross-encoder model as show on Fig. 2. This model aims to predict the probability of a pair of examples belonging to the same class. While maintaining the overall structure of the nearest neighbors classifier, the distance function is modified as demonstrated below:

$$h = BERT(u, e_i) \in \mathbb{R}^d \tag{3}$$

$$S(u, e_i) = \sigma(W * h + b) \in [0, 1] \tag{4}$$

A pair of examples (u, e_i) is fed to deep similarity function $S(u, e_i)$ as one long input: [CLS] ; u ; [SEP] ; e_i ; [SEP] effectively removing the step of compressing texts into embedding-vectors.

2.1 OOS Detection

To detect out-of-scope examples, we employ an uncertainty-based strategy. Given an input utterance u and a predicted class C_i, we compare the model's confidence $p(c = C_i|u)$ with a specified threshold $T \in [0, 1]$ and if $p(c = C_i|u) < T$ then u is considered to be an OOS example. In kNN-based methods we use maximum similarity score among all the training examples as the model's confidence (see Eq. 2).

3 Proposed Method

In the original paper strong capabilities of similarity function S were achieved by training an underlying cross-encoder to predict the pairwise similarity of

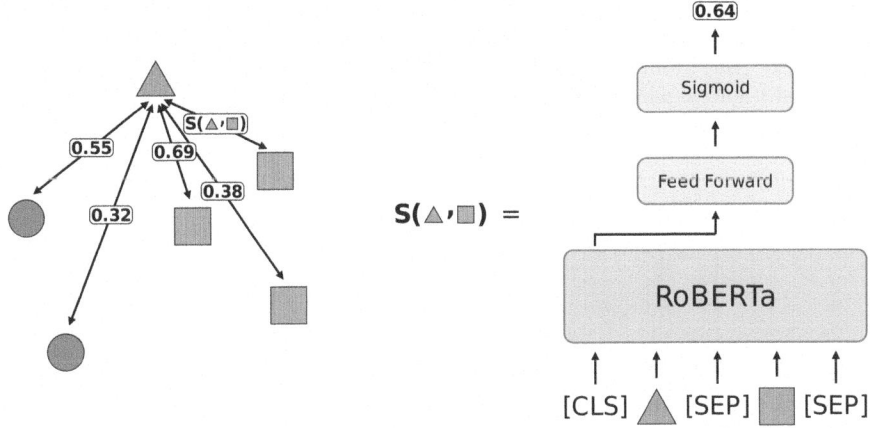

Fig. 2. Discriminative nearest neighbors prediction pipeline: green triangle represents an input utterance and other objects represent labeled utterances from the training set. (Color figure online)

examples from the training set. However, this approach necessitates training on large batch sizes for squared-size data, making it impractical for the average user. As an alternative, we skip the fine-tuning process and instead focus on pre-training a deep intent-similarity function that can effectively differentiate between unseen intents.

3.1 Pre-training

Since similarity prediction is a binary classification task, our subsequent investigations will be primarily focused on transferring knowledge from tasks that share a binary format.

Natural Language Inference involves determining whether a hypothesis can be logically inferred from a premise. It has proven to be a successful pre-training task for robust sentence embedders such as SBERT [20], SimCSE [7], InferSent [4], and has also improved the quality of the original DNNC [24] model. Typically, NLI is approached as a three-way classification task, with *entailment*, *contradiction*, and *neutral* classes, but for our purposes we use a dataset provided by DNNC paper, where last two classes are merged into one *non_entailment* class to align with the binary classification format. This dataset consists of three NLI datasets (3NLICombined): MNLI [23], SNLI [1], and WNLI [22], comprising over 300,000+ *entailment* pairs and 600,000+ *non_entailment* pairs.

Paraphrase Detection task aims to determine whether a pair of sentences has the same semantic meaning. Here, the true label of a text pair is independent of their order, making this task more suitable for similarity prediction. One major downside of paraphrase detection is the scarcity of large annotated datasets. We

Table 1. Examples of text pair from original MRPC dataset and from it's augmented version. Numbers prepended after the label are NLI model confidence scores. (a) and (b) are text pairs from original dataset; (c) – augmented hard negative pair; (d) – augmented negative pair with low entailment score.

	Text's pair	Label
(a)	He said the foodservice pie business doesn't fit the company's long-term growth strategy.	paraphrase
	The foodservice pie business does not fit our long-term growth strategy.	
(b)	Security lights have also been installed and police have swept the grounds for booby traps.	non_paraphrase
	Security lights have also been installed on a barn near the front gate.	
(c)	Associated Press Writer Jay Reeves in Birmingham, Ala., contributed to this story.	non_paraphrase (0.8)
	An Associated Press report is included.	
(d)	The Institute said dioxin levels in the environment have fallen by as much as 76 percent	non_paraphrase (0.1)
	It has a margin of error of plus or minus three to four percentage point	

utilize the small MRPC [6] dataset, which consists of 2,753 positive and 1,323 negative annotated pairs.

Another possible pre-training strategy is a consecutive knowledge transfer, which simulates pre-training and fine-tuning stages. Initially, model trains on a large 3NLICombined dataset, similar to the standard pre-training, and then it fine-tunes on a smaller paraphrasing dataset.

3.2   Paraphrases Augmentation

The size of the paraphrasing dataset can significantly impact the generalization abilities of the deep function S. To mitigate this issue, we have augmented the paraphrasing dataset and both increased the number of training examples and modified the portion of negative pairs.

In order to predict the class of the input utterance u, we compute the distance between u and all the training examples $e_i \in E$. During this process, our binary classifier S compares the input with K (shot) training examples from the same class, as well as with $(N-1)*K$ examples from different classes. This implies that an optimal proportion during the validation step would consist of one positive pair for every $(N-1)$ negative pairs.

Paraphrase Clustering. Paraphrasing is an equivalence relation, so we can divide our paraphrasing dataset into equivalence classes, where each class represents a cluster of utterances with similar meaning. Assuming that all the formed clusters have distinct meanings, pairs from different clusters can be categorized as non-paraphrases.

Hard Negatives Mining. Many studies in the field of contrastive learning [7,8,10,21] have demonstrated the importance of including hard negative examples in order to establish high-quality boundaries between clusters in the feature space. In Table 1, in row (d), you can see a negative augmented pair where the

Table 2. Evaluation datasets statistics: N - number of classes; In-Scope/OOS - are the number of in-scope and out-of-scope examples accordingly.

	N	Train In-Scope	Dev. In-Scope/OOS	Test In-Scope/OOS
CLINC150-Banking-Domain	15	1500	300/100	450/1000
CLINC150	150	15000	3000/100	4500/1000
BANKING77-OOS	77	5905	1506/200	2000/1000

texts differ in both meaning and subject matter, which makes it easy to predict whether they are paraphrases or not.

To identify the most challenging negative pairs, we employed a RoBERTa classifier trained on the NLI classification task. For each pair of generated non-paraphrases, we calculated the probability of logical entailment. This enabled us to find examples that are topically and logically closer to each other, while still being labeled as *non_paraphrase*.

When constructing negative pairs, we initially assumed that clusters would possess distinct meanings. However, this assumption proved to be incorrect, leading to some pairs being wrongly labeled as *non_paraphrase*. These misclassified examples can exhibit considerably high entailment scores and to avoid utilizing such mislabeled instances as hard negatives, we filtered out all pairs with entailment scores exceeding 80%.

4   Experimental Settings

We will employ a 5-shot setting for all our experiments.

4.1   Evaluation Datasets

For multi-domain intent detection, we utilize the CLINC150 [15] dataset, which supports 150 intents across 10 different domains, each with 15 labels. Additionally, we conducted experiments to evaluate the ability to distinguish between 77 similar intents within the single domain of BANKING77-OOS [3,25] dataset. Similarly to the original paper, we evaluated our approach using the banking domain subset from the CLINC150 dataset (referred to as CLINC150-Banking-Domain). During our evaluation, we only employed out-of-domain out-of-scope examples [25], which do not belong to any of the domains supported by dataset (Table 2).

4.2   Evaluation Metrics

According to [15,24] we use in-scope accuracy (Acc_{in}), out-of-scope recall (R_{oos}) and out-of-scope precision (P_{oos}) for all evaluations. OOS recall and precision are defined as standard recall and precision for binary classification, where positive class consists of out-of-scope examples, and negative class is a combination of all the in-scope classes.

Table 3. Test set metrics of our approach with different variations of pre-training task and augmentation proportions of MRPC dataset.

CLINC150-Banking-Domain

	Paraphrase pre-training				Consecutive pre-training			
	Acc_{in}	R_{oos}	P_{oos}	$F1_{oos}$	Acc_{in}	R_{oos}	P_{oos}	$F1_{oos}$
No aug.	78.7 ± 1.9	92.6 ± 1.2	96.2 ± 0.6	94.3 ± 0.8	76.9 ± 2.6	91.5 ± 0.8	94.9 ± 0.6	84.2 ± 1.3
1 vs 10	71.0 ± 2.3	96.4 ± 0.5	91.7 ± 0.8	94.0 ± 0.4	82.5 ± 2.3	93.8 ± 0.8	97.0 ± 0.4	95.4 ± 0.5
1 vs 100	73.4 ± 2.1	92.1 ± 1.1	94.0 ± 0.6	93.0 ± 0.7	82.6 ± 1.7	95.8 ± 0.5	96.2 ± 0.5	96.0 ± 0.4
1 vs 1000	67.3 ± 0.0	81.4 ± 0.0	92.6 ± 0.0	86.6 ± 0.0	82.2 ± 1.4	95.0 ± 0.8	96.4 ± 0.3	95.7 ± 0.5

BANKING77-OOS

	Paraphrase pre-training				Consecutive pre-training			
	Acc_{in}	R_{oos}	P_{oos}	$F1_{oos}$	Acc_{in}	R_{oos}	P_{oos}	$F1_{oos}$
No aug.	39.7 ± 1.0	94.1 ± 0.8	68.4 ± 0.4	79.2 ± 0.4	36.8 ± 1.2	94.3 ± 3.8	65.6 ± 1.0	77.4 ± 2.0
1 vs 10	39.9 ± 0.8	93.7 ± 0.3	65.7 ± 0.5	77.2 ± 0.4	42.9 ± 1.7	95.8 ± 0.7	71.3 ± 1.1	81.7 ± 0.7
1 vs 100	41.8 ± 1.3	93.4 ± 0.7	66.1 ± 0.4	77.5 ± 0.5	44.5 ± 1.7	95.2 ± 0.6	73.7 ± 1.1	83.1 ± 0.8
1 vs 1000	39.6 ± 1.4	95.2 ± 0.6	65.6 ± 0.7	77.7 ± 0.4	46.1 ± 1.9	93.1 ± 0.6	77.4 ± 1.6	84.5 ± 1.1

CLINC150

	Paraphrase pre-training				Consecutive pre-training			
	Acc_{in}	R_{oos}	P_{oos}	$F1_{oos}$	Acc_{in}	R_{oos}	P_{oos}	$F1_{oos}$
No aug.	55.0 ± 0.5	85.8 ± 0.9	39.5 ± 0.4	54.1 ± 0.5	49.6 ± 1.6	80.2 ± 9.2	33.2 ± 1.7	46.9 ± 3.2
1 vs 10	55.0 ± 0.2	83.8 ± 0.9	38.0 ± 0.4	52.3 ± 0.5	64.0 ± 1.1	83.0 ± 2.2	45.0 ± 0.5	58.3 ± 0.4
1 vs 100	50.4 ± 0.5	82.0 ± 0.9	34.9 ± 0.3	48.9 ± 0.4	64.7 ± 1.4	83.8 ± 1.9	46.5 ± 0.9	59.8 ± 0.7
1 vs 1000	54.9 ± 0.2	84.7 ± 0.4	39.2 ± 0.3	53.6 ± 0.4	60.8 ± 1.5	84.7 ± 1.3	42.3 ± 1.0	56.4 ± 0.8

Threshold Selection. Confidence threshold T controls the probability of OOS detection. In order to neglect the influence of this hyperparameter during the evaluation on test set, we choose T from $[0.0, 0.1, ..., 1.0]$, which maximizes the $Acc_{in} + R_{oos}$ on a development set.

Few-Shot Averaging. The choice of few training examples has a significant impact on the overall predictions [14]. To address this problem, we employ a random selection process where we choose *shot* examples from each intent-class of the original training set. This process is repeated 5 times for CLINC150 and BANKING77-OOS and 10 times for CLINC150-Banking-Domain to generate different few-shot training sets. Each model is then individually trained on each of these few-shot training sets, and the average and standard deviation of metrics on the test set are reported. For kNN-based methods, the few-shot training sets are utilized for predictions.

4.3   Model Configurations

We use base configuration of RoBERTa model in all our experiments, due to it's high scores on various binary classification tasks within the GLUE [22] benchmark. RoBERTa was also used in original paper and showed superior results when compared to BERT and SBERT. During the pre-training phase, we utilize the binary cross-entropy loss function. We set batch size to 256 for the setting,

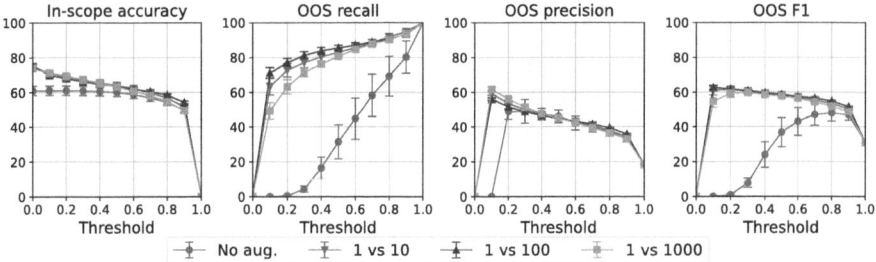

Fig. 3. Test set results on CLINC150 dataset for consecutive pre-training on 3NLICombined and augmented MRPC dataset. Higher area under curve means more robustness to threshold selection.

where MRPC dataset was augmented in proportions of one positive pair to 1000 negative. For all the other settings a batch size of 32 is used.

Models Compared. We include models provided by the original paper [24] and, if necessary, train them on our own data.

The following models were not fine-tuned on target data:

- TF-IDF-kNN – kNN baseline with TF-IDF vectors.
- Emb-kNN-vanilla – kNN-based approach with embeddings produced by [CLS] token of SRoBERTa-NLI-base [20] encoder. SRoBERTa model was trained with NLI dataset in style of siamese-networks.

We also compare our approach with strong fine-tuned models:

- RoBERTa – transformer-encoder with linear classification head. It was pre-trained with a large corpus of open-domain data, which allows it to faster adapt to specific domains.
- Emb-kNN – same as Emb-kNN-vanilla, but SRoBERTa was additionally fine-tuned on pairs of examples from the target dataset.
- DNNC – state of the art approach for few-shot intent classification and out-of-scope detection based on a cross-encoder RoBERTa model. Has high demands on GPU memory during the fine-tuning phase.

5 Experimental Results

5.1 Pre-training Results

As demonstrated in Sect. 3.2 optimal proportions of negative and positive pairs can vary depending on the dataset. Therefore, we compared all pre-training tasks with different proportions of augmented pairs on three intent classification datasets (see Table 3). Knowledge transfer from solely NLI yielded the worst results due to it's directional nature, achieving 42.2% in-scope accuracy, 78.8%

Table 4. Test set metrics of knn-based tuning-free models.

CLINC150-Banking-Domain

	Acc_{in}	R_{oos}	P_{oos}	$F1_{oos}$
TF-IDF-kNN	59.2 ± 3.3	61.0 ± 2.7	$\mathbf{97.7 \pm 0.6}$	75.1 ± 2.0
Emb-kNN-vanilla	64.4 ± 2.7	89.5 ± 1.1	95.2 ± 0.5	92.2 ± 0.5
TDNNC (our)	$\mathbf{82.6 \pm 1.7}$	$\mathbf{95.8 \pm 0.5}$	96.2 ± 0.5	$\mathbf{96.0 \pm 0.4}$

BANKING77-OOS

	Acc_{in}	R_{oos}	P_{oos}	$F1_{oos}$
TF-IDF-kNN	43.8 ± 1.7	55.2 ± 2.2	63.0 ± 2.2	58.8 ± 1.5
Emb-kNN-vanilla	$\mathbf{55.0 \pm 1.6}$	$\mathbf{96.5 \pm 0.3}$	$\mathbf{86.3 \pm 1.6}$	$\mathbf{91.1 \pm 0.9}$
TDNNC (our)	44.5 ± 1.7	95.2 ± 0.6	73.7 ± 1.1	83.1 ± 0.8

CLINC150

	Acc_{in}	R_{oos}	P_{oos}	$F1_{oos}$
TF-IDF-kNN	35.3 ± 0.8	71.6 ± 2.0	24.4 ± 0.3	36.4 ± 0.5
Emb-kNN-vanilla	54.2 ± 0.4	$\mathbf{87.5 \pm 0.6}$	39.6 ± 0.3	54.5 ± 0.4
TDNNC (our)	$\mathbf{64.7 \pm 1.4}$	83.8 ± 1.9	$\mathbf{46.5 \pm 0.9}$	$\mathbf{59.8 \pm 0.7}$

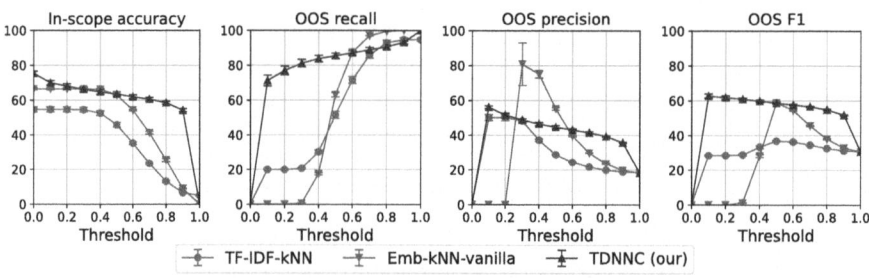

Fig. 4. Comparison of tuning-free methods on test subset of CLINC150. Higher area under curve means more robustness to threshold selection.

OOS recall and 31.6% OOS precision on CLINC150 dataset. Consecutive pre-training on 3NLICombined dataset and augmented MRPC dataset with one positive pair for every 100 negative produced one of the best overall results, primarily due to the significant improvements in OOS detection quality.

On Fig. 3 you can see, that augmentation significantly increased OOS recall. When TDNNC was trained on augmented version of paraphrasing dataset, the number of negative pairs increased and consequently, the cross-encoder model was less likely to predict that out-of-scope example is similar to some in-scope examples. The reduction in number of such mistakes resulted in much higher OOS recall metric.

5.2   Comparison with Existing Solutions

In further comparisons we will be using a version of our model that was pre-trained firstly with 3NLICombined dataset and then trained with augmented MRPC dataset with 1 positive pair to 100 negative.

Tuning-Free Methods. When compared to methods that do not require fine-tuning (see Table 4), our approach attains the highest in-scope classification and OOS detection metrics on both CLINC150-Banking-Domain and the entire CLINC150 dataset, with a margin of 10.5% in-scope accuracy and 5.3% OOS f1 on the last one. TF-IDF-kNN and Emb-kNN-vanilla embeddings struggle to distinguish OOS examples from clusters of in-scope examples. As a result both models generate a lot of low-confidence predictions, making it harder to detect OOS examples with uncertainty-based strategy. Although Emb-kNN-vanilla model exhibits relatively high in-domain performance (see Fig. 4), lack of robustness to threshold selection leads to a significant drop in in-scope accuracy, resulting in lower overall metrics. These facts indicate that binary discriminators are superior to standard approaches in terms of out-of-scope detection accuracy.

Fine-Tuned Methods. Fine-tuned approaches attain higher in-domain accuracy due to higher quality of fine-tuned embeddings, as shown in Table 5. On Fig. 5 RoBERTa classifier exhibits a rapid decline in in-scope accuracy and has a low area under OOS F1 curve, which means that despite it's high initial in-scope accuracy it struggles with handling outliers.

Table 5. Test set metrics of our approach and fine-tuned models.

CLINC150-Banking-Domain

	In-scope accuracy	OOS recall	OOS precision	OOS f1
RoBERTa	81.1 ± 1.8	88.3 ± 3.3	92.8 ± 0.8	90.5 ± 1.9
Emb-kNN	79.6 ± 2.5	92.2 ± 1.9	92.2 ± 0.8	92.2 ± 0.8
DNNC	**87.0 ± 1.5**	**96.1 ± 0.9**	96.1 ± 0.5	**96.1 ± 0.4**
TDNNC (our)	82.6 ± 1.7	95.8 ± 0.5	**96.2 ± 0.5**	96.0 ± 0.4

BANKING77-OOS

	In-scope accuracy	OOS recall	OOS precision	OOS f1
RoBERTa	60.9 ± 1.0	95.2 ± 1.3	67.6 ± 1.8	79.0 ± 1.2
Emb-kNN	58.3 ± 2.3	86.8 ± 2.3	64.8 ± 1.9	74.2 ± 1.8
DNNC	**66.5 ± 1.4**	**97.7 ± 1.3**	**74.6 ± 1.4**	**84.6 ± 0.7**
TDNNC (our)	44.5 ± 1.7	95.2 ± 0.6	73.7 ± 1.1	83.1 ± 0.8

CLINC150

	In-scope accuracy	OOS recall	OOS precision	OOS f1
RoBERTa	81.7 ± 1.1	**91.4 ± 1.9**	20.8 ± 1.4	33.9 ± 1.8
Emb-kNN	77.1 ± 0.4	88.6 ± 1.0	52.3 ± 0.3	65.8 ± 0.4
DNNC	**84.3 ± 0.7**	90.6 ± 2.1	**63.9 ± 1.3**	**74.9 ± 1.3**
TDNNC (our)	64.7 ± 1.4	83.8 ± 1.9	46.5 ± 0.9	59.8 ± 0.7

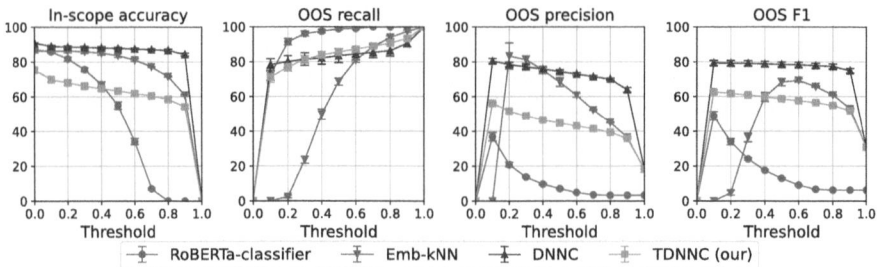

Fig. 5. Comparison between fine-tuned models and our approach on CLINC150 dataset. Higher area under curve means more robustness to threshold selection.

While our method lacks in-scope accuracy, due to the fact it was not fine-tuned on the target data, it shows one of the highest OOS recall and precision scores on all datasets. Furthermore, both our model and the original DNNC proved to be more robust to threshold selection than standard kNN-based methods and transformer-based classifiers.

6 Conclusion

In this paper, we presented a training-free modification of DNNC architecture for intent classification and out-of-scope detection. Our approach uses consecutive knowledge transfer from natural language inference and paraphrasing tasks, to develop a robust deep similarity function capable of generalizing to unseen intents. Additionally, we enhanced the classification quality by expanding the paraphrasing dataset with negative pairs we got from clustering and selecting hard negatives with high entailment scores. As a result, our final model achieved the highest in-scope classification and OOS detection scores on CLINC150-Banking-Domain and CLINC150 datasets compared to alternative tuning-free methods.

For future work, we aim to optimize the computational complexity of our approach. We have already tried to increase the prediction speed by replacing a RoBERTa model with a distilled BERT [12], but observed significant decrease in OOS accuracy. Another potential optimization involves utilizing a faster tuning-free model to filter candidates, similar to the strategy employed in retrieval models [17].

It is also possible to apply the proposed method to other text classification tasks like topic classification or speech function classification [18,19], by adjusting the pre-training strategy of cross-encoder model.

Acknowledgments. This work was supported by a grant for research centers in the field of artificial intelligence, provided by the Analytical Center for the Government of the Russian Federation in accordance with the subsidy agreement (agreement identifier 000000D730321P5Q0002) and the agreement with the Moscow Institute of Physics and Technology dated November 1, 2021 No. 70-2021-00138.

A   Appendix

See Table 6 and Figs. 6, 7.

Table 6. Hyperparameter settings used for fine-tuning

	CLINC150-Banking-Domain			BANKING77-OOS			CLINC150		
	batch size	epochs	lr	batch size	epochs	lr	batch size	epochs	lr
RoBERTa	15	25	5e−5	30	25	5e−5	50	25	5e−5
Emb-kNN	200	35	1e−5	200	15	2e−5	200	7	2e−5
DNNC	370	15	1e−5	600	7	2e−5	900	7	2e−5

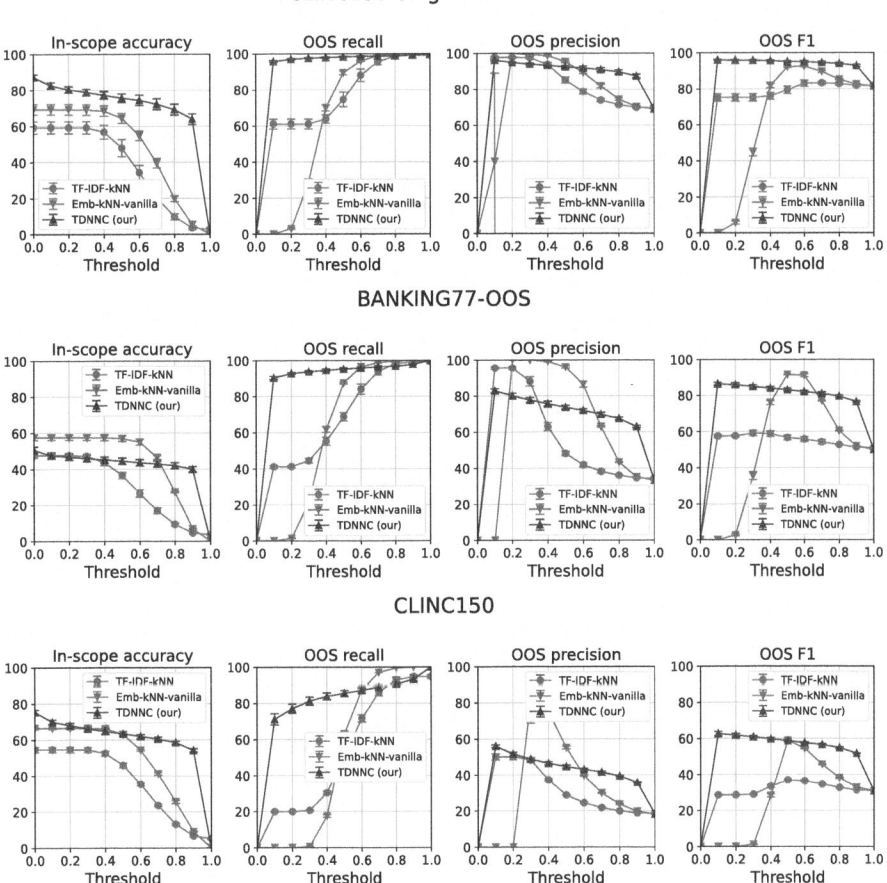

Fig. 6. Comparison of kNN-based methods on test sets of 3 different intent classification datasets. Higher area under curve means more robustness to threshold selection.

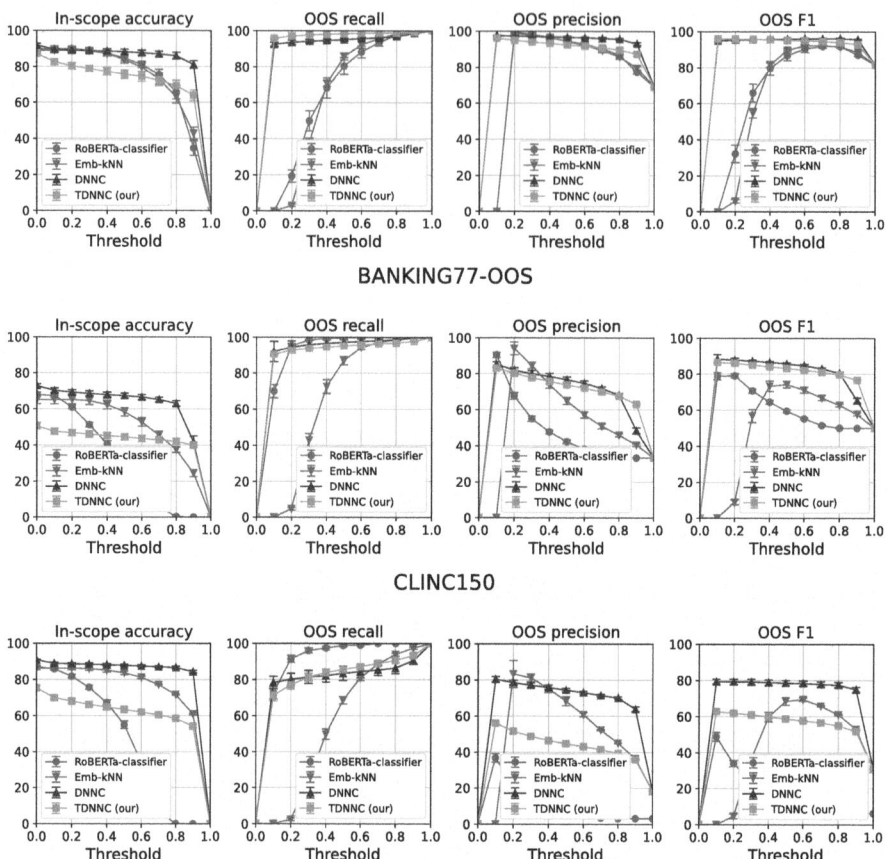

Fig. 7. Comparison between fine-tuned baselines and our approach on test sets of 3 different intent classification datasets. Higher area under curve means more robustness to threshold selection.

References

1. Bowman, S.R., Angeli, G., Potts, C., Manning, C.D.: A large annotated corpus for learning natural language inference. CoRR abs/1508.05326 (2015). http://arxiv.org/abs/1508.05326
2. Burtsev, M., et al.: Deeppavlov: an open source library for conversational AI. In: NIPS (2018). https://openreview.net/pdf?id=BJzyCF6Vn7
3. Casanueva, I., Temcinas, T., Gerz, D., Henderson, M., Vulic, I.: Efficient intent detection with dual sentence encoders. CoRR abs/2003.04807 (2020). https://arxiv.org/abs/2003.04807
4. Conneau, A., Kiela, D., Schwenk, H., Barrault, L., Bordes, A.: Supervised learning of universal sentence representations from natural language inference data. In: Proceedings of the 2017 Conference on Empirical Methods in Natural

Language Processing, pp. 670–680. Association for Computational Linguistics, Copenhagen, Denmark (2017). https://doi.org/10.18653/v1/D17-1070, https://aclanthology.org/D17-1070

5. Devlin, J., Chang, M., Lee, K., Toutanova, K.: BERT: pre-training of deep bidirectional transformers for language understanding. CoRR abs/1810.04805 (2018). http://arxiv.org/abs/1810.04805

6. Dolan, W.B., Brockett, C.: Automatically constructing a corpus of sentential paraphrases. In: Proceedings of the Third International Workshop on Paraphrasing (IWP2005) (2005). https://aclanthology.org/I05-5002

7. Gao, T., Yao, X., Chen, D.: SimCSE: simple contrastive learning of sentence embeddings (2022)

8. Gunel, B., Du, J., Conneau, A., Stoyanov, V.: Supervised contrastive learning for pre-trained language model fine-tuning (2021)

9. Hendrycks, D., Lee, K., Mazeika, M.: Using pre-training can improve model robustness and uncertainty (2019)

10. Iskender, B., Xu, Z., Kornblith, S., Chu, E.H., Khademi, M.: Improving dense contrastive learning with dense negative pairs (2023)

11. Karpov, D., Konovalov, V.: Knowledge transfer between tasks and languages in the multi-task encoder-agnostic transformer-based models. In: Computational Linguistics and Intellectual Technologies, vol. 2023 (2023). https://doi.org/10.28995/2075-7182-2023-22-200-214, https://www.dialog-21.ru/media/5902/karpovdpluskonovalovv002.pdf

12. Kolesnikova, A., Kuratov, Y., Konovalov, V., Mikhail, M.: Knowledge distillation of Russian language models with reduction of vocabulary. In: Computational Linguistics and Intellectual Technologies. RSUH (2022). https://doi.org/10.28995/2075-7182-2022-21-295-310, https://www.dialog-21.ru/media/5770/kolesnikovaaplusetal036.pdf

13. Konovalov, V., Gulyaev, P., Sorokin, A., Kuratov, Y., Burtsev, M.: Exploring the BERT cross-lingual transfer for reading comprehension. In: Computational Linguistics and Intellectual Technologies, pp. 445–453 (2020). https://doi.org/10.28995/2075-7182-2020-19-445-453, http://www.dialog-21.ru/media/5100/konovalovvpplusetal-118.pdf

14. Konovalov, V., Melamud, O., Artstein, R., Dagan, I.: Collecting better training data using biased agent policies in negotiation dialogues. In: Proceedings of WOCHAT, the Second Workshop on Chatbots and Conversational Agent Technologies. Zerotype, Los Angeles (2016). http://workshop.colips.org/wochat/documents/RP-270.pdf

15. Larson, S., et al.: An evaluation dataset for intent classification and out-of-scope prediction. CoRR abs/1909.02027 (2019). http://arxiv.org/abs/1909.02027

16. Liu, Y., et al.: RoBERTa: a robustly optimized BERT pretraining approach. CoRR abs/1907.11692 (2019). http://arxiv.org/abs/1907.11692

17. Nie, Y., Wang, S., Bansal, M.: Revealing the importance of semantic retrieval for machine reading at scale. CoRR abs/1909.08041 (2019). http://arxiv.org/abs/1909.08041

18. Ostyakova, L., Molchanova, M., Petukhova, K., Smilga, N., Kornev, D., Burtsev, M.: Corpus with speech function annotation: challenges, advantages, and limitations. In: Computational Linguistics and Intellectual Technologies, pp. 1129–1139 (2022)

19. Ostyakova, L., PetukhovaO, K., Smilga, V., ZharikovaO, D.: Linguistic annotation generation with chatGPT: a synthetic dataset of speech functions for discourse

annotation of casual conversations. In: Proceedings of the International Conference "Dialogue, vol. 2023 (2023)

20. Reimers, N., Gurevych, I.: Sentence-BERT: sentence embeddings using Siamese BERT-networks. CoRR abs/1908.10084 (2019). http://arxiv.org/abs/1908.10084

21. Schroff, F., Kalenichenko, D., Philbin, J.: FaceNet: a unified embedding for face recognition and clustering. In: 2015 IEEE Conference on Computer Vision and Pattern Recognition (CVPR). IEEE (2015). https://doi.org/10.1109/cvpr.2015.7298682

22. Wang, A., Singh, A., Michael, J., Hill, F., Levy, O., Bowman, S.R.: GLUE: a multi-task benchmark and analysis platform for natural language understanding. CoRR abs/1804.07461 (2018). http://arxiv.org/abs/1804.07461

23. Williams, A., Nangia, N., Bowman, S.R.: A broad-coverage challenge corpus for sentence understanding through inference. CoRR abs/1704.05426 (2017). http://arxiv.org/abs/1704.05426

24. Zhang, J., et al.: Discriminative nearest neighbor few-shot intent detection by transferring natural language inference. CoRR abs/2010.13009 (2020). https://arxiv.org/abs/2010.13009

25. Zhang, J.G., et al.: Are pretrained transformers robust in intent classification? A missing ingredient in evaluation of out-of-scope intent detection. In: The 4th Workshop on NLP for Conversational AI. ACL 2022 (2022)

Prompt-Tuning for Targeted Sentiment Analysis in Russian

Yuliana Solomatina[✉] and Natalia Loukachevitch

Lomonosov Moscow State University, Moscow, Russia
julianasolomatina@yandex.ru

Abstract. This paper describes prompt-based methods for targeted sentiment classification of the Russian news texts, proposed during the RuSentNE-2023 competition. This task is challenging in two following aspects: first, it is required to extract sentiment towards specific entities, not of the overall text; second, the sentiments in the news texts are often implicit, which means they are harder to detect. In the present study, several strategies of prompt-tuning BERT-like models were explored. The best result was achieved when incorporating external knowledge into the prompt verbalizer. We demonstrate that prompt-tuning methods help achieve high results while being less computationally intensive than other fine-tuning and ensembling strategies.

Keywords: targeted sentiment analysis · prompt tuning · named entities

1 Introduction

Sentiment analysis is one of the most popular areas of research in natural language processing. The task of sentiment analysis is to determine the author's attitude towards the information they convey. This application is used for various purposes, including analyzing reviews on recommendation services to automatically identify users' positive, neutral, or negative opinions; monitoring reputation of large companies; identifying users' political preferences on social networks, and many others. There are two basic approaches to solving sentiment analysis tasks: one is based on vocabularies and rules [1,13], and the other one is based on machine learning [2]. The most effective sentiment analysis systems are those created using machine learning methods. With this approach, the data is first annotated, then significant features are identified, and finally, the most appropriate classification algorithm is chosen. Machine learning methods are much better at handling difficulties such as recognizing irony or sarcasm, resolving ambiguity in sentiment lexicon, and being oriented towards a specific domain. The best results in solving this task, as well as many other natural language processing tasks, are demonstrated by pre-trained language models based on attention mechanism [27,28], particularly BERT [3].

© The Author(s), under exclusive license to Springer Nature Switzerland AG 2024
D. I. Ignatov et al. (Eds.): AIST 2023, CCIS 1905, pp. 111–124, 2024.
https://doi.org/10.1007/978-3-031-67008-4_9

Speaking of the latest approaches for classification tasks in NLP, it is important to mention the "pre-train and prompt" paradigm, which has outperformed traditional fine-tuning in many tasks. In this approach, a textual prompt is used to reformulate the downstream task in such a way that it resembles the model's pre-training task [19,24]. Prompt-tuning can be implemented along with fine-tuning or instead of it (the latter strategy is usually used for large-scaled models).

In the current study, we aimed to evaluate the prompt-based learning approach for targeted sentiment analysis in Russian. Models were developed and tested for predicting sentiment towards extracted named entities based on the RuSentNE dataset [20]. Pre-trained encoder ruRoBERTa [14] was used for this purpose. BERT-style models are state-of-the-art in most NLP tasks related to text data analysis and also give good results on unbalanced datasets. The experiments were based on a question-answering approach - the task was formulated as a question in natural language, which was fed into the model, and the expected output was a response that could be unambiguously compared to the class label (negative, positive, or neutral sentiment). The evaluation showed that tuning both the prompt and the model's weights demonstrates higher scores, and more complex prompt-tuning strategies also help improve the result.

2 Related Work

Before deep learning, the task of targeted sentiment analysis was handled using systems based on vocabularies and rules, as well as classical machine learning methods. For instance, in [1], the authors proposed an algorithm based on the SentiWordNet sentiment vocabulary, and in the work [2], the algorithm proposed by the authors took into account a whole set of linguistic features. All studies were mainly conducted on the English language material, and according to the results of the SemEval-2014 competition, the best result was an F-score of 63.3 [15].

The invention of transformer models had a significant impact on research in the field of targeted sentiment analysis. Most recent studies are based on fine-tuning the BERT model. For example, in the study [5], the authors propose fine-tuning all the weights of the model for a specific task. Later, more complex approaches to targeted sentiment analysis were proposed, for example, those based on the integration of external knowledge. In [11], the authors added an additional layer of embeddings to the BERT model, encoding information from external linguistic resources (vocabularies and knowledge graphs). In [9], the authors developed this idea by adding several layers of external knowledge, achieving a state-of-the-art result for the English language (F-score 83.1). Another approach is based on the idea of prompting the model with a question to the target entity as input and obtaining an output that corresponds to the class label [6].

The next important stage in NLP came with the idea of training not only (and not necessarily) the weights of the language model, but also prompts for it

[17]. In the field of aspect-based sentiment analysis, the SentiPrompt strategy was proposed [18], the main idea of which is to embed information from linguistic resources into the text of the trained prompt.

All of the studies mentioned above were conducted on English language data. There are significantly fewer studies on Russian language material using modern approaches. For example, in [8] authors tested several deep learning models (CNN, LSTM, BiLSTM) and BERT model on Russian sentiment evaluation datasets. For the BERT-based model, the targeted sentiment analysis task was formulated as a question-answering task and as an natural-language inference task. The latter approach demonstrated the best results. In [25], the authors describe the distance supervision approach for automatic annotation of sentiment in texts, which significantly improved the classification results of neutral networks. In [23], authors proposed an attention-mechanism-based model that separately processed the sentence embedding and the target word embedding.

The prompt-tuning-based models, trained on Russian texts, is also an under-researched subject, and most existing papers describe prompt-tuning only in relation to GPT-like models while solving text generation tasks [7,10]. In 2023, the RuSentNE-2023 competition was organized for sentiment analysis of named entities in news texts, which is expected to increase scientific interest in this area. The models described in the present paper were developed as a result of the RuSentNE-2023 competition participation. The ruRoBERTa model for classification was chosen as the baseline model during the preliminary evaluation stage of the current study. As compared with ruBERT, ruRoBERTa uses the BPE tokenizer, which helps process rare tokens more effectively. This is essential for the news data since there are many mentions of named entities. Moreover, ruRoBERTA was pre-trained on a much larger dataset, almost as large as the one used for ruGPT-3 [26].

3   Methods

The models developed in the current study were based on the prompting approach, which has been widely used for various NLP tasks and has proven to be more effective than the standard "pre-training and fine-tuning" paradigm. With this approach, a template is provided as the model's input, which represents a piece of text formulated in such a way that the downstream task can be reduced to a language modeling task. The template consists of a *prompt, context,* and *target,* i.e. the language element that the model is expected to predict. This method was applied to the targeted sentiment classification in the study [6]. The authors attempted to feed the BERT model with a question about the target entity and obtain an output that corresponds to the class label. The question was formulated as follows: *What do you think of the TARGET of it?*

Despite the fact that using a manual prompt works good for downstream tasks, it is still unclear how to formulate the prompt in the most optimal way. Experiments show that even replacing one word in the template can lead to a significant change in quality metrics [19]. Moreover, we cannot claim that the most well-formulated prompt in terms of natural language is the most optimal prompt for the model.

To maintain the advantage of using prompts while eliminating its drawbacks, the authors of the paper [19] proposed a method called *prompt tuning*. Instead of first formulating the template in natural language and then encoding them into embeddings, the authors suggest learning the prompt via gradient descent. Such prompts were called *soft prompts*, opposed to *discrete (manual) prompts*. The approach was tested on BERT and GPT and demonstrated higher quality metrics than fine-tuning.

In the current study, we applied prompt-tuning methods to the targeted sentiment analysis task. First, we tested **manual prompts** for BERT model along with some techniques to handle data imbalance (class weights and augmentation). We used the following input format:

$$[CLS]\langle sentence\rangle[SEP]\langle prompt\rangle.$$

The prompt is first initialized with manually chosen tokens. In the prompt-tuning tasks, it is also important to set a *verbalizer* - a list of words that the model should predict in the language modeling block. The predicted word is then expected to be easily mapped to a class label. For the sentiment analysis tasks, the verbalizer only usually contains class names: positive, negative, and neutral.

The example of a tunable prompt is presented below:

What:soft is the attitude:soft towards:soft X in the sentence:soft X promoted Y? [MASK]

In this example, *soft* indicates tunable tokens. Depending on a strategy, either all tokens are being tuned during training or only some of them, while others remain fixed. [MASK] indicates the masked class label, which is predicted with accordance to the verbalizer: [*positive, negative, neutral*].

We implemented several prompt-tuning strategies (they are explained in more detail in Sect. 5):

- **Mixed Template**: creating a prompt template, in which some tokens are fixed and others are tunable. They can be differentianted via special placeholders, for example, { *"soft": "What is the attitude"*}{ *"text": "positive, negative or neutral?"*}.
- **Prompt Tuning with Rules**: learning subprompts and then aggregating them to get the class prediction [10]. The template contains several masks while the verbalizer contains a broader description. For instance, a sequence { *"mask"*}{ *"mask"*}{ *"mask"*}{ *"mask"*} in the template corresponds to the sequence [*"implicitly", "or", "explicitly", "expressed"*] in the verbalizer;
- **Knowledgeable Verbalizer**: extending the list of verbalizers with the words from the Russian sentiment lexicons [12] - RuSentiLex [21] and RuSentiFrames [22]. It allows the model to predict not necessarilly the class name but also some word associated with it. For example, *gift, love, thankful - positive*; *painful, crime, harm - negative*.

4   Dataset

4.1   Dataset for the RuSentNE-2023 Competition

The RuSentNE-2023 dataset contains news texts in Russian, extracted from Wikinews. The task of the competition was to predict sentiment (positive, negative, or neutral) towards a given entity in a single sentence. The dataset was provided in the following format: *sentence, entity, entity_tag, entity_pos_start_rel, entity_pos_end_rel, label*. An example from the dataset is presented below.

Example 1. **Sentence**: After the departure of Radiy Khabirov, the administration significantly weakened, and its influence sharply decreased compared to the government apparatus," the interlocutor told Kommersant.
Entity: Radiy Khabirov
Entity tag: PERSON
Entity pos start rel: 24
Entity pos end rel: 37
Label: 1

The dataset consisted of 12 245 pairs of sentences and entities. Table 1 presents the data distribution. It shows that the data is significantly imbalanced, which should be taken into consideration during model training.

Table 1. Data distribution in the RuSentNE-2023 dataset.

Class	Number of entries
Neutral	9580
Negative	1472
Positive	1193

The dataset was split into train, validation and test sets. The proportions of the split are presented in Table 2.

Table 2. Sizes of train, validation, and test sets for RuSentNE-2023.

Type	Size
Train	6637
Validation	2845
Test	1947

Participants of the competition could evaluate their model on the Codalab platform by uploading an archive with a file containing the model's predictions in the format of a one-dimensional table.

4.2   Evaluation Metrics

In the current study, cross-validation was performed on three fixed splits (with a ratio of 70/30 between the train and validation sets sizes). Additionally, all models were evaluated on the split provided by the organizers of the RuSentNE-2023 competition.

The metric used for model performance evaluation was the F1-score. During cross-validation on the three splits, the F1-score was averaged across all three classes. When testing on the splits from the RuSentNE-2023 competition, the metric was averaged only across the positive and negative classes, it was the official metric of the competition. This is important for this specific task, as the main difficulty lies in extracting opinions. The neutral class, largely due to its quantitative prevalence, is classified with high precision and recall and is not of significant interest for the task at hand.

5   Experiments

5.1   Implementation Details

All the experiments were based on fine-tuning pre-trained ruRoBERTa-large model with the following hyperparameters:

- number of epochs: 5
- batch size: 16
- optimizer: AdamW
- loss function: cross-entropy

We used the largest batch size possible due to the computational limitations. The number of epochs was chosen with accordance to [9], where the original RoBERTa model was fine-tuned for a similar task.

None of the model's weights were frozen during training, which means that both fine-tuning and prompt-tuning were implemented. The preliminary study showed that tuning only prompt tokens does not work well in this particular case, while tuning the whole model along with its prompt works better than vanilla fine-tuning. For all the prompt-tuning experiments, the OpenPrompt package was used [4].

5.2   Manual Prompts

The first block of experiments aimed to evaluate the influence of the following two factors:

- prompt type: target word or question about the target word (similar to [6]);
- class imbalance handling method (class weight calculation or augmentation).

First, the text is tokenized and padded to the maximum sequence length of 200. Then, the tokens are processed by the pre-trained model, and the [CLS] token embedding is extracted at the output, which, after regularization via a dropout layer, is fed to a fully connected layer that performs the classification.

Experiment 1. In the first experiment, the model was fed with the concatenation of the target word and the sentence. No methods to combat class imbalance were used.

Experiment 2. In the second experiment, the same data input method as in the first one was used, but weights were calculated for each class to handle data imbalance in the cross-entropy loss function (higher weights were given to smaller classes to increase loss value during training).

Experiment 3. The prompt was a question with the following formulation: *Как относятся к X?* '(How do they feel about X?), where X is the target word. No methods to combat class imbalance were used.

Experiment 4. In this experiment, the input data format is the same as in the previous one, but class weight calculation was used to handle data imbalance, as in Experiment 2.

Experiments 5, 6. In these experiments, data augmentation was implemented for the two prompt types (target word and question). To achieve greater diversity of augmented sentences, two different augmentation strategies were used. To avoid overfitting, each sentence was augmented only once. To do this, sentences belonging to the negative or positive class were extracted from the dataset, and this subset was split in half, with one half using back-translation (into English and back) and the other replacing 30% of tokens with the most similar ones based on cosine distance using the contextual embeddings of the BERT model.

5.3   Prompt-Tuning with Mixed Template

The main idea of this experiment is to use both discrete and soft prompt tokens. The template was formulated as follows:
{ "soft": "Какое отношение выражено к *(What is the attitude towards)*" }
{ "placeholder": "text_a" } { "soft": "в предложении *(in the sentence)*" }
{ "placeholder": "text_b"}: { "text": "негативное, положительное или нейтральное *(negative, positive or neutral)*" } { "soft": "Отношение в данном предложении к *(In this sentence, the attitude towards)*" }{ "placeholder": "text_a"}{ "mask"}. Here and below:

- soft: discrete tokens, initialized with the word embeddings and tuned during the training
- text: soft tokens
- text_a: target word
- text_b: sentence
- mask: the word that is predicted during masked language modelling

5.4 Prompt-Tuning with Rules

The idea of this experiment was to create a template where not only the verbalizer of the true class is masked (as in basic prompt tuning), but also some "auxiliary" tokens (*subprompts*) that provide a more detailed description of the task (this strategy was proposed in [16]). In general, it helps encode information, specific for the particular class. In order to evaluate how the lexical content of the prompt tokens affect the result, two templates were compared.

Template 1. {"text": "В предложении *(In the sentence)*"} {"placeholder": "text_b"} {"mask"} {"mask"} {"mask"} {"mask"} {"mask"} {"mask"} {"text": "относится к *(feels about)*"} {"placeholder": "text_a"}

The verbalizer:
{
0: ["автор", "или", "другой", "участник", "ситуации", "нейтрально"],
["author", "or", "another", "participant", "of the situation", "neutral"]
1: ["автор", "или", "другой", "участник", "ситуации", "отрицательно"],
["author", "or", "another", "participant", "of the situation", "negative"]
2: ["автор", "или", "другой", "участник", "ситуации", "положительно"]
["author", "or", "another", "participant", "of the situation", "positive"]
}

Template 2. {"text": "Какое отношение *(What is the attitude)*"} {"mask"} {"mask"} {"mask"} {"mask"} {"text": " к *(towards)*"} {"placeholder": "text_a"} {"text": "в данном новостном тексте *(in the given news text)*"} {"placeholder": "text_b"}{"text": "?"}{"mask"} The verbalizer:
{
0: ["имплицитно", "или", "эксплицитно", "выражено", "нейтральное"],
["implicitly", "or", "explicitly", "expressed", "neutral"]
1: ["имплицитно", "или", "эксплицитно", "выражено", "отрицательное"],
["implicitly", "or", "explicitly", "expressed", "negative"]
2: ["имплицитно", "или", "эксплицитно", "выражено", "положительное"]
["implicitly", "or", "explicitly", "expressed", "positive"]
}

Both templates contain several masked tokens, the logits of each masked token are processed into label logits, which are then combined into a single label logits of the whole task via conjunctive normal form (as proposed in [16]).

In the first template, we emphasise presence of the source of the expressed sentiment in the sentence, while the second template implies that the sentiment can be expressed either explicitly (i.e. by means of sentiment words) or implicitly (i.e. by mentioning a (dis-)advantageous fact). This presumably helps "explain" the task to the model more accurately, since each prompt points out one of the two main aspects of the targeted sentiment analysis. After tuning the prompt tokens and the model's weights, the class label is assigned using logical rules (e.g., based on conjunctive normal form) and concatenation of the predictions of the subprompts.

5.5   Knowledgeable Verbalizer

The idea of this experiment is as follows: in the classic task setup, the model should predict the class name ("positive", "negative" or "neutral"). However, the list of verbalizers can be extended using linguistic resources. For this purpose, Russian sentiment lexicons RuSentiFrames [22] and RuSentiLex [21] were applied.

To deal with words which are ambiguous in terms of their sentiment, we tested two strategies:

First Strategy

- RuSentiLex: if the sentiment is present in at least one of the word's meanings, it is added to the corresponding list. If the sentiment can be both negative and positive depending on the meaning, the word is added to both lists.
- RuSentiFrames: for the predicates which are absent in RuSentiLex, check if sentiment is present in the frame and then add it to the corresponding list. If there are several possible sentiments, the word is added to both lists.

This strategy ensures high completeness for both verbalizer lists, but it allows for overlapping.

Second Strategy

- RuSentiLex: if the sentiment is present in at least one of the word senses, it is added to the corresponding list. If the sentiment can be both negative and positive depending on the meaning, the word is only added to the positive list.
- RuSentiFrames: for the predicates which are absent in RuSentiLex, check if their sentiment is present in the frame, and then add it to the corresponding list. If there are several possible sentiment, the word is only added to the positive list.

In this case, we avoid overlapping of classes and give priority to the positive sentiment class (since in all previous experiments the F1-score for this class was always the lowest). For the neutral class, the verbalizer remains unchanged. The sizes of verbalizer lists for both strategies are presented in Table 3.

Table 3. Extended verbalizer sizes.

Polarity	First strategy	Second strategy
Neutral	1	1
Positive	6350	8425
Negative	12634	12485

6   Results and Discussion

6.1   Cross-validation

The results of the experiments are provided in Table 4. In general, manually prompted models demonstrate lower metrics' values than prompt-tuned models. Techniques aimed at handling data imbalance do not significantly improve models' performance, although some of them can be rather time-consuming (such as augmentation). Apart from that, we assume that the perfect prompt that the model needs to perform the given task is unlikely a grammatically or semantically natural linguistic expression (the prompt formulated as a question explains the task more accurately from the human perspective, than the target word alone, but this does not lead to performance leap).

Table 4. Experimental results on cross-validation.

Prompt Type	Model	$F1_{avg}$	$F1(pos, neg)_{avg}$
Manual	target + sentence	0.651	0.522
	target + sentence + class weights	0.658	0.550
	question + sentence	0.649	0.550
	question + sentence + class weights	0.650	0.547
	target + sentence + aug	0.672	0.566
	question + sentence + aug	0.671	0.560
Soft	Mixed Template	0.687	0.580
	Prompt-tuning with rules (1)	0.727	0.630
	Prompt-tuning with rules (2)	0.697	0.635
	Knowledgeable verbalizer (1)	**0.73**	**0.66**
	Knowledgeable verbalizer (2)	0.72	0.65

When the prompt contains soft tokens, it leads to better performance (up to 0.02–0.09). Sophisticated techniques, such as prompt-tuning with rules, also contribute to an improvement (0.04–0.05 as compared with the Mixed Template strategy). However, the lexical content of the initialized prompt does not influence the final result. Since such tokens are tuned during training, many semantic aspects, though important for the natural language, are neglected.

Incorporating external knowledge into the verbalizer leads to the highest results ($F1_{avg} = 0.73$, $F1(pos, neg)_{avg} = 0.66$.). It also shows that the fuller verbalizer lists have better impact on the result than the disjoint ones, while emphasising positive sentiments does not increase classification quality for this class.

After the quantitative analysis of the results, it was of particular interest to look at the predicted tokens. Generally, analyzing soft tokens doesn't make much sense because, when projected onto natural language words, they tend to represent a sequence of word parts or entirely non-existent words in the language. However, for the specific task, it was important to see how the predicted tokens relate to information from vocabularies.

From the model block responsible for masked language modeling (MLM), logits were extracted and then transformed into tokens. It is precisely the logits of the tokens at the output of the MLM block that are projected onto class probabilities, which means they have a substantial influence on the final prediction. After extracting these tokens, we highlighted those that are present in the vocabulary but absent in the sentence (substrings were also considered) and evaluated their impact on prediction. The statistics is demonstrated in Table 5.

Table 5. Analysis of the predictions' statistics

Average number of vocabulary tokens in prediction	20
Maximum number of vocabulary tokens in prediction	281
Proportion of correct predictions for the neutral class given that at least 1 vocabulary token appears in the prediction	0.50
Proportion of correct predictions for the positive class given that at least 1 vocabulary token appears in the prediction	0.58
Proportion of correct predictions for the negative class given that at least 1 vocabulary token appears in the prediction	0.67
Correlation between prediction correctness and number of vocabulary tokens for positive and negative classes (using biserial criterion)	0.01 (p_value = 0.7)

Thus, it is evident that for positive and negative classes, a large proportion of correctly predicted class labels are associated with the examples in which the prediction text contains vocabulary tokens. For the negative class, this proportion is even higher than for the positive one due to the fact that its verbalizer contains more vocabulary words. Additionally, no correlation was detected between the number of vocabulary tokens and the prediction correctness.

6.2   RuSentNE 2023 Post-evaluation

As was mentioned above, the model that demonstrated the highest results both in cross-validation and the RuSentNE-2023 competition's validation set was the model with an extended verbalizer formed according to the first strategy described in 5.5. This model was used to evaluate the quality on the RuSentNE-2023 competition's test set during the post-evaluation stage of the competition. The results are presented in Table 6[1].

This result can be compared to the third best result of the conducted competition, lagging behind the first place by 4.51 according to the $F1(\text{pos}, \text{neg})_{\text{avg}}$

---

[1] These results are as of July 2023.

Table 6. Evaluating on the RuSentNE-2023 test set

RuSentNE-2023 participants	$F1_{avg}$	$F1(pos, neg)_{avg}$
1st place: MTS [29]	74.11	66.67
2nd place: cookies [30]	74.29	66.64
Knowledgeable verbalizer (1)	70.94	62.16

metric and by 3.17 according to the $F1_{avg}$ metric. However, the models from the top of the leaderboard leveraged ensembling methods, which are time-consuming and computationally expensive [29,30]. Moreover, these models were trained on additional datasets, while the model presented in the current study gained competitively high scores without dataset extension.

7   Conclusion

In this paper, we proposed a prompt-tuning method based on the knowledgeable verbalizer for targeted sentiment analysis of the Russian news texts. Specifically, we tested different learning strategies for both discrete and soft prompts as well as several technique to handle data imbalance. It turned out that prompt-tuning along with fine-tuning surpasses vanilla fine-tuning and manual prompting. Furthermore, it is relatively cheap in time and computational resources, as compared with the ensembling methods, which occupied the top of the leaderboard during the RuSentNE-2023 competition. Among prompt-tuning strategies, extending verbalizer lists via external linguistic resources performs the best, which is probably because it allows the MLM-block of the model to predict a wider range of tokens. All in all, it is quite clear that the present task, while being relatively new and under-researched, poses a challenge and presents an open field for future studies.

Acknowledgement. This work was supported by the Russian Science Foundation (grant No. 21-71-30003).

References

1. Baccianella, S., Esuli, A., Sebastiani, F.: SentiWordNet 3.0: an enhanced lexical resource for sentiment analysis and opinion mining. In: Proceedings of the Seventh International Conference on Language Resources and Evaluation (LREC'10), Valletta, Malta. European Language Resources Association (ELRA) (2010)
2. Biber, D., Finegan, E.: Styles of stance in English: lexical and grammatical marking of evidentiality and affect. Text-Interdisciplinary J. Study Discourse **9**, 93–124 (1989)
3. Devlin, J., Chang, M., Lee, K., Toutanova, K.: BERT: pre-training of deep bidirectional transformers for language understanding. In: Proceedings of NAACL-HLT, pp. 4171–4186 (2019)

4. Ding, N., et al.: OpenPrompt: an open-source framework for prompt-learning. In: Proceedings of the 60th Annual Meeting of the Association for Computational Linguistics: System Demonstrations, pp. 105–113 (2022)

5. Du, C., Sun, H., Wang, J., Qi, Q., Liao, J.: Adversarial and domain-aware BERT for cross-domain sentiment analysis. In: Proceedings of the 58th Annual Meeting of the Association for Computational Linguistics, pp. 4019–4028 (2020)

6. Gao, Z., Feng, A., Song, X., Wu, X.: Target-dependent sentiment classification With BERT. IEEE Access **7**, 154290–154299 (2019)

7. Goloviznina, V., Fishcheva, I., Peskisheva, T., Kotelnikov, E.: Aspect-based argument generation in Russian. In: Proceedings of the International Conference Dialogue (2023)

8. Golubev, A., Loukachevitch, N.: Improving results on Russian sentiment datasets. In: Filchenkov, A., Kauttonen, J., Pivovarova, L. (eds.) AINL 2020. CCIS, vol. 1292, pp. 109–121. Springer, Cham (2020). https://doi.org/10.1007/978-3-030-59082-6_8

9. Hamborg, F., Donnay, K.: NewsMTSC: a dataset for (multi-)target-dependent sentiment classification in political news articles. In: Proceedings of the 16th Conference of the European Chapter of the Association for Computational Linguistics: Main Volume, pp. 1663–1675, Online. Association for Computational Linguistics (2021)

10. Han, X., Zhao, W., Ding, N., Liu, Z., Sun, M.: PTR: Prompt Tuning with Rules for Text Classification. ArXiv, abs/2105.11259 (2021)

11. Hosseinia, M., Dragut, E., Mukherjee, A.: Stance prediction for contemporary issues: data and experiments. In: Proceedings of the Eighth International Workshop on Natural Language Processing for Social Media, pp. 32–40 (2020)

12. Hu, S., et al.: Knowledgeable prompt-tuning: incorporating knowledge into prompt verbalizer for text classification. In: Proceedings of the 60th Annual Meeting of the Association for Computational Linguistics (Volume 1: Long Papers), pp. 2225–2240 (2022)

13. Hutto, C., Gilbert, E.: VADER: a parsimonious rule-based model for sentiment analysis of social media text. In: Proceedings of the International AAAI Conference on Web and Social Media, 8(1), pp. 216–225 (2014). https://doi.org/10.1609/icwsm.v8i1.14550

14. HuggingFace ruRoBERTa-large https://huggingface.co/ai-forever/ruRoberta-large. Accessed 17 Jul 2023

15. Kiritchenko, S., Zhu, X., Cherry, C., Mohammad, S.: NRC-Canada-2014: detecting aspects and sentiment in customer reviews. In: Proceedings of the 8th International Workshop on Semantic Evaluation (SemEval 2014), pp. 437–442 (2014)

16. Konodyuk, N.: Prompt tuning for text detoxification. In: Computational Linguistics and Intellectual Technologies, pp. 1098–1105 (2022)

17. Lester, B., Al-Rfou, R., Constant, N.: The Power of Scale for Parameter-Efficient Prompt Tuning. ArXiv, abs/2104.08691 (2021)

18. Li, C., et al.: SentiPrompt: Sentiment Knowledge Enhanced Prompt-Tuning for Aspect-Based Sentiment Analysis. ArXiv, abs/2109.08306 (2021)

19. Liu, X., et al.: GPT Understands, Too. ArXiv, abs/2103.10385 (2021)

20. Loukachevitch N., et al.: NEREL: A Russian Dataset with Nested Named Entities, Relations and Events. In: Proceedings of RANLP-2021, pp. 880–889 (2021)

21. Loukachevitch, N., Levchik, A.: Creating a General Russian sentiment Lexicon. In: Proceedings of the Tenth International Conference on Language Resources and Evaluation (LREC'16), pp. 1171–1176 (2016)

22. Loukachevitch, N.V., Rusnachenko, N.: Sentiment frames for attitude extraction in Russian. In: Computational Linguistics and Intellectual Technologies, pp. 541–552 (2020)
23. Naumov, A., Rybka, R., Sboev, A., Selivanov, A., Gryaznov: Neural-network method for determining text author's sentiment to an aspect specified by the named entity. In: CEUR Workshop Proceedings, pp. 134–143 (2020)
24. Radford, A., Wu, J., Child, R., Luan, D., Amodei, D., Sutskever, I.: Language models are unsupervised multitask learners. OpenAI Blog **1**(8), 9 (2019)
25. Rusnachenko, N., Loukachevitch, N., Tutubalina, E.: Distant supervision for sentiment attitude extraction. In: Proceedings of the International Conference on Recent Advances in Natural Language Processing (RANLP 2019), pp. 1022–1030 (2019)
26. ruT5, ruRoBERTa, ruBERT: how we trained a series of models for Russian *(translated from Russian)*https://habr.com/ru/companies/sberbank/articles/567776/. Accessed 6 Sep 2023
27. Sun, C., Huang, L., Qiu, X.: Utilizing BERT for aspect-based sentiment analysis via constructing auxiliary sentence. North American Chapter of the Association for Computational Linguistics, pp. 380–385 (2019)
28. Yin, D., Meng, T., Chang, K.: SentiBERT: a transferable transformer-based architecture for compositional sentiment semantics. In: Annual Meeting of the Association for Computational Linguistics, pp. 3695–3706 (2020)
29. Podberezko, P., Kaznacheev, A., Abdullayeva, P., Kabaev, A.: HAlf-MAsked Model for Named Entity Sentiment analysis. In: Proceedings of the International Conference "Dialogue (2023)
30. Glazkova, A.: Fine-tuning text classification models for named entity oriented sentiment analysis of russian texts. In: Proceedings of the International Conference "Dialogue (2023)

On the Way to Controllable Text Summarization in Russian

Alena Dremina[1] and Maria Tikhonova[1,2(✉)]

[1] HSE University, Moscow, Russia
m_tikhonova@mail.ru
[2] SberDevices, Moscow, Russia

Abstract. This paper is devoted to controllable text summarization in Russian. We explore the effectiveness of a mixture-of-experts architecture for the Russian language. Namely, we adapt the HydraSum method, which involves multiple decoders, for Russian. We compare the performance of this approach with fine-tuning pre-trained language models and show that HydraSum is applicable for text summarization in Russian, yielding good-quality and stylistically diverse summaries.

Keywords: Language Modeling · Text Summarization · Text Generation

1 Introduction

This paper focuses on automatic text summarization, which belongs to the *Natural Language Generation (NLG)* group of tasks. Text summarization is a fundamental task in NLG, as quickly getting information from various sources is critical in the modern information-saturated world. Modern summarization models, driven mainly by neural networks, can produce accurate and concise texts, automatically capturing multiple essential features, such as the output length, specificity, abstractiveness, etc. The development of summarization models led to new areas of research. One such area is controllable text generation, which has become crucial in NLG in recent years due to its significant practical value in real-world applications. The importance of controllable text generation lies in the ability of users to control the output text in accordance with their needs. Thus, the generated text appears to be more appropriate for specific context. Existing summarization methods lack user control over the generated text, providing summaries that are not stylistically diverse. To address this issue, HydraSum architecture [1] was introduced. The main idea of the approach is to use a transformer-based model as an encoder and multi-decoder architecture, where different decoders capture different style features from the training data.

In this paper, we adapt the HydraSum method for a controllable text summarization task for Russian. The contribution of this paper is the following: we explore the potential of controllable text summarization for Russian by adapting the HydraSum method and comparing it to several summarization models. We

© The Author(s), under exclusive license to Springer Nature Switzerland AG 2024
D. I. Ignatov et al. (Eds.): AIST 2023, CCIS 1905, pp. 125–135, 2024.
https://doi.org/10.1007/978-3-031-67008-4_10

show that this architecture is applicable to the Russian language and can be used as an approach to controllable text summarization.

The paper is structured as follows: the Sect. 2 presents the overview of the related work; the Sect. 3 describes the methodology of the study including the description of the dataset and its preprocessing steps, model architecture, training, and inference processes; Sect. 4 describes the results, and finally Sect. 5 concludes the paper.

2   Related Work

In this section, we progressively examine the literature relevant to our study, starting with automatic summarization task in general and then moving forward to more specific topics.

The goal of automatic text summarization is to generate a concise and coherent paragraph of a long text capturing its essential information. In recent years, automatic text summarization has become crucial in many applications, including media monitoring (condensing huge amount of information into smaller parts), automatic document processing such as summarization of long documents, information retrieval, and question-answering systems. Various approaches for this task have been proposed. Most of them fall into two categories: extractive and abstractive summarization.

2.1   Extractive Summarization

The core idea of extractive summarization is to generate a summary by selecting and reordering the most important sentences or phrases directly from the original text. In such a formulation, the goal is to determine which sentences to consider important and how to combine them into a coherent text. For extractive summarization, ranking methods can be used, which score the sentences, pick the most significant ones, and place them in the right order. Such methods include *term frequency-inverse document frequency (TF-IDF)* [2] approach, graph-based methods [3–5], topic-based methods relying on topic representations such as the *Latent Dirichlet Allocation (LDA)* [6–8], machine learning-based methods, such as SVM [9], methods based on feedforward neural network [10], convolution neural network [11], RNN or GRU [12], and transformer based models [13,14].

However, despite the popularity of this approach, relatively simple implementation, and high accuracy of generated texts, extractive summarization has some obvious drawbacks. Its primary disadvantage is that texts generated within this approach must still be more human-like. Due to the wrong linking between sentences, summaries can be semantically poor and incohesive [15]. Moreover, if the input text consists of multiple topics, the generated summary may not capture all of them correctly.

2.2   Abstractive Summarization

Abstractive summarization partly handles these problems and produces more human-like texts. Its main idea is to generate a new text, which summarizes the input text, fully reflecting its key ideas, without taking the exact sentences from the source text.

Most methods of abstractive summarization are based on *Sequence-to-Sequence (seq2seq)* models [16,17], which represent an encoder-decoder architecture and are being trained on text pairs: original text - it's summary. First seq2seq models, however, showed poor quality for long texts, which are common it text summarization, due to the vanishing gradient problem. The appearance of the Transformer architecture with the underlying Attention mechanism [18] was a genuine breakthrough in NLP in general and in text summarization in particular. Self-attention mechanism enables the Transformer model to selectively focus on relevant parts of the input sequence [19], which proved very fruitful in text processing. Almost all of the latest NLP models are based on the Transformer architecture: *Generative Pre-Training (GPT)* [20], BERT [21], GPT-2 [22], DistilBERT [23], BART [24], T5 [25], PEGASUS [26], and GPT-3 [27]. These language models have been successfully used for abstractive text summarization either in the few-/zero-shot setting or by fine-tuning a pre-trained model.

In recent years methods of controllable text generation, in general, and controllable text summarization, in particular, are gaining popularity. These methods introduce some kind of control into the text generation process so that output satisfies the desired characteristics. One of the latest approaches to controllable summarization is HydraSum [1] which showed quite promising results for the English language.

2.3   Summarization in Russian

Automatic summarization has made significant progress with the advent of deep learning techniques.

The research [33–35] devoted to text summarization in Russian explores the abilities of mBART [30], ruGPT3[1], mT5 [31], and ruT5[2] for extractive [32] and abstractive summarization. The comparative analysis of these models' performance [36] has revealed that ruGPT3 most abstractive summaries, which, however, contain a lot of mistakes and often are not completed; mBart generates the least abstractive summaries and tends to repeat parts of the original text; ruT5 models have provided the best summaries, with a high degree of abstractiveness and little number of mistakes.

The main datasets, which can be used for text summarization in Russian include XSum [14], Gazeta [33], MLSUM [38], CNN/Dailymail [39], Xl-Sum [35], and Wikilingua [40].

Thus, automatic text summarization for Russian constantly develops with the appearance of new Russian-based Transformer models. However, the progress in

---

[1] https://huggingface.co/ai-forever/rugpt3large_based_on_gpt2.

[2] https://huggingface.co/ai-forever/ruT5-large.

this area still needs to catch up to the one for the English language. There are still questions to be explored and room for improvement. One such question is controllable text summarization, which is able to provide stylistically diverse summaries. This paper addresses this issue and for this purpose adapts Hydra-Sum architecture to the Russian language.

3   HydraSum Method

In this work, we explore one of the latest approaches to controllable text summarization called HydraSum [1]. The authors suggest a new mixture-of-experts architecture with multiple decoders based on the pre-trained language model. They choose Facebook's Bart-large as a base model. In HydraSum, the base model's architecture is extended to k decoders (the authors experiment with 2 and 3-decoder architecture). Each decoder captures some stylistic feature of the input text. The lower layers of decoders have shared parameters, which is done to minimize the number of additional parameters introduced into the model architecture. The top layers of the different decoders are trained independently. Thus, each decoder can specialize and learn distinct representations to suit the specific task while still benefiting from the shared parameters from the lower layers.

Another important part of HydraSum is a gating mechanism (g), which is a weighted sum of the k decoders' output. It dynamically determines how much each decoder's output contributes to the overall result, enabling flexibility in decision-making based on the weighted contributions. After utilizing the gating mechanism, the outputs from the shared layers are fed into a feedforward layer followed by a softmax activation function, which outputs the overall next-token output probability is computed using Eq. 1.

$$P(y_i|x, y < i) = \sum_{j=1:k} g_i^j \times P_{\phi_j}(y_i|x, y < i) \tag{1}$$

Thus, this process assigns weights to the outputs, determining their relative importance.

The authors provided three inference strategies:

1. sampling from individual decoders, where one decoder is more extractive, as it learns to copy parts of the input text, whereas the other is abstractive, as it learns how to paraphrase sentences;
2. a mixture using the model's gating mechanism;
3. mixture using a manually specified gating mechanism.

The generation process is controllable in the sense that the model automatically separates distinct summary styles across different decoders in process of training. This in turn introduces flexibility during inference as one can sample from either individual decoders or their mixtures to vary summary features.

To evaluate the HydraSum performance several metrics were chosen: Quality (ROUGE [37]), Abstractiveness, Specificity, Length, and Readability. The original paper shows that for English, all HydraSum strategies achieve comparable or even better results than the baseline model (the BART model fine-tuned for the summarization task). Moreover, a mixture of decoders shows better results than individual decoders.

4 Experimental Setup

4.1 Models

HydraSum architecture shows good results in English, outperforming the pre-trained language model fine-tuned for the task. In our work, we adapt Hydra-Sum architecture to the Russian language and explore whether this approach is suitable for other languages besides English. For adaptation we use mBART[3] model from HuggingFace library. mBART [30] is a multilingual language model pre-trained on massive multilingual text corpora, including Russian.

We choose the 2-decoder architecture of HydraSum. The number of decoder blocks in mBART equals 12. Following the original approach, we set the number of shared layers to 8. The authors also experiment with the number of shared layers, setting it to 6 and 10; however, it does not significantly influence the result. The model is trained for three epochs, with the batch equal to 8 and the learning rate 1e−5. We use cross-entropy loss as the training.

At the inference stage, we perform sampling from both decoders separately (in the experiments, we call them *Decoder 0* and *Decoder 1*) and sample from a mixture of decoders (to which we refer to as *Mixture of decoders*) to examine the differences in results between these strategies.

As a baseline, we fine-tune several pre-trained language models: ruT5-base, mBART and ruGPT3Small. The models are fine-tuned on the Gazeta dataset. Fine-tuning hyperparameters are presented in Table 1. The code is available in our repository[4].

Table 1. Fine-tuning hyperparameters.

ruT5-base	ruGPT3Small	mBART
Max input length = 1024	Max output length = 128	Max input length = 1024
Max output length = 128	Max input length = 1024	Max output length = 128
Learning rate = 2e−5	Learning rate = 1e−5	Learning rate = 1e−5
Batch size = 16	Batch size = 16	Batch size = 3
Epochs = 3	Epochs = 3	Epochs = 3
no repeat ngram size = 4	no repeat ngram size = 4	no repeat ngram size = 4

---

[3] https://huggingface.co/facebook/mbart-large-cc25.
[4] https://github.com/AlyDr/ControllableTextSum.

4.2 Data

In our experiments, we use the Gazeta dataset[5], one of the most popular datasets for summarization task in Russian. The dataset consists of the news articles and their summaries from Gazeta.ru[6]. It also includes article titles, dates, and URLs. The dataset is split by date into the train, validation, and test parts with 52 400, 5 265, and 5 770 examples in each part, respectively. Detailed dataset statistics are presented in Table 2.

Table 2. Gazeta dataset statistics.

	Train		Validation		Test	
	Text	Summary	Text	Summary	Text	Summary
Min words	28	15	191	18	357	18
Max words	1 500	85	1 500	85	1 498	85
Avg. words	766.5	48.8	772.4	54.5	750.3	53.2
Avg. Sentences	37.2	2.7	38.5	3.0	37.0	2.9
Unique words	611 829	148 073	167 612	42 104	175 369	44 169
Unique lemmas	282 867	63 351	70 210	19 698	75 214	20 637

4.3 Metrics

For the evaluation, we generate three summaries with the HydraSum approach (one for each strategy) and one summary from each fine-tuned model. We evaluate generated summaries with ROUGE metrics [37], a standard text summarization metric, which is measured between the generated and reference summaries. Apart from ROUGE, following the original methodology, we compute several summarization-relevant metrics: abstractiveness, specificity, length, and readability.

Abstractiveness, proposed in [48], is measured by two aspects: coverage, which counts the proportion of words presented in both input text and in summary; density, which counts the average longest continuous extract copied from the input text. Specificity is measured with a Speciteller tool [49].

The length of summaries is measured by an absolute length (number of words) and by compression rate, computed as the number of words in the generated summary divided by the number of words in the input text.

Finally, we compute readability score [50] adapted for Russian language to measure the quality of the generated texts.

---

[5] https://huggingface.co/datasets/IlyaGusev/gazeta.
[6] https://www.gazeta.ru/.

5 Results

Results of fine-tuned pre-trained language models and HydraSum architecture are presented in Table 3 and examples of generated summaries can be found in Fig. 1. The analysis reveals a significant difference between the performance of two individual HydraSum decoders. It can be seen that Decoder 0, although it generates longer summaries, has lower coverage and density than Decoder 1. Moreover, it can be observed that Decoder 1 is more extractive than Decoder 0, as it shows bigger coverage. However, the most extractive summaries are produced by ruT5-base and mBART, and the most abstractive summary appears to be the reference one, showing low coverage and density. As for the mixture of decoders, it produces more abstractive summaries than each decoder individually.

Input Text	Prediction
Сооснователь Microsoft Билл Гейтс в интервью Bloomberg ответил на вопрос о том, можно ли считать главу Tesla и SpaceX Илона Маска «новым Стивом Джобсом», учитывая прогресс, которого добились его компании. Однако филантроп считает, что такое сравнение нельзя назвать уместным. ... По его словам, между двумя новаторами есть ряд ключевых отличий. «Илон — это скорее инженер-практик. Стив же был гением дизайна, маркетинга и подбора персонала. ...Между тем Стив Джобс и Билл Гейтс на протяжении 30 лет были непримиримыми соперниками. Ранее «Газета.Ru» писала о непростой истории их взаимоотношений. Теперь же, после смерти Джобса, кажется, что Гейтс нашел себе нового достойного противника в лице Илона Маска. В том же интервью Bloomberg филантроп заявил, что «крайне уважает» Маска за электроавтомобили Tesla... Ранее на этой неделе два миллиардера вновь поссорились из-за электромобилей. В своем блоге Билл Гейтс порассуждал на тему будущего этого вида транспорта. ... он выразил сомнения касательно применения этой технологии для грузовых машин, перемещающихся на дальние расстояния...на что глава Tesla заявил, что Гейтс «не имеет понятия о том, о чем говорит». ..."	**Reference.** Илона Маска нельзя назвать новым Стивом Джобсом, так как это «грубое упрощение», заявил основатель Microsoft Билл Гейтс в интервью Bloomberg. Филантроп, который был лично знаком с обоими визионерами, пояснил, что их невозможно сравнивать между собой, так как Маск является инженером-практиком, а Джобс был «гением» сразу в нескольких сферах бизнеса. **mBART.** Сооснователь Microsoft Билл Гейтс заявил, что не считает Илона Маска «новым Стивом Джобсом». По словам филантропа, между двумя новаторами есть ряд ключевых отличий. Он также отметил, что Маск был гением дизайна, маркетинга и подбора персонала. **RuT5-base.** Сооснователь Microsoft Билл Гейтс ответил на вопрос о том, можно ли считать главу Tesla и SpaceX Илона Маска «новым Стивом Джобсом», учитывая прогресс, которого добились его компании. По его словам, между двумя новаторами есть ряд ключевых отличий. **RuGPT3-small.** Глава Tesla Илон Маск в интервью Bloomberg назвал основателя Microsoft Стивена Джобса «новым Илоном Маском» из-за того, что тот «был гением дизайна и подбора персонала». **D0.** Билл Гейтс считает сравнение Илона Маска и Стива Джобса неуместным. По его словам, между двумя новаторами есть ряд отличий: Маск - инженер-практик, а Джобс - гений дизайна и маркетинга. **D1.** Сооснователь Microsoft Билл Гейтс в интервью Bloomberg назвал Илона Маска «новым Стивом Джобсом». Билл Гейтс и Илон Маск поспорили из-за электрокаров." **Mixture of decoders.** Билл Гейтс, сооснователь Microsoft, в интервью Bloomberg отметил различия между Илоном Маском и Стивом Джобсом, назвав их разными типами инноваторов. Между тем, Маск и Гейтс стали новыми соперниками после смерти Джобса. Во время спора о будущем электромобилей Маск отметил, что Гейтс "не имеет понятия о том, о чем говорит".

Fig. 1. Examples of summaries generated by different models.

Regarding specificity, here, the results are pretty close to each other except for Decoder 0, which shows the lowest specificity score of 0.57. The highest result of 0.68 is achieved by a mixture of decoders and the ruT5-base model.

One of the interesting findings is that summaries generated by HydraSum decoders, individually or in a mixture, provide information from the input text,

Table 3. Results of summarization experiments. *Decoder 0* stays for sampling from the first decoder. *Decoder 1* stays for sampling from the second decoder. *Mixture of decoders* stays for sampling from both decoders.

	abstractiveness		specificity	length		readability	ROUGE		
	coverage	density		absolute length	compression ratio		R1	R2	RL
Reference	0.71	1.52	0.64	**49.0**	**0.12**	44.0	-		
ruT5-base	**1.0**	**22.2**	**0.68**	37.0	0.09	59.0	30.18	11.88	27.31
ruGPT3- small	0.77	1.61	0.63	25.0	0.06	58.0	17.93	3.79	15.47
mBART	0.95	5.68	0.61	35.0	0.09	59.0	**30.97**	**12.61**	**28.36**
Decoder 0	0.82	3.65	0.57	28.0	0.07	55.0	29.45	10.72	26.89
Decoder 1	0.92	4.36	0.64	21.0	0.05	**78.0**	29.83	10.79	27,05
Mixture of decoders	0.76	2.48	**0.68**	**49.0**	**0.12**	61.0	30.12	11.14	27.68

which is not reflected in the summaries generated by fine-tuned models. However, this should be more thoroughly analyzed further.

As for ROUGE, HydraSum decoders show results comparable with the fine-tuned models by this metric. This demonstrates that additional decoders do not influence the quality of text generation.

An important finding of this study is that without any initial difference in training between individual decoders of HydraSum architecture, they tend to capture different stylistic features in Russian-language texts as well as they do in English texts due to the randomly assigned weights by the gate (g), which differently contribute to the final output probabilities. This observation proves that the HydraSum approach is applicable to the Russian language.

6   Conclusion

In this paper, we attempted to adapt the HydraSum approach for controllable text summarization in Russian. We used the HydraSum with the mBART model and compared it with several fine-tuned baselines.

In the experiments, the new approach proved to be applicable to the Russian language showing promising results on qualitative and style-specific metrics.

As a part of future research, it is planned to train this model with more decoders and on a more extensive dataset to try to capture more stylistically diverse features. Moreover, we plan to use HydraSum with other models. In addition, we plan to manually specify gate (g) during the inference stage to explore the architecture's capabilities for the Russian language further.

References

1. Goyal, T., Rajani, N., Liu, W., Kryściński, W.: HydraSum: disentangling style features in text summarization with multi-decoder models. In: Proceedings of the 2022 Conference on Empirical Methods in Natural Language Processing, pp. 464–479 (2022)
2. Rossiello, G., Basile, P., Semeraro, G.: Centroid-based text summarization through compositionality of word embeddings. MultiLing **2017**, 12 (2017)

3. Mallick, C., Das, A.K., Dutta, M., Das, A.K., Sarkar, A.: Graph-based text summarization using modified TextRank. Soft Comput. Data Anal. **137** (2019)
4. Uçkan, T., Karcı, A.: Extractive multi-document text summarization based on graph independent sets. Egypt. Inf. J. **21**(3), 145–157 (2020)
5. Van Lierde, H., Chow, T.W.: Query-oriented text summarization based on hypergraph transversals. Inf. Process. Manag. **56**(4), 1317–1338 (2019)
6. Haghighi, A., Vanderwende, L.: Exploring content models for multi-document summarization. In: Proceedings of Human Language Technologies: The 2009 Annual Conference of the North American Chapter of the Association for Computational Linguistics, pp. 362–370 (2009)
7. Chang, Y.L., Chien, J.T.: Latent Dirichlet learning for document summarization. In: 2009 IEEE International Conference on Acoustics, Speech and Signal Processing, pp. 1689–1692. IEEE (2009)
8. Belwal, R.C., Rai, S., Gupta, A.: Text summarization using topic-based vector space model and semantic measure. Inf. Process. Manag. **58**(3), 102536 (2021)
9. Leskovec, J., Milic-Frayling, N., Grobelnik, M.: Extracting summary sentences based on the document semantic graph (2005)
10. Svore, K., Vanderwende, L., Burges, C.: Enhancing single-document summarization by combining RankNet and third-party sources. In: Proceedings of the 2007 Joint Conference on Empirical Methods in Natural Language Processing and Computational Natural Language Learning (EMNLP-CoNLL), pp. 448–457 (2007)
11. Cao, Z., Li, W., Li, S., Wei, F., Li, Y.: AttSum: joint learning of focusing and summarization with neural attention. In: Proceedings of COLING 2016, the 26th International Conference on Computational Linguistics: Technical Papers, pp. 547–556 (2016)
12. Nallapati, R., Zhai, F., Zhou, B.: SummaRuNNer: a recurrent neural network based sequence model for extractive summarization of documents. In: Proceedings of the AAAI Conference on Artificial Intelligence, vol. 31, no. 1 (2017)
13. Liu, Y., Lapata, M.: Text summarization with pretrained encoders. In: Proceedings of the 2019 Conference on Empirical Methods in Natural Language Processing and the 9th International Joint Conference on Natural Language Processing (EMNLP-IJCNLP), p. 3721. Association for Computational Linguistics (2019)
14. Narayan, S., Cohen, S.B., Lapata, M.: Don't give Me the details, just the summary! Topic-aware convolutional neural networks for extreme summarization. In: Proceedings of the 2018 Conference on Empirical Methods in Natural Language Processing. Association for Computational Linguistics (2018)
15. Moratanch, N., Chitrakala, S.: A survey on abstractive text summarization. In: 2016 International Conference on Circuit, Power and Computing Technologies (ICCPCT), pp. 1–7. IEEE (2016)
16. Sutskever, I., Vinyals, O., Le, Q.V.: Sequence to sequence learning with neural networks. In: Advances in Neural Information Processing Systems, vol. 27 (2014)
17. Cho, K., et al.: Learning phrase representations using RNN encoder-decoder for statistical machine translation. In: Proceedings of the 2014 Conference on Empirical Methods in Natural Language Processing (EMNLP), p. 1724. Association for Computational Linguistics (2014)
18. Bahdanau, D., Cho, K., Bengio, Y.: Neural machine translation by jointly learning to align and translate. arXiv preprint arXiv:1409.0473 (2014)
19. Vaswani, A., et al.: Attention is all you need. In: Advances in Neural Information Processing Systems, vol. 30 (2017)
20. Radford, A., Narasimhan, K., Salimans, T., Sutskever, I.: Improving language understanding by generative pre-training (2018)

21. Kenton, J.D.M.W.C., Toutanova, L.K.: BERT: pre-training of deep bidirectional transformers for language understanding. In: Proceedings of NAACL-HLT, pp. 4171–4186 (2019)
22. Radford, A., Wu, J., Child, R., Luan, D., Amodei, D., Sutskever, I.: Language models are unsupervised multitask learners. OpenAI Blog **1**(8), 9 (2019)
23. Sanh, V., Debut, L., Chaumond, J., Wolf, T.: DistilBERT, a distilled version of BERT: smaller, faster, cheaper and lighter. arXiv preprint arXiv:1910.01108 (2019)
24. Lewis, M., et al.: BART: denoising sequence-to-sequence pre-training for natural language generation, translation, and comprehension. In: Proceedings of the 58th Annual Meeting of the Association for Computational Linguistics, pp. 7871–7880 (2020)
25. Raffel, C., et al.: Exploring the limits of transfer learning with a unified text-to-text transformer. J. Mach. Learn. Res. **21**(1), 5485–5551 (2020)
26. Zhang, J., Zhao, Y., Saleh, M., Liu, P.: PEGASUS: pre-training with extracted gap-sentences for abstractive summarization. In: International Conference on Machine Learning, pp. 11328–11339. PMLR (2020)
27. Brown, T., et al.: Language models are few-shot learners. In: Advances in Neural Information Processing Systems, vol. 33, pp. 1877–1901 (2020)
28. OpenAI.: GPT-4 Technical Report (2023)
29. Joshi, P., Santy, S., Budhiraja, A., Bali, K., Choudhury, M.: The state and fate of linguistic diversity and inclusion in the NLP world. In: Proceedings of the 58th Annual Meeting of the Association for Computational Linguistics, pp. 6282–6293 (2020)
30. Liu, Y., et al.: Multilingual denoising pre-training for neural machine translation. Trans. Assoc. Comput. Linguist. **8**, 726–742 (2020)
31. Xu, P., et al.: MEGATRON-CNTRL: controllable story generation with external knowledge using large-scale language models. In: Proceedings of the 2020 Conference on Empirical Methods in Natural Language Processing (EMNLP), pp. 2831–2845 (2020)
32. Polyakova, I., Pogoreltsev, S.: Extractive Russian text summarization as greedy sentence sequence continuation search with probabilities from pretrained language models. VV Golenkov-Editor-in-chief, 303 (2021)
33. Gusev, I.: Dataset for automatic summarization of Russian news. In: Filchenkov, A., Kauttonen, J., Pivovarova, L. (eds.) AINL 2020. CCIS, vol. 1292, pp. 122–134. Springer, Cham (2020). https://doi.org/10.1007/978-3-030-59082-6_9
34. Alexandr, N., Irina, O., Tatyana, K., Inessa, K., Arina, P.: Fine-tuning GPT-3 for Russian text summarization. In: Silhavy, R., Silhavy, P., Prokopova, Z. (eds.) CoMeSySo 2021. LNNS, vol. 231, pp. 748–757. Springer, Cham (2021). https://doi.org/10.1007/978-3-030-90321-3_61
35. Hasan, T., et al.: XL-sum: large-scale multilingual abstractive summarization for 44 languages. In: Annual Meeting of the Association of Computational Linguistics and International Joint Conference on Natural Language Processing 2021, pp. 4693–4703. Association for Computational Linguistics (ACL) (2021)
36. Goloviznina, V., Kotelnikov, E.: Automatic summarization of russian texts: comparison of extractive and abstractive methods. In: Proceedings of the International Conference "Dialogue 2022" (2022)
37. Chin-Yew, L.: ROUGE: a package for automatic evaluation of summaries. In: Proceedings of the Workshop on Text Summarization Branches Out (2004)

38. Scialom, T., Dray, P.A., Lamprier, S., Piwowarski, B., Staiano, J.: MLSUM: The multilingual summarization corpus. In: Proceedings of the 2020 Conference on Empirical Methods in Natural Language Processing (EMNLP), pp. 8051–8067 (2020)
39. Hermann, K.M., et al.: Teaching machines to read and comprehend. In: Advances in Neural Information Processing Systems, p. 28 (2015)
40. Ladhak, F., Durmus, E., Cardie, C., Mckeown, K.: WikiLingua: a new benchmark dataset for cross-lingual abstractive summarization. In: Findings of the Association for Computational Linguistics: EMNLP 2020, pp. 4034–4048 (2020)
41. Nan, F., et al.: Entity-level factual consistency of abstractive text summarization. In: Proceedings of the 16th Conference of the European Chapter of the Association for Computational Linguistics: Main Volume, pp. 2727–2733 (2021)
42. Li, X.L., Liang, P.: Prefix-tuning: optimizing continuous prompts for generation. In: Proceedings of the 59th Annual Meeting of the Association for Computational Linguistics and the 11th International Joint Conference on Natural Language Processing (Volume 1: Long Papers), pp. 4582–4597 (2021)
43. Keskar, N.S., McCann, B., Varshney, L.R., Xiong, C., Socher, R.: CTRL: a conditional transformer language model for controllable generation. arXiv preprint arXiv:1909.05858 (2019)
44. Zeng, Y., Nie, J.Y.: Generalized conditioned dialogue generation based on pretrained language model. arXiv preprint arXiv:2010.11140 (2020)
45. Fan, A., Lewis, M., Dauphin, Y.: Hierarchical neural story generation. In: Proceedings of the 56th Annual Meeting of the Association for Computational Linguistics (Volume 1: Long Papers). Association for Computational Linguistics (2018)
46. Krause, B., et al.: GeDi: generative discriminator guided sequence generation. In: Findings of the Association for Computational Linguistics: EMNLP 2021, pp. 4929–4952 (2021)
47. Mireshghallah, F., Goyal, K., Berg-Kirkpatrick, T.: Mix and match: learning-free controllable text generationusing energy language models. In: Proceedings of the 60th Annual Meeting of the Association for Computational Linguistics (Volume 1: Long Papers), pp. 401–415 (2022)
48. Grusky, M., Naaman, M., Artzi, Y.: NEWSROOM: a dataset of 1.3 million summaries with diverse extractive strategies. In: Proceedings of the 2018 Conference of the North American Chapter of the Association for Computational Linguistics: Human Language Technologies, Volume 1 (Long Papers). Association for Computational Linguistics (2018)
49. Li, J., Nenkova, A.: Fast and accurate prediction of sentence specificity. In: Proceedings of the AAAI Conference on Artificial Intelligence, vol. 29, no. 1 (2015)
50. Solnyshkina, M., Ivanov, V., Solovyev, V.: Readability formula for Russian texts: a modified version. In: Batyrshin, I., Martínez-Villaseñor, M.L., Ponce Espinosa, H.E. (eds.) MICAI 2018. LNCS (LNAI), vol. 11289, pp. 132–145. Springer, Cham (2018). https://doi.org/10.1007/978-3-030-04497-8_11
51. Jin, D., Jin, Z., Hu, Z., Vechtomova, O., Mihalcea, R.: Deep learning for text style transfer: a survey. Comput. Linguist. **48**(1), 155–205 (2022)

Document-Level Relation Extraction in Russian

Alexey V. Yandutov$^{(\boxtimes)}$ and Natalia V. Loukachevitch

Lomonosov Moscow State University, Moscow, Russia
wynand.enterprises@gmail.com

Abstract. We explore automatic relation extraction at the document level for Russian texts, considering nested entities. Using state-of-the-art models, we conducted first-of-its-kind experiments on Russian texts and made necessary model adjustments to implement experiments on the NEREL dataset, which is annotated with nested named entities and document-level relations between them. The results validate the efficiency of the proposed techniques in relation extraction from Russian texts, including nested entities.

Keywords: Document level relation extraction · Nested named entities · Russian language

1 Introduction

The present era of rapid advancements in artificial intelligence (AI) marks the prominence of Natural Language Processing (NLP), a field that is particularly affected by the burgeoning amounts of digitally-stored data. With the increasing demand for automatic extraction of information from text, vital to numerous NLP applications such as question-answer systems, information retrieval, and augmentation of knowledge databases, the role of automatic extraction of relationships or relations between entities within a text, a process termed Relation Extraction (RE), is increasingly pivotal.

Entities within textual contexts often establish distinct relationships that can be harnessed in a multitude of applications, including the analysis of news articles, scholarly publications, and social media posts. RE thereby has far-reaching implications, significantly impacting industries beyond academia, particularly the business sector.

Despite substantial strides since the 1990s, RE as a field is complex and underdeveloped relative to Named Entity Recognition (NER). Significant improvement in RE has been observed with the introduction of transformer architectures, but traditional RE methods, largely confined to intra-sentence relationships, fall short when it comes to untangling complex relations that span sentences in a document. This highlights the need for extraction at the document level, a need that is particularly pressing given that most existing research in the area is largely focused on English texts.

© The Author(s), under exclusive license to Springer Nature Switzerland AG 2024
D. I. Ignatov et al. (Eds.): AIST 2023, CCIS 1905, pp. 136–148, 2024.
https://doi.org/10.1007/978-3-031-67008-4_11

Our work addresses this gap, embarking on the exploration of document-level relation extraction in Russian texts, a domain that remains largely uncharted. Until recently, available datasets for relation extraction in the Russian language were scarce. The advent of the NEREL dataset, annotated with nested named entities and their relationships, offers an unprecedented opportunity to delve into relation extraction methodologies, encompassing nested named entities and document-level relationships.

Within this framework, we investigate the automatic extraction of relations from Russian texts at the document level using advanced transformer models. The long-tail nature of these texts, often exceeding 1024 tokens, necessitates modifications to existing models to ensure accurate and efficient processing.

Document-level RE is inherently computationally intensive and poses challenges with regard to resource allocation. To address this, our research includes strategies to balance the quantity of negative relationships in the training set, thus optimizing memory utilization and accelerating model convergence, while maintaining the high quality of the models.

Through these experiments, our research aims to further the capabilities of RE in Russian language texts and to provide a robust foundation for subsequent research in the field.

2 Related Work

In sentence-level relation extraction approaches, the primary focus lies on extracting relations between two entities within a sentence. However, many real-world relations can only be extracted by reading across several sentences, leading to the more complex task of document-level relation extraction. For this reason, document-level relation extraction still attracts the attention of many researchers [10]. The document-level relation extraction cannot be reduced to cross-sentence relation-extraction. The main modern approaches to document-level relation extraction include graph-based models and transformer-based models.

One of the earliest works was the paper by Yao et al. [16], where the authors proposed the DocRED dataset and as a basic solution, adapted CNN, LSTM, BiLSTM, context-aware LSTM [7] models for document-level relation extraction. Graph-based models are widely used in relation extraction due to their efficiency and ability to justify predictions. In the study [4], a model that combines representations learned across various textual areas in the document and across the hierarchy of sub-relations was proposed. In the work [1], a graph-based model with directed edges (EoG) was proposed for document-level relation extraction. Explicit graphical justification can bridge the gap between entities appearing in different sentences, mitigating the dependence on long distances, and achieving promising performance.

Contrary to this, given the transformer architecture which can implicitly model long-distance dependencies, some researchers directly apply pre-trained language models without creating document graphs. In the work [12], a two-step approach to training in DocRED was proposed using BERT vector representations. In the study [18], a novel transformer-based model (ATLOP) with an adaptive threshold and a localised context pool based on BERT was proposed.

Subsequent approaches such as DocUNET [17], based on semantic segmentation, were proposed in line with the idea of segmentation in computer vision. At present, the state-of-the-art approach in the DocRED dataset is the DREEAM model proposed in the paper [6], which uses an attention mechanism to guide the model to the most relevant statements in the document.

Significantly, there has been a lack of experimentation with document-level relation extraction in the Russian language. Although there are several Russian datasets annotated with relations [2,3,8], there is a noticeable absence of research focus on this area. In 2020, the RuReBus evaluation event was organized, specifically dedicated to the advancement of relation extraction in Russian [3]. This event produced promising outcomes with the R-BERT model, which showed superior results [13]. But in recent RuReBus and RuRed datasets, relations were annotated on the sentence level. In particular, the NEREL dataset [5] is the first Russian resource that includes cross-sentence relation annotations suitable for document-level relation extraction research. In [5] some initial experiments in cross-sentence relation extraction based on NEREL were described, which showed complexity and low performance of extracting such relations.

While the DeepPavlov framework[1] incorporates model for document-level RE (ATLOP model [18]), known for its adaptive thresholding and localized context pooling features, it is worth noting that our study considers more recent advancements. Specifically, the DREEAM model (2023) [6], which we include, can be seen as an extension or enhancement of the ATLOP model. This allows us to explore newer approaches in document-level relation extraction. Also it is important to highlight that DeepPavlov's model does not offer a straightforward way to be retrained on new datasets. Additionally, the relationships predicted by DeepPavlov are based on the RuRed dataset [2], which differs from the NEREL dataset used in our study and mainly contains relation annotations on the sentence level. This presents limitations in terms of adaptability and scope.

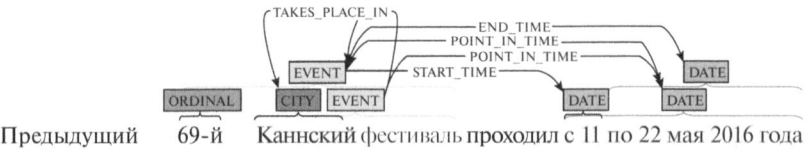

Fig. 1. Example of nested entities and relations between them in a sentence "The previous 69th Cannes Film Festival took place from 11 to 22 May 2016".

---

[1] https://docs.deeppavlov.ai/en/master/features/models/re.html.

3   Dataset

Supervised models for RE require adequate amounts of annotated data for their training. It is time-consuming to manually label large-scale training data.

NEREL [5] is a dataset in Russian for named entity recognition and relation extraction based on Russian Wikinews documents. NEREL is much larger than the existing Russian datasets: today it contains 933 documents annotated with 29 entity and 49 relations types. NEREL also contains an event annotation involving named objects and their roles in the events.

An important distinction of NEREL from the previous datasets is the annotation of nested named entities (up to 6 levels of nestedness), as well as relations within nested entities and at sentence and the document level. NEREL allows the development of new models that can extract relations between nested named entities as well as relations at both the sentence and document levels. Figure 1 shows an example of nested entities and relations between them on sentence-level.

All the annotated relations can be subdivided into cross-sentence (24% of the total number of relations) in which the subject and object are in different sentences and within sentence relations (76% of the total number of relations).

Figure 2 shows an example of cross-sentence relations between entities. In NEREL there are also such relations as *alternative_name* and *abbreviation*, which allow connecting mentions of the same entities in the text. In Fig. 2 such relations are established between "Konstantin Ryabinov" and "Kuzia UO" mentions. A special document-level representation can be generated from the available NEREL annotation, which lists all mentioned entities and relations between them on the document level.

The average text length is 264 words. However, the maximum length is significantly larger at 1991 words, notably more than the standard DOCRED dataset for English, where the text length does not exceed 1024 tokens. This necessitates handling long sequences. Moreover, this task demonstrates a quadratic dependency on the number of entities in a document. It is crucial to note the large number of entities in a document: the 99th percentile equals 89 entities, while the 95th percentile equals 69 entities.

4   Approaches

A document can contain multiple mentions of an entity and its relations. The task of the document-level relation extraction is as follows: having a list of extracted entities (not mentions) to determine relations between them.

The DocRED dataset is one of the most popular and widely used benchmarks for document-level relation extraction. To identify the most effective models, we analysed the state-of-the-art models for DocRED and ReDocRED [11] listed on https://paperswithcode.com/ in February-March 2023. From this list, we selected different top-performing approaches that not only demonstrated superior performance, but also had publicly available working code and could be

> **Субъект**: «Гражданская оборона»
> **Объект**: Константин Рябинов
>
> [0] В Санкт-Петербурге в возрасте 55 лет умер музыкант и поэт *Константин Рябинов*, известный как *Кузя УО*. [1] Он был одним из основателей панк-группы *«Гражданская оборона»* ... [5] *Кузя УО* играл на гитаре и бас-гитаре в студийных записях и концертных выступлениях ...
>
> **Отношение**: FOUNDED_BY
> **Подтверждающие предложения**: [0, 1]

Fig. 2. Example of document level relation in text: "[0] In St. Petersburg, musician and poet Konstantin Ryabinov, known as Kuzia UO, has died at the age of 55. [1] He was one of the founders of the punk band "Grazhdanskaya Oborona" ... [5] Kuzia UO played guitar and bass in studio recordings and concert performances ...".

adopted to reproduce on a Russian dataset. In addition, we considered it essential to establish a robust baseline for our experiments. To this end, we adopted the model recommended as a baseline by the authors of the DocRED paper. This baseline serves as our initial point of comparison, enabling us to evaluate the performance of the other models and iteratively improve our approaches.

4.1   BiLSTM and Context-Aware LSTM

For document-level relation extraction, the creators of the DocRED dataset proposed an adaptation of BiLSTM and context-aware LSTM models as a baseline solution. This implementation is described as follows:

- Document Encoding: BiLSTM is deployed to transcribe a document of n words into a sequence of corresponding hidden state vectors, denoted as h_i, with i spanning from 1 to n.
- Input Features: For every word within the document, the input features are constituted by the concatenation of GloVe word embeddings, entity-type vectors, and coreference vectors.
- Entity Representation: The representation of each named entity mention is determined as the mean of the hidden states for the words encompassed within that particular mention.
- Relation Prediction: The prediction of relations between pairs of entities is formulated as a multi-class classification task, employing a bilinear function for this purpose.

4.2   DocuNET

DocuNET [17] models relation extraction by predicting entity-level relation matrices akin to semantic segmentation in computer vision. The proposed model is constituted of:

Encoder Module. This module utilises a pre-trained language model to yield vector representations of tokens. To manage extensive documents, a dynamic window is used, and entity vector representations are derived through Logsum exp pooling.

Relation Matrix Calculation Module Using entity vector representations, an entity relation matrix is computed, employing both similarity-based and context-based features for entity pairs.

U-shaped Segmentation Module Employing a U-Net architecture, this module treats the entity-level relation matrix as a D-channel input image. It integrates both local and global information in predicting entity relationships via two down-sampling and up-sampling blocks with skip connections.

Classification Module Using a bidirectional linear function, this module predicts relationship probabilities between entity pairs. A balanced softmax method, inspired by circle loss in computer vision, addresses imbalanced distribution of relations.

4.3   DREEAM

Research indicates that evidence sentences, defined as those containing information about the relationship between a pair of entities, can aid DocRE systems in concentrating on the relevant text, thereby improving relation extraction. DREEAM [6] enhances the ATLOP [18] model by directing attention modules to assign higher weights to evidence-containing sentences without introducing any additional learnable parameters. Nevertheless, evidence prediction (ER) in DocRE encounters two principal challenges: high memory consumption and limited availability of evidence annotations. The model can be used for both supervised learning and self-training to compensate for the lack of manually annotated evidence, employing the same architecture with various training signals. A self-training ER process is proposed, comprising four main stages:

– Training a teacher model on human-annotated data, with labeled relations and supporting sentences;
– Applying the trained teacher model for evidence prediction on weakly-labeled data;
– Training a student model on weakly-labeled data with ER from earlier automatically generated evidence;
– Fine-tuning the student model on manually annotated data for model refinement.

Inference Procedure At this stage, it computes the importance of sentences based on an attention threshold, selecting sentences with importance exceeding a predefined threshold as evidence. A pseudo-document is constructed from the sentences deemed important. A single-parameter overlay layer, represented by τ, which represents the threshold value, is used to aggregate predictions from the pseudo-document and the entire document. Each triplet relation is chosen as the final prediction only if the sum of its aggregated values across the entire document and pseudo-documents exceeds τ. The τ parameter is tuned to minimize binary cross-entropy losses in relation extraction on the validation set.

Experiments on DocRED demonstrate that DREEAM currently achieves state-of-the-art results on the tasks of relation extraction and evidence detection, including through weakly-supervised learning on data obtained as a result of distant relation prediction on a large dataset.

5   Experiments

5.1   Data Processing

Initially, NEREL was annotated in the format of the brat annotation tool [9], in which mentions of entities and relations between them were established. For the inference models, we converted the NEREL dataset into the DocRED format, which lists entities, relations between them and supporting sentences. In adapting to the requirements of our task, we converted the entity mentions from character-based indices in the text to token-based indices in the sentence, subsequently converting this data into the DocRED format.

Initially, for baseline experiments with BiLSTM we excluded large documents exceeding 512 words from the training dataset, as tokenized using the Natasha library. This action affected less than 16 out of the total 746 documents. These particular documents were not only lengthy but also possessed a high volume of entities, causing a significant surge in GPU memory usage in certain models. For experiments involving transformer models, only a few documents were removed that had an extremely high number of entities, constituting the 95th percentile of this number across the entire dataset.

The authors of the original dataset provided a split of 80:10:10 for training, validation, and test sets. The metrics reported in the results tables have been calculated based on the validation set (also referred to as the dev set) over numerous runs. Furthermore, we have verified that the performance on the test set yields similar metric values.

We should also note the sequence of entities in the relationships in the data. In the DocRED dataset, the head and tail entities are sorted based on their order of appearance in the text, with the head entity always preceding the tail entity. However, in the NEREL dataset, the order is determined by the semantic relation, which might introduce extra complexities for the model when discerning which entity is the head and which is the tail. We conducted our experiments without altering the order of entities in these relationships.

BiLSTM and Context-Aware LSTM. To construct a basic solution for this task, models from the DocRED paper were employed.

- Distilled GLOVE embeddings, specifically the navec_news_v1_1B_250K_300d_100q from the Navec library of the Natasha project, were adopted as pretrained vector representations for words. These representations were selected because:

a) Their dimensionality is 100, similar to that used for DocRED, which required fewer modifications in the original model. Moreover, in comparison to the vectors from RusVectores, these vectors take up about 10 times less space (50 MB).

b) These embeddings were trained on news datasets: lenta, ria taiga_fontanka, buriy news, buriy webhose, ods_gazeta, ods_interfax, which align well with the nature of the NEREL dataset.

- Metadata were also prepared: mappings of entity and relation titles to indices (ner2id.json, rel2id.json), a word embedding dictionary (vocab.json).

- The model implementation required a parameter for the maximum number of relations in a document. An estimate of the upper bound was taken as 93^2, representing the square of the maximum number of entities in a document (excluding mentions) in the training set after preliminary filtering.

- One of the main problems encountered in solving this task is the high computational resource requirements of the algorithms. Specifically, a significant amount of memory is required for storing and processing information about entities and their relations at the document level. Moreover, the complexity of the algorithms often has an order of n^2, where n is the number of entities in the document. This stems from the necessity to determine the presence and type of relation between every pair of entities, leading to an exponential growth in the amount of computations as the number of entities increases.

Furthermore, in the dataset under consideration, the number of negative relations can exceed the number of positive ones by up to 80 times, leading to a severe imbalance during training.

One approach to undersampling negative relations involves limiting the number of negative relations in a document to a certain boundary, for instance, negative_bound = K * {number of positive examples}, where K is the undersampling coefficient, K = [15, 30, 50]. This method enables a reduction in the number of negative relations involved in training without a significant loss in model quality, as well as lowering memory consumption and accelerating algorithm convergence.

Undersampling negative relations can reduce memory load and expedite algorithm convergence, which in turn can shorten training time and simplify the process of model tuning.

Experiments demonstrate that randomly eliminating negative relations from the training set leads to reduced memory usage and faster algorithm convergence (Fig. 3), without greatly impacting the final quality of the models. This indicates that methods for undersampling negative relations can be beneficial in the context of document-level relation extraction, as they optimise the training process and enhance the efficient utilisation of computational resources.

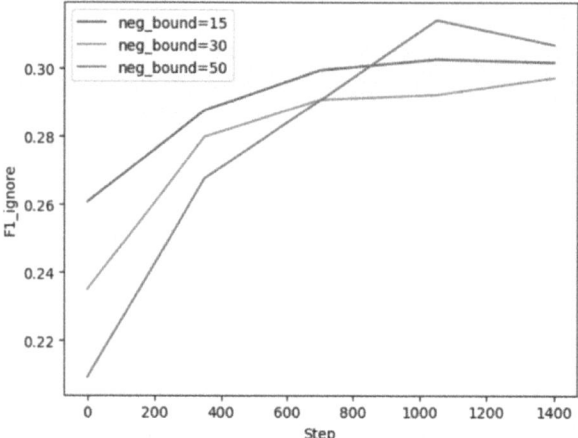

Fig. 3. F1_ignore score progression during training for different undersampling rates and their effect on convergence speed.

5.2 DREEAM and DocuNET Model

Handling Long Sequences. The document-level relation extraction task involves processing the entire text. However, transformer neural network architectures, such as BERT and RoBERTa, have limitations on the length of the input text, with the maximum number of tokens capped at 512.

In the applied models, this issue is addressed by dividing the input text into chunks, each of which is processed separately by the encoder. In the model implementations, documents containing more than 512 tokens (but fewer than 1024 tokens) are split into two parts: one for the first 512 tokens and another for the last ones. In other words, tokens at the very beginning/end of the document are encoded only once as they appear only in the first/second half, whereas tokens between them may be encoded twice as they can appear in both halves. The final vector representation of each token is calculated as the average of its vector representations in the first and second halves.

In the current implementation, the maximum number of chunks is 2, implying that the input text should not exceed 1024 tokens, which suffices for the DocRED dataset. However, in the NEREL dataset, some documents are longer, leading to incomplete processing of such documents. Despite this, the length on the filtered dataset does not exceed 1500 tokens.

Moreover, the model does not implement a mechanism for truncating long texts to 1024 tokens. This restriction precluded resolving the problem of long texts in such a way and resulted in errors. It is important to note that truncating documents necessitates additional adjustments to the training data (document-level relations), which adds complexity to the implementation of this feature.

To apply these models to the NEREL dataset, we made changes to the software implementation of the model. We added the capability to include additional chunks, which allowed for more accurate processing of sequences up to 1536 tokens. This required alterations in handling long sequences and adjusting the attention indices of input tokens for proper chunk division. No significant changes were noticed in the resulting metrics when comparing the results before and after the fix. This can be attributed to the relatively small number of documents exceeding 1024 tokens in length. However, this enhancement helped avoid critical errors on long texts.

Incorporating Nested Entities

NEREL contains nested entities, for which relationships with other entities are also annotated. Working with texts with nested entities can introduce additional problems for many models due to the peculiarities of input processing and entity overlaps. Therefore, they are often ignored, leading to information loss and lower metrics.

To account for nested entities, we have chosen models that use entity markers for entity representation. According to work [15], this helps avoid the described issues. Specifically, for the BERT encoder, the entity classes are denoted by specific tokens, while in the RoBERTa model, distinct symbols are used. This strategy facilitates these models' ability to extract document-level relations, which is particularly beneficial when handling nested entities, a capability evident in the outcome of our model predictions.

Employed Encoder Variations

The transformer models mentioned in this study make use of several pretrained transformer models tailored for the Russian language. These function as encoders and include models such as xlm-RoBERTa-large, DeepPavlov/ruBert, ruBert-base, ruBert-large, and ruRoBERTa-large, the last three being offerings from Sber.

Our selection of these models was influenced by their performance on the Russian SuperGLUE benchmark. We considered the suitability of their architecture and their demands on GPU resources. Although we also attempted to incorporate RuLeanALBERT into our system, it proved somewhat incompatible with the current software implementation of our models. Therefore, despite its potential, its application was sidelined pending further optimization work.

It is also worth noting that NEREL does not contain weakly labelled data, which means an approach using student-teacher learning methods cannot be applied in this case.

For evaluation, we used F1 and F1_ignore as the evaluation metrics. F1_ignore denotes F1 score excluding relational facts shared by the training and development/test sets. Our experimental outcomes, as detailed in Table 1, reveal a variable performance across models, contingent on the encoders in use. Notably, the DREEAM + Inference-stage Fusion model [14], in combination with the ruRoBERTa-large encoder, delivered the highest F1 scores. These findings underscore the critical role that the encoder selection plays in determining the ultimate effectiveness of a model in performing document-level relation extraction (Table 2).

The code for experiments, along with additional supplementary material, is available on GitHub. Interested readers can access the repository at https://github.com/iden-alex/ru_relation_extraction.

Table 1. Document-level relation extraction results on NEREL

Model	Encoder	F1_ignore	F1
BILSTM	Russian Navec glove vectors	31.42	32.24
ContextAware	Russian Navec glove vectors	30.88	31.73
DocuNET	DeepPavlov/ruBert	53.55	54.04
DocuNET	ruBert-base	53.92	54.4
DocuNET	ruBert-large	52.15	52.64
DocuNET	xlm-Roberta-large	57.22	57.67
DocuNET	ruRoberta-large	53.21	53.72
DREEAM	ruBert-base	51.7	52.78
DREEAM	ruBert-large	51.77	52.78
DREEAM	xlm-Roberta-large	62.126	62.99
DREEAM	ruRoberta-large	67.07	67.73
DREEAM + Inference-stage Fusion	ruRoberta-large	**67.87**	**68.53**

Table 2. Evidence sentence prediction results on NEREL

Model	Encoder	F1_evidence
BILSTM	Russian Navec glove vectors	–
ContextAware	Russian Navec glove vectors	–
DocuNET	DeepPavlov/ruBert	–
DocuNET	ruBert-base	–
DocuNET	ruBert-large	–
DocuNET	xlm-Roberta-large	–
DocuNET	ruRoberta-large	–
DREEAM	ruBert-base	55.73
DREEAM	ruBert-large	56.13
DREEAM	xlm-Roberta-large	58.98
DREEAM	ruRoberta-large	58.74
DREEAM + Inference-stage Fusion	ruRoberta-large	–

6   Conclusion

We have undertaken the first exploration of the task of document-level relation extraction for the Russian language. Using state-of-the-art models that achieve the best results on the main English dataset for doc-level RE, a series of novel experiments were conducted with Russian texts at the document level. To enable the accurate execution of these experiments, the model was refined to process texts exceeding 1024 tokens in length.

The task of document-level relation extraction is computationally demanding, necessitating substantial computational resources. Consequently, the research included the application of undersampling the number of negative relations in the training sample for certain models. This approach optimizes memory usage and expedites convergence, without significantly deteriorating the final quality of the models.

As a result of the conducted research and experiments, we demonstrated that the proposed methods and models can successfully extract relationships from Russian texts at the document level, inclusive of nested entities. This pioneering work in the field of Natural Language Processing signifies a considerable contribution to the advancement of research on document-level relation extraction for underrepresented languages, specifically Russian.

References

1. Christopoulou, F., Miwa, M., Ananiadou, S.: Connecting the dots: Document-level neural relation extraction with edge-oriented graphs. CoRR **abs/1909.00228** (2019). http://arxiv.org/abs/1909.00228
2. Gordeev D.I, Davletov A.A., R.A.A.G.R.G.G.A.: Relation extraction dataset for the Russian (2020)
3. Ivanin, V., et al.: Rurebus-2020 shared task: Russian relation extraction for business, pp. 416–431 (01 2020). https://doi.org/10.28995/2075-7182-2020-19-416-431
4. Jia, R., Wong, C., Poon, H.: Document-level n-ary relation extraction with multiscale representation learning. CoRR **abs/1904.02347** (2019). http://arxiv.org/abs/1904.02347
5. Loukachevitch, N.V., et al.: NEREL: A Russian dataset with nested named entities and relations. CoRR **abs/2108.13112** (2021). https://arxiv.org/abs/2108.13112
6. Ma, Y., Wang, A., Okazaki, N.: Dreeam: Guiding attention with evidence for improving document-level relation extraction (2023)
7. Sorokin, D., Gurevych, I.: Context-aware representations for knowledge base relation extraction. In: Proceedings of the 2017 Conference on Empirical Methods in Natural Language Processing, pp. 1784–1789. Association for Computational Linguistics, Copenhagen, Denmark (Sep 2017). https://doi.org/10.18653/v1/D17-1188, https://aclanthology.org/D17-1188
8. Starostin, A., et al.: Factrueval 2016: evaluation of named entity recognition and fact extraction systems for Russian (2016)
9. Stenetorp, P., Pyysalo, S., Topić, G., Ohta, T., Ananiadou, S., Tsujii, J.: brat: a web-based tool for NLP-assisted text annotation. In: Proceedings of the Demonstrations at the 13th Conference of the European Chapter of the Association for Computational Linguistics, pp. 102–107. Association for Computational Linguistics, Avignon, France (Apr 2012). https://www.aclweb.org/anthology/E12-2021

10. Tan, Q., Xu, L., Bing, L., Ng, H.T., Aljunied, S.M.: Revisiting docred – addressing the false negative problem in relation extraction (2022)
11. Tan, Q., Xu, L., Bing, L., Ng, H.T., Aljunied, S.M.: Revisiting docred – addressing the false negative problem in relation extraction (2023)
12. Wang, H., Focke, C., Sylvester, R., Mishra, N., Wang, W.Y.: Fine-tune BERT for docred with two-step process. CoRR abs/1909.11898 (2019). http://arxiv.org/abs/1909.11898
13. Wu, S., He, Y.: Enriching pre-trained language model with entity information for relation classification. CoRR abs/1905.08284 (2019). http://arxiv.org/abs/1905.08284
14. Xie, Y., Shen, J., Li, S., Mao, Y., Han, J.: Eider: Empowering document-level relation extraction with efficient evidence extraction and inference-stage fusion (2022)
15. Yandutov, A.V., Loukachevitch, N.V.: Approaches to relation extraction for nested named entities. Lobachevskii J. Math. **44**, 249-258 (2023). https://link.springer.com/article/10.1134/S1995080223010456
16. Yao, Y., et al.: DocRED: A large-scale document-level relation extraction dataset. In: Proceedings of the 57th Annual Meeting of the Association for Computational Linguistics. pp. 764–777. Association for Computational Linguistics, Florence, Italy (Jul 2019). https://doi.org/10.18653/v1/P19-1074, https://www.aclweb.org/anthology/P19-1074
17. Zhang, N., et al.: Document-level relation extraction as semantic segmentation. In: Zhou, Z.H. (ed.) Proceedings of the Thirtieth International Joint Conference on Artificial Intelligence, IJCAI-21, pp. 3999–4006. International Joint Conferences on Artificial Intelligence Organization (8 2021). https://doi.org/10.24963/ijcai.2021/551, https://doi.org/10.24963/ijcai.2021/551, main Track
18. Zhou, W., Huang, K., Ma, T., Huang, J.: Document-level relation extraction with adaptive thresholding and localized context pooling. CoRR abs/2010.11304 (2020). https://arxiv.org/abs/2010.11304

The Battle of Information Representations: Comparing Sentiment and Semantic Features for Forecasting Market Trends

Andrei Zaichenko[1] , Aleksei Kazakov[2], Elizaveta Kovtun[2] ,
and Semen Budennyy[2,3](✉)

[1] Higher School of Economics University, Moscow, Russia
[2] Sber AI Lab, Moscow, Russia
budennyysemen@gmail.com
[3] Artificial Intelligence Research Institute (AIRI), Moscow, Russia

Abstract. The study of the stock market with the attraction of machine learning approaches is a major direction for revealing hidden market regularities. This knowledge contributes to a profound understanding of financial market dynamics and getting behavioural insights, which could hardly be discovered with traditional analytical methods. Stock prices are inherently interrelated with world events and social perception. Thus, in constructing the model for stock price prediction, the critical stage is to incorporate such information on the outside world, reflected through news and social media posts. To accommodate this, researchers leverage the implicit or explicit knowledge representations: (1) sentiments extracted from the texts or (2) raw text embeddings. However, there is too little research attention to the direct comparison of these approaches in terms of the influence on the predictive power of financial models. In this paper, we aim to close this gap and figure out whether the semantic features in the form of contextual embeddings are more valuable than sentiment attributes for forecasting market trends. We consider the corpus of Twitter posts related to the largest companies by capitalization from NASDAQ and their close prices. To start, we demonstrate the connection of tweet sentiments with the volatility of companies' stock prices. Convinced of the existing relationship, we train Temporal Fusion Transformer models for price prediction supplemented with either tweet sentiments or tweet embeddings. Our results show that in the substantially prevailing number of cases, the use of sentiment features leads to higher metrics. Noteworthy, the conclusions are justifiable within the considered scenario involving Twitter posts and stocks of the biggest tech companies.

Keywords: stock market · embedding · time series · event study · NLP

1 Introduction

With the advancements in machine learning and natural language processing (NLP), researchers explore the leveraging of various features extracted from

© The Author(s), under exclusive license to Springer Nature Switzerland AG 2024
D. I. Ignatov et al. (Eds.): AIST 2023, CCIS 1905, pp. 149–163, 2024.
https://doi.org/10.1007/978-3-031-67008-4_12

textual data for predicting stock prices. Such NLP technique as sentiment analysis allows the extraction of the sentiment or emotion from a piece of text, which can be used to infer the overall sentiment of the market towards a particular company or stock. Information representation in the context of financial markets is proposed to be viewed as explicit or implicit.

The implicit representation involves extracting sentiment polarity directly from text (by a pre-trained NLP classifier) and then using this information to assess the expected reaction of a signal. This typically involves classifying the text into positive or negative categories based on its perceived sentiment towards market and using this classification as a feature in a further machine learning model to predict price movements. This approach benefits from computational ease and straightforward implementation. However, it comes with the drawback of potentially losing important semantic features and contextual information in the text. Additionally, the implicit approach relies on the accuracy of the used sentiment analysis algorithm, and inaccuracies in its operation can introduce distortions into the final predictive model.

The explicit approach involves creating an embedding from the text directly and then using this embedding to predict the reaction for signal change [29]. An embedding is a mathematical representation of a text in a lower-dimensional space, which can capture the meaning and context of the word or phrase. By creating an embedding from text, this approach is able to retain more of the semantic links contained in the text, which may be useful for predicting financial market trends. However, this approach may be more complex and time-consuming to implement, as it involves creating and training an embedding model on the text data.

After analyzing two ways of representing information related to the stock market, we can conclude that each of them has its own strengths and weaknesses. In this work, our aim is to compare the performance of the stock price prediction model built on the basis of either implicit or explicit knowledge representations.

2   Contribution

In this research, our goal is to explore the effectiveness of the explicit embedding vector approach and compare it with the more established implicit sentiment solution. The main contributions achieved during the study of this topic are as follows:

- Demonstrated the statistical dependence between stock price volatility and Twitter post sentiments.
- Proved intrinsic linkage between two kinds of representations by showing that embeddings hold information on sentiments.
- Discovered the superiority of the binary sentiment extraction approach over the usage of embedding vectors in the prevailing number of cases.

3   Related Work

One of the main streams of the research is dedicated to studying the usage of purely technical analysis trading indicators and historical data in combination with statistical methods for stock price prediction with machine learning. Many researchers employ the GRU-based models in this tasks [2, 11, 17] while the others explore the transformer architecture in this field [21, 30].

With the development of NLP methods, the number of works aiming to predict stock price trends and volatility proposed a combination of financial news and social media data is increasing [16]. Clinical trial announcements were used as a source of sentiment in pursuit of predicting pharma stock market price changes by [5]. Another research used Valence Aware Dictionary and Sentiment Reasoning (VADER) [14] for sentiment analysis [15]. [18] proposed a novel Deep Learning Transformer Encoder Attention (TEA) model. [10] considered such under-explored content as Environmental, Social, and Corporate Governance (ESG) news flow for volatility forecasting. [3] analyzed the impact of sentiment and attention variables on the stock market volatility by adding search engine and information consumption data on top of widely used social media and news texts.

Sentiment scores are commonly used in stock price prediction studies as they are easy to compute and provide a simple metric to gauge market sentiment [3, 10, 13, 15, 16, 18]. On the other hand, embedding vectors are dense numerical representations of words or phrases that capture the semantic meaning of the text. They are created by mapping words or phrases to high-dimensional vectors in a way that similar words or phrases are located close to each other in this vector space using techniques like Word2vec [22] or GloVe [24], BERT [8] and GPT [23], and have been shown to be effective in capturing complex relationships between words and phrases.

There are several works that concentrate on the usage of text semantics in the context of stock price prediction. [29] proposed a Multi-head Attention Fusion Network to exploit aspect-level semantic information from texts to enhance prediction. [20] developed a Spatial-temporal attention-based convolutional network. The authors converted news articles into vector embeddings and used them as a feature in their model. They noted that in the case of the utilization of the preprocessed text features, latent information in the text is lost because the relationships between the text and stock price are not considered. The main conclusion derived from these papers can be formulated as follows - the effectiveness of exploiting and fusing semantic aspect-level textual information leads to an improved performance upon the baselines. This fact means that the topic needs further investigation and refining.

The main advantage of embeddings over sentiment analysis is that they capture more complex relationships in the data. Despite the promising results of using embedding vectors as a feature [6, 7], there is still a research gap in the comparison of the effectiveness of using embedding vectors versus sentiment scores in stock price prediction. Moreover, researchers tend to experiment with different datasets and methodologies, making it difficult to draw meaningful comparisons.

4   Methods

In this section, we highlight the key steps of our pipeline for constructing the model for stock price prediction. In particular, we focus on the phase in which we use different techniques for getting textual information representations that served as part of the model input. We emphasize the prediction quality depends on the formed feature space and provide a framework for fair comparison and selection of the optimal configuration.

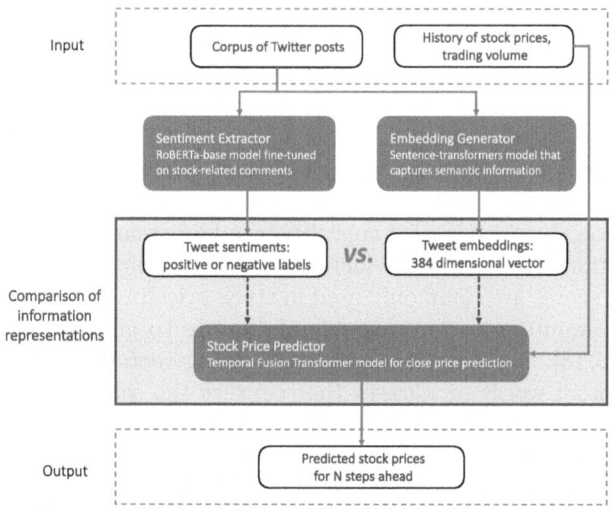

Fig. 1. Pipeline for comparison of utilizing either text sentiments or embeddings within the stock price prediction problem statement. There are three main steps in the scheme: (1) preprocessing of the collected Twitter and financial datasets, (2) obtaining either tweet sentiments or tweet embeddings, and (3) making prediction of stock close prices for N steps ahead.

An overall scheme of processes to make a stock close price prediction is described in Fig. 1. It consists of three main steps:

- **Data retrieval and preprocessing.** As the initial data, we take a corpus of Twitter posts related to the selected group of companies and the historical values of their stock prices and trading volumes. Tweets preprocessing steps were described in the Github repository [32]. Financial data are modified through the volume feature being smoothed. Finally, a business day resampling is applied to get our attributes at the same time frequency.
- **Model input formation.** Depending on the experiment, we pass the Twitter dataset either through the sentiment extractor or embedding generator to get the part of the input, which is responsible for the outside world information. Calendar holidays, trading volumes, and preceding close prices constitute the rest part.

– **Model prediction.** Data samples described with the composed feature space are fed into the TFT model. After a training process, a N steps ahead prediction for the close price is made.

4.1 Data Preparation

For accurate price prediction over the next N steps, we combine Twitter and financial data. Tweets contribute sentiment and semantic information. We incorporate historical financial data into our model. Additionally, we include a formula for measuring price volatility to support in-depth market analysis.

Data. For Twitter data, we use the dataset from Kaggle published in 2020 [9]. This dataset contains tweets related to 5 companies: Amazon, Apple, Google, Microsoft, and Tesla from 2015 to 2020. Raw data contains over 3 million tweets and information on the tweet author, post date, tweet text body, and the number of comments, likes, and retweets. For the preprocessing steps, a simple spam and duplicate tweet reduction was performed. All the preprocessing steps in more detail are described in the GitHub repository [32]. The financial data were collected with yfinance python library [31] using Yahoo! Finance's API for the same time period as the tweet data. Adjusted Close, High, Low, Open, Close, and Volume features were used.

Sentiment Information. The initial dataset with Twitter posts related to the particular group of companies did not contain information about the sentiment of the tweets. First, we had to extract sentiment polarity from the tweets in order to use them as a feature in a predictive model for comparison purposes. The following pre-trained models were applied to extract sentiment data from the tweets:

– FinBERT sentiment [1] built by further training the BERT language model on a large financial corpus and thereby adopting it for financial sentiment classification. The sentiment model is fine-tuned on 10000 manually annotated sentences collected from analyst reports about S&P500 firms;
– RoBERTa-base model fine-tuned on the Stocktwits dataset [28], which contains 3.2 million comments with the user labelled tags: "Bullish" or "Bearish".

Based on a manual comparative analysis of 300 tweets, we selected the latter option for our binary sentiment-based solution due to its superior performance, aligning with our research objectives and providing transparency in our methodology.

As a result of the sentiment extraction procedure, we get either 1 or 0 as labels for each tweet. Label 1 is associated with positive sentiment, while label 0 - with negative sentiment. However, before feeding obtained information into the predictive model, we need to group sentiment labels by their relation to company and date. Then, we suggest two types of scores, which reflect the ultimate sentiment of the group of tweets:

$$Sentiment\ Score\ 1 = \frac{negative}{negative + positive},$$
$$Sentiment\ Score\ 2 = \frac{negative}{positive}, \tag{1}$$

*SentimentScore*2 provided us with better metrics on the validation dataset in the task of price prediction. Thus, it is selected as the primary score for the final pipeline and is denoted *Sentiment Score* in the subsequent discussion. Although, we included the behavior comparison of *Sentiment Score* 1 and *Sentiment Score* 2 in Sect. 5.2.

To achieve a more pronounced sentiment score trend and reduce highly fluctuating values, we applied a three-week smoothing technique. We compared the results of using a simple rolling mean with the Exponential Weighted Moving Average (EWMA) method implemented in the pandas library [25]. We chose the EWMA method due to its faster response to downtrends, primarily because it incorporates a weight decay parameter, giving more weight to recent data points than to older ones.

Additionally, we addressed the issue of the stock market operating only on business days, leaving gaps on weekends. Since sentiment data is available daily, we opted to convert the daily sentiment frequency to a business day frequency. This approach prevents the introduction of extraneous, information-poor values during weekends, with sentiment scores achieved on weekends being aggregated to the following business day.

Semantic Information. We use embeddings, generated by the "all-MiniLM-L6-v2" sentence transformer model [26], which efficiently maps tweets into a dense vector space. We chose this model for its strong performance, driven by pre-training on diverse data to capture intricate linguistic patterns. The 384-dimensional embeddings strike a balance between representation capacity and computational efficiency in our task. To construct the representation for a single trading day, we compute the average of all vectors associated with that day. However, similar to the sentiment approach, we encounter the challenge of missing market prices during weekends. To address this, we concatenate the weekend embeddings vectors with the vectors related to the next business day.

Financial Information. The predictions for future stock prices are made on the basis of historical price values. We consider open, close, high, and low prices of stocks for a particular day. Open price is a selling price of a stock at the time the exchange opens. Close price is the last price during a trading session. High and low prices are the maximum and minimum prices during a session, correspondingly. In addition, we leverage trading volume as a feature, which shows a number of shares that have been bought or sold during the trading day.

Volatility. In the following discussion, we mean Average True Range measure under the volatility concept. Average True Range is defined according to the formula:

$$ATR_t = \frac{n-1}{n}ATR_{t-1} + \frac{1}{n}\Big(max(H_t, C_{t-1}) - min(L_t, C_{t-1})\Big), \qquad (2)$$

where t - given day; ATR_t - Average True Range in day t; n - considered number of periods; H_t - highest price in day t; L_t - lowest price in day t; C_{t-1} - close price in day $t-1$.

4.2   Predictive Model Parameters

When selecting the architecture for price prediction, we experimented with multiple models available in the Darts Python library [12]. The Temporal Fusion Transformer [19] implementation demonstrated the best performance, and we chose it as the core predictive model for our final pipeline. Unlike a univariate approach that necessitates creating distinct models for each company, our approach involves training multiple time-series simultaneously, streamlining both the training and forecasting processes.

For the input history length of the model, we selected the previous three business weeks (15 d). This choice was determined through grid-search, and you can find detailed results on our GitHub repository [32]. Opting for a shorter lookback window introduces limitations, as sentiment reactions to events require time, and excessively short observation windows may lead to information loss. In our main experiments, we set the output prediction length to 3 d. We allocated 80% of the data for training and reserved 20% for testing purposes.

A common target variable in all experiments is close price. To test an effect of different input compositions, three groups were identified:

- HLOV - High, Low, Open, Volume attributes. In this case, we test out the model performance only working with market data as input.
- HLOVS - the same as above, but with adding *Sentiment Score* attribute.
- HLOVE - market data and embedding vectors as model input

4.3   Loss Function

In order to prevent overfitting, a custom loss function is introduced. It is built on top of one of the most popular loss functions for regression tasks - Mean Squared Error (MSE) loss, but with the addition of a directional component. MSE does not consider whether the direction of the predicted close price is correct, just focusing on the difference between true and predicted prices. However, the direction of price movement is an essential factor in the financial world. Multiple values from 10 to 10^4 of α were tested in the loss function. Lower values of α did not punish the directional errors enough to make a difference in comparison with standard MSE loss, while having α set to 10^4 only prolonged the loss convergence time. In the end, we come up with the following loss function:

$$DMSE = \frac{1}{n} \sum_{i=1}^{n} \alpha(x_i - y_i)^2, \quad \text{where} \quad \begin{cases} \alpha = 1, & \text{if } (x_i - x_{i-1})(y_i - y_{i-1}) \geq 0, \\ \alpha = 10^3, & \text{otherwise,} \end{cases}$$

(3)

where $DMSE$ - Directional Mean Squared Error; n - number of data points; x_i - true values; y_i - predicted values.

5   Results

In this section, we share our study's findings on the connections between Twitter sentiments, stock prices, and volatility. Additionally, we present metrics obtained from our predictive model development using binary sentiment labels and sentence embeddings.

5.1   Daily Return Analysis

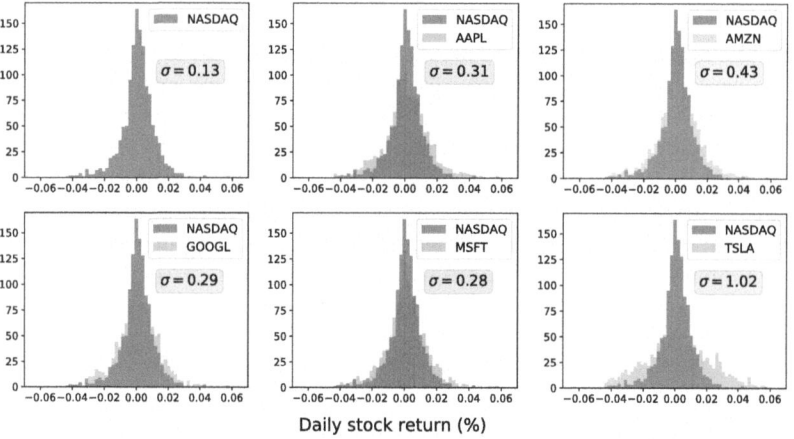

Fig. 2. Histograms of daily stock return values for the companies during a five-year period. NASDAQ daily stock returns are given for comparison purposes. Tesla is the most volatile among the analyzed five companies.

Histograms of daily stock return values for all five companies in a five-year period are presented in Fig. 2. Here σ is calculated as the sum of squares of deviations of daily return values from 0. As we can observe, Tesla is a clear outlier among five other companies having much greater daily return oscillations. One can infer that this stock has a higher risk/reward ratio. This shows that no matter that all the described stocks are traded inside the same NASDAQ index, there is still evidence of them behaving in a different way caused by the other factors.

5.2   Relation Between Tweet Sentiments and Stock Market

Before the construction of the predictive model on the basis of Twitter data, our goal is to explore the direct connection between price variables and Twitter sentiments. Figure 3 demonstrates stock prices as well as sentiment scores calculated in both ways for Apple and Amazon. Visually, we can observe a great amount of resemblance between price and sentiment variables, especially in the case of Amazon.

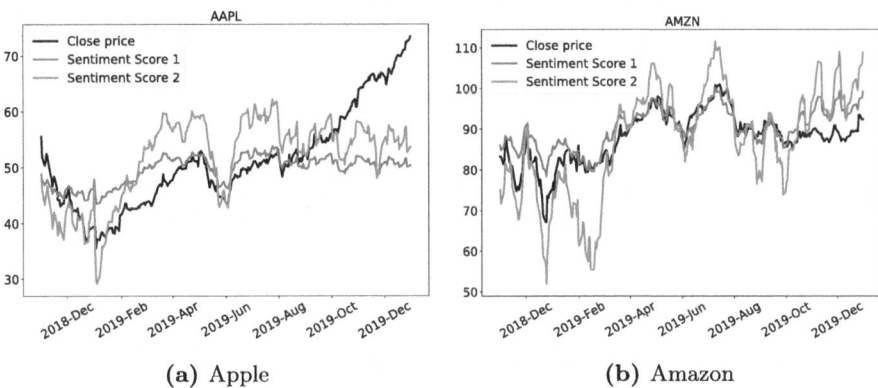

(a) Apple   **(b)** Amazon

Fig. 3. Sentiment scores and closing prices comparison with a constant multiplier applied to sentiment scores for visualization purposes. Left graph is denoted to the moving of Apple prices and sentiments, while Amazon patterns are demonstrated on the right.

When comparing stock volatility with EWMA smoothed sentiment scores in Figs. 4 and 5, we notice a significant similarity in the shape of the volatility curve and Twitter sentiment scores. Notably, Amazon exhibits a stronger correlation between these values. It's noteworthy to observe a time lag between public reaction and stock volatility, particularly for Apple, where volatility tends to precede sentiment response. In contrast, Amazon's sentiment changes are more synchronized with volatility, making it a more influential predictor of price movement. Examining scatter plots in Figs. 4 and 5, we identify a clear directional pattern in sentiment-volatility dependence.

To conduct a more comprehensive correlation analysis between different variables, we consider Google company and calculate pairwise Spearman correlation coefficients between close price, volume, volatility, and sentiment score. The results are given in Table 1. We applied three-week EWMA smoothing to volume and sentiment score in order to get a more meaningful correlation with the stock close price. After comparing with non-smoothed calculations, we found out that smoothing indeed helped to improve correlation indicators. It happens because we smoothed out short-term fluctuations and consequently captured the underlying trends.

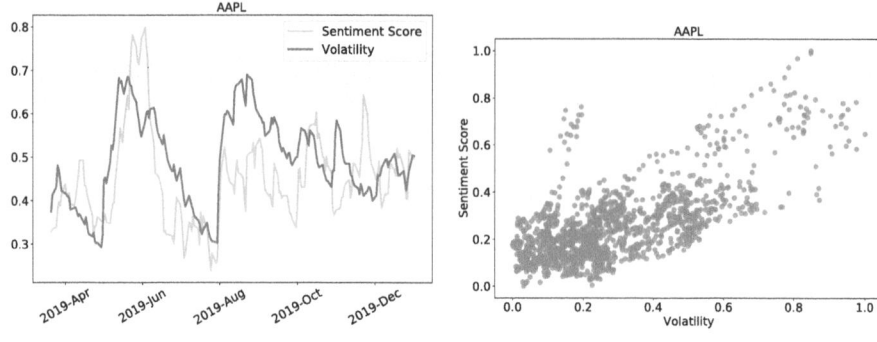

(a) Sentiment score and volatility behavior

(b) Scatter plot for sentiment score and volatility

Fig. 4. Sentiment score and volatility comparison for Apple with Min-Max scaling applied to both features. There is a comparison of behaviour of sentiment scores and volatility overtime on the left side. We can observe a time lag of sentiment reaction to volatility. The right side is denoted to the scatter plot of sentiment score and volatility to discover hidden linear-like patterns.

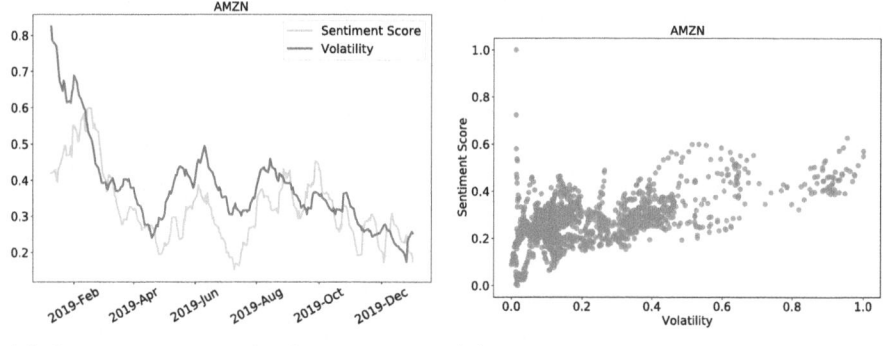

(a) Sentiment score and volatility behavior

(b) Scatter plot for sentiment score and volatility

Fig. 5. Sentiment score and volatility comparison for Amazon with Min-Max scaling applied to both features. There is a comparison of behaviour of sentiment scores and volatility overtime on the left side. We can observe quite synchronized directional movements. The right side is denoted to the scatter plot of sentiment score and volatility to explore hidden linear-like patterns in more detail.

Table 1. Spearman correlation between different variables with and without the application of a three-week window smoothing to volatility and sentiment score.

Using EWMA	Close	Volume	Volatility	Sentiment Score
Close	1.000	−0.464	0.626	0.401
Volume	−0.464	1.000	0.295	0.085
Volatility	0.626	0.295	1.000	0.508
Sentiment Score	0.401	0.084	0.508	1.000

5.3   Connection of Sentiments and Embeddings

After exploring of relationships between sentiment scores and financial indicators, we became interested in the interdependency of other entities, namely tweets' sentiments and their embeddings. As mentioned, tweet sentiments were extracted with a pre-trained RoBERTa-base model, and tweet embeddings were generated by sentence transformers. We hypothesized the presence of an underlying connection between these two representations. To check this, we fitted a linear regression model on vector embeddings, for which corresponding sentiment scores were set as targets. The prediction quality was estimated with R^2 metric. The results of such experiment are given in Table 2. For a better understanding of the magnitude of the effect that embedding encompasses sentiment knowledge, we generate vectors of random variables with a dimension equal to the size of embedding vectors. Then, we try to predict sentiment scores from obtained pointless vectors.

Table 2. R^2 values for each company achieved after fitting a linear regression model on embeddings as features and sentiment scores as targets. R^2 scores for the case when predictions are made from randomly generatedvectors are given for comparison purpose.

R^2 for *Sentiment Score* prediction	Apple	Amazon	Google	Microsoft	Tesla
from embeddings	0.800	0.821	0.778	0.804	0.891
from random vectors	0.344	0.352	0.358	0.381	0.396

For the majority of companies, R^2 values are greater than 0.8. Thus, there is a strong relationship between the vector embeddings and the sentiment scores. High metrics for the linear regression model indicate that the feature variable is a strong predictor of the response variable.

5.4   Price Prediction Results and Sensitivity Analysis

When choosing price prediction models to include in the final pipeline, several candidates were tested. Moreover, we experimented with different horizons. All details and results of intermediate experiments are given in the GitHub repository [32]. For the main experiments, we took the model and forecast horizon that proved themselves in the best way during the selection phase. We stopped at the TFT model with the forecast horizon equal to 3.

Table 3. Metrics for three-day forecast horizon considering five companies. The predictions are made with TFT model using feature sets including either sentiments (HLOVS) along with financial data or embeddings (HLOVE, HLOVE2).

	Model	MAPE	MAE	R^2	RMSE	MSE	SMAPE
Apple	**tft_HLOVS**	1.7602	**0.8932**	**0.9648**	**1.2014**	**1.4434**	1.7502
	tft_HLOVE	**1.7557**	0.8947	**0.9648**	1.2023	1.4455	**1.7492**
	tft_HLOVE2	3.9357	1.6854	0.7368	2.1665	4.6938	3.8641
Amazon	**tft_HLOVS**	**1.0544**	**0.5605**	**0.8535**	**0.7469**	**0.5579**	**1.0526**
	tft_HLOVE	1.1442	0.6078	0.8334	0.7965	0.6344	1.1408
	tft_HLOVE2	1.9503	1.0404	0.4551	1.5635	2.4446	1.9184
Google	**tft_HLOVS**	1.6830	0.9044	**0.8326**	1.2160	**1.4786**	1.6807
	tft_HLOVE	**1.6824**	**0.9039**	0.8323	**1.2171**	1.4814	**1.6781**
	tft_HLOVE2	2.9347	2.3175	0.4988	2.6873	2.9715	3.3174
Microsoft	**tft_HLOVS**	**1.2687**	**0.8267**	**0.9626**	**1.0643**	**1.1327**	**1.2721**
	tft_HLOVE	1.3059	0.8503	0.9615	1.0798	1.1660	1.3086
	tft_HLOVE2	3.9615	1.6081	0.7412	3.5431	2.9846	3.0965
Tesla	**tft_HLOVS**	**3.5182**	**1.3833**	**0.8923**	**1.9810**	**3.9242**	**3.5386**
	tft_HLOVE	3.5717	1.4020	0.8915	1.9880	3.9523	3.5994
	tft_HLOVE2	4.5540	1.9531	0.5735	2.3948	5.7354	4.4676

The aim of the experiments is to compare metrics for a three-day prediction period when using sentence embedding or sentiments as features. We basically compare the effect by considering two feature sets: HLOVS and HLOVE, which are described in 4.2. The metrics are performed in Table 3. As we wanted to make a generalizable conclusion, we examined embeddings that are generated by one more language model. We considered Microsoft model - all-mpnet-base-v2 [27]. It has twice the number of dimensions compared with the approach mentioned in 4.1. The feature set with embeddings obtained with this model is denoted as HLOVE2 in Table 3. However, the accuracy of closing price prediction dropped significantly for such embeddings.

If we look at prediction qualities for Apple company in Table 3, we observe that for some metrics embedding approach shows better performance. However, the difference from the sentiment approach is not substantial, and HLOVS set gives better results for 4 out of 6 metrics. For Microsoft, sentiment representations clearly perform better. Another company that has nearly similar results for both feature sets is Google, and the difference in metrics is negligible -

only 0.036% for Mean Absolute Percentage Error (MAPE). For Amazon and Tesla sentiments again prove to be an optimal feature for all metrics. To sum up, although embeddings seem to contain more information, sentiment solution remains a strong baseline that is hard to beat.

If we analyze MAPE metric, models with embedding feature show better results only in two cases out of five - for Apple and Google. However, this prevalence is minor. During the analysis of volatility and close price correlations with sentiment score, these companies had one of the lowest scores among the considered five. It might be the reason why Twitter sentiments do not have such a performance-enhancing impact for these two companies. For Amazon, Google and Tesla sentiment score method outperforms embedding vector solutions.

6   Conclusions

This study provides evidence that information obtained from Twitter posts serves as a strong indicator for stock price movements. In particular, Twitter sentiment scores are highly correlated with price volatility and improve the performance of predictive models for financial markets. We conclude that sentiments provide valuable insights on events around and contribute to better capturing underlying market dynamics.

The main goal of this paper is to investigate whether sentence embeddings yield better results in stock price prediction compared to the binary sentiment analysis approach. In the majority of conducted experiments, the sentiment approach outperforms the embedding vectors method. This fact might be counterintuitive because embeddings seem to encompass more valuable contextual information. However, sentiments tend to represent information in a more concise way, bringing less noise into the prediction model. Nevertheless, the embedding approach still has an advantage that it does not require an additional model for sentiment extraction and the consequent quality verification of that procedure. It is important to note the limitation of the conducted research. We made a comparison analysis only within restricted use case of Twitter posts and stocks of top companies from NASDAQ.

7   Statement on Computational Resources and Environmental Impact

We used a NVIDIA GeForce RTX 3080 Ti GPU to train the models, extract the sentiment score and make embedding vectors from the tweets using BERT model [26]. Two NVIDIA A100 80 GB PCIe GPUs were used for testing out the MPNet [27] approach and inferencing. This work contributed totally 6.12 kg equivalent CO_2 emissions. The carbon emissions information was generated using the open-source library *eco2AI*[1] [4].

---

[1] Source code for *eco2AI* is available at https://github.com/sb-ai-lab/Eco2AI.

References

1. Araci, D.: Finbert: financial sentiment analysis with pre-trained language models. CoRR abs/1908.10063 (2019). http://arxiv.org/abs/1908.10063
2. Aseeri, A.O.: Effective short-term forecasts of Saudi stock price trends using technical indicators and large-scale multivariate time series. PeerJ Comput. Sci. **9**, e1205 (2023). https://doi.org/10.7717/peerj-cs.1205
3. Audrino, F., Sigrist, F., Ballinari, D.: The impact of sentiment and attention measures on stock market volatility. Int. J. Forecast. **36**(2), 334–357 (2020). https://doi.org/10.1016/j.ijforecast.2019.05.010
4. Budennyy, S.A., et al.: eco2AI: carbon emissions tracking of machine learning models as the first step towards sustainable AI. Doklady Math. **106**(S1), S118–S128 (2022). https://doi.org/10.1134/S1064562422060230
5. Budennyy, S., Kazakov, A., Kovtun, E., Zhukov, L.: New drugs and stock market: how to predict pharma market reaction to clinical trial announcements (2022). https://doi.org/10.48550/ARXIV.2208.07248
6. Chandola, D., Mehta, A., Singh, S., Tikkiwal, V.A., Agrawal, H.: Forecasting directional movement of stock prices using deep learning. Ann. Data Sci. (2022). https://doi.org/10.1007/s40745-022-00432-6
7. Chen, J., Chen, T., Shen, M., Shi, Y., Wang, D., Zhang, X.: Gated three-tower transformer for text-driven stock market prediction. Multimed. Tools Appl. **81**(21), 30093–30119 (2022). https://doi.org/10.1007/s11042-022-11908-1
8. Devlin, J., Chang, M.W., Lee, K., Toutanova, K.: Bert: Pre-training of deep bidirectional transformers for language understanding (2018). https://doi.org/10.48550/ARXIV.1810.04805
9. Dogan, M., Metin, O., Tek, E., Yumusak, S., Oztoprak, K.: Speculator and influencer evaluation in stock market by using social media. In: 2020 IEEE International Conference on Big Data (Big Data), pp. 4559–4566 (12 2020). https://doi.org/10.1109/BigData50022.2020.9378170
10. Guo, T., Jamet, N., Betrix, V., Piquet, L.A., Hauptmann, E.: Esg2risk: a deep learning framework from esg news to stock volatility prediction (2020). https://doi.org/10.48550/ARXIV.2005.02527
11. Gupta, U., Bhattacharjee, V., Bishnu, P.S.: Stocknet-gru based stock index prediction. Expert Syst. Appl. **207**, 117986 (2022). https://doi.org/10.1016/j.eswa.2022.117986
12. Herzen, J., et al.: Darts: user-friendly modern machine learning for time series. J. Mach. Learn. Res. **23**(124), 1–6 (2022). http://jmlr.org/papers/v23/21-1177.html
13. Hu, Z., Liu, W., Bian, J., Liu, X., Liu, T.Y.: Listening to chaotic whispers: A deep learning framework for news-oriented stock trend prediction. In: Proceedings of the Eleventh ACM International Conference on Web Search and Data Mining, pp. 261–269. WSDM '18, Association for Computing Machinery, New York, NY, USA (2018). https://doi.org/10.1145/3159652.3159690
14. Hutto, C., Gilbert, E.: Vader: a parsimonious rule-based model for sentiment analysis of social media text. In: Proceedings of the International AAAI Conference on Web and Social Media, vol. 8, pp. 216–225 (2014)
15. Kamal, S., Sharma, S., Kumar, V., Alshazly, H., Hussein, H.S., Martinetz, T.: Trading stocks based on financial news using attention mechanism. Mathematics **10**(12) (2022). https://doi.org/10.3390/math10122001
16. Khan, W., Ghazanfar, M.A., Azam, M.A., Karami, A., Alyoubi, K.H., Alfakeeh, A.S.: Stock market prediction using machine learning classifiers and social media,

news. J. Ambient. Intell. Humaniz. Comput. **13**(7), 3433–3456 (2022). https://doi.org/10.1007/s12652-020-01839-w

17. Li, C., Qian, G.: Stock price prediction using a frequency decomposition based GRU transformer neural network. Appl. Sci. **13**(1) (2023). https://doi.org/10.3390/app13010222
18. Li, Y., Lv, S., Liu, X., Zhang, Q.: Incorporating transformers and attention networks for stock movement prediction. Complexity **2022**, 7739087 (2022). https://doi.org/10.1155/2022/7739087
19. Lim, B., Arik, S.O., Loeff, N., Pfister, T.: Temporal fusion transformers for interpretable multi-horizon time series forecasting (2019). https://doi.org/10.48550/ARXIV.1912.09363
20. Lin, C.T., Wang, Y.K., Huang, P.L., Shi, Y., Chang, Y.C.: Spatial-temporal attention-based convolutional network with text and numerical information for stock price prediction. Neural Comput. Appl. **34**(17), 14387–14395 (2022). https://doi.org/10.1007/s00521-022-07234-0
21. Lin, F., Li, P., Lin, Y., Chen, Z., You, H., Feng, S.: Kernel-based hybrid interpretable transformer for high-frequency stock movement prediction. In: 2022 IEEE International Conference on Data Mining (ICDM), pp. 241–250 (2022). https://doi.org/10.1109/ICDM54844.2022.00034
22. Mikolov, T., Chen, K., Corrado, G., Dean, J.: Efficient estimation of word representations in vector space (2013). https://doi.org/10.48550/ARXIV.1301.3781
23. Muennighoff, N.: SGPT: GPT sentence embeddings for semantic search (2022). https://doi.org/10.48550/ARXIV.2202.08904
24. Pennington, J., Socher, R., Manning, C.: GloVe: global vectors for word representation. In: Proceedings of the 2014 Conference on Empirical Methods in Natural Language Processing (EMNLP), pp. 1532–1543. Association for Computational Linguistics, Doha, Qatar (Oct 2014). https://doi.org/10.3115/v1/D14-1162
25. Reback, J., McKinney, W., et al.: pandas: data analysis and manipulation library for python (2023). https://pandas.pydata.org
26. Sentence transformer model all-minilm-l6-v2. https://huggingface.co/sentence-transformers/all-MiniLM-L6-v2. Accessed 10 Dec 2022
27. Sentence transformer model all-mpnet-base-v2. https://huggingface.co/microsoft/mpnet-base. Accessed Jan 15 2023
28. Sentiment inferencing model for stock related commments. https://huggingface.co/zhayunduo/roberta-base-stocktwits-finetuned. Accessed 10 Dec 2022
29. Tang, N., Shen, Y., Yao, J.: Learning to fuse multiple semantic aspects from rich texts for stock price prediction. In: Cheng, R., Mamoulis, N., Sun, Y., Huang, X. (eds.) Web Information Systems Engineering - WISE 2019, pp. 65–81. Springer International Publishing, Cham (2019)
30. Wang, C., Chen, Y., Zhang, S., Zhang, Q.: Stock market index prediction using deep transformer model. Expert Syst. Appl. **208**, 118128 (2022). https://doi.org/10.1016/j.eswa.2022.118128
31. Python library yfinance. https://pypi.org/project/yfinance/. Accessed Jan 10 2022
32. Zaichenko, A., Kazakov, A., Kovtun, E., Budennyy, S.: Comparing sentiment and semantic features for forecasting market trends. https://github.com/azadata/Sentiment-Semantic-Paper (2023)

Computer Vision

Interactive Image Segmentation with Superpixel Propagation

Hrach Ayunts[1] ⓘ, Varduhi Yeghiazaryan[3](✉) ⓘ, Shant Navasardyan[1] ⓘ,
and Humphrey Shi[2,4,5,6] ⓘ

[1] Picsart AI Research (PAIR), Yerevan, Armenia
{hrach.ayunts,shant.navasardyan}@picsart.com
[2] Picsart AI Research (PAIR), Miami, USA
humphrey.shi@picsart.com
[3] American University of Armenia, Yerevan, Armenia
vyeghiazaryan@aua.am
[4] University of Oregon, Eugene, USA
[5] University of Illinois Urbana-Champaign, Champaign, USA
[6] Georgia Institute of Technology, Atlanta, USA

Abstract. Interactive tools play an important role in solving image segmentation problems. In this paper, we present a new interactive segmentation framework for high-accuracy segmentation. The main interaction of the user is to provide clicks inside the object of interest and control the mask-growing process with a slider. We use propagation on superpixels for region growth. To do large-scale evaluation we automate the user interactions and compare our method with state-of-the-art approaches on a few datasets with detailed annotations. Our method consistently outperforms the competitors on high accuracies and certain classes of images. We also do experiments with human annotators to show how more time-consuming naive approaches are compared to our method. Moreover, in contrast to state-of-the-art deep learning methods that stop improving segmentation accuracy beyond 20–100 clicks, our algorithm guarantees accuracy improvement after every iteration.

Keywords: Image segmentation · interactive segmentation · superpixels · front propagation

1 Introduction

Image segmentation is one of the classic problems in computer vision with a very wide range of applications and has attracted decades of research. Recently, deep learning (DL) approaches have significantly improved solutions for semantic [3,16,22,32] and instance [9,21,31] segmentation tasks. However, DL algorithms are usually trained on a huge amount of carefully annotated data, the collection of which requires massive human effort, and is time-consuming and expensive. To reduce the cost of data annotations, some algorithms [10,11,14,28] propose

ⓒ The Author(s), under exclusive license to Springer Nature Switzerland AG 2024
D. I. Ignatov et al. (Eds.): AIST 2023, CCIS 1905, pp. 167–179, 2024.
https://doi.org/10.1007/978-3-031-67008-4_13

semi-automatic matting and segmentation with human interactions. Hence user-guided interactive segmentation becomes an important problem in computer vision, solutions of which can also be useful in such fields as medical imaging, e-commerce, creative editing, etc.

The task of interactive segmentation assumes the partitioning of an image into semantically correct regions using some form of guidance from a user while allowing iterative refinement. User input usually comes in the form of bounding boxes, scribbles, or clicks [10,29]. Others [7] rely on contours to speed up the selection of the regions of interest by the user in comparison with click-based interactions.

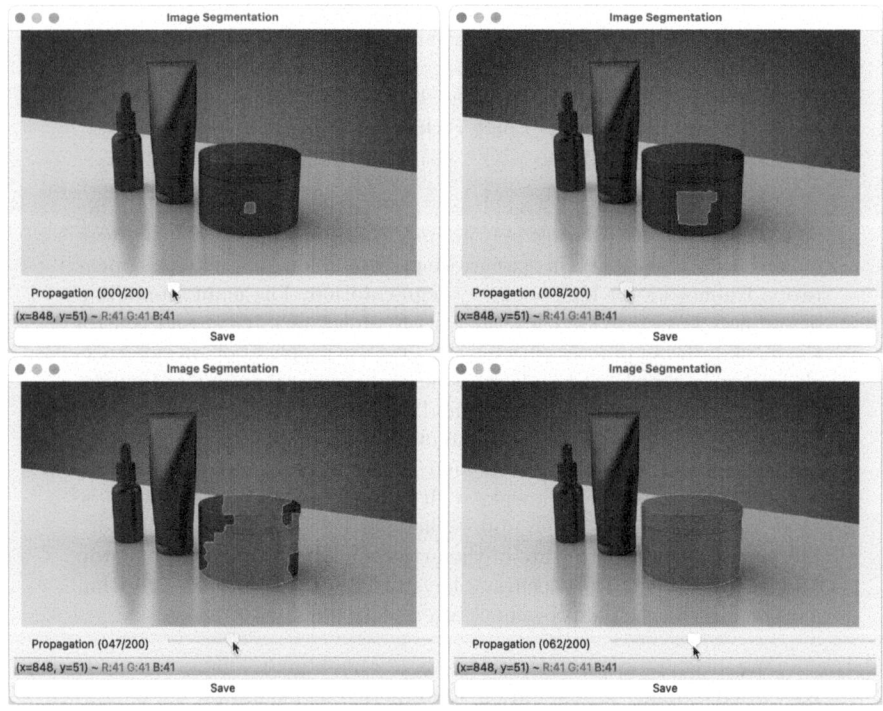

Fig. 1. Screenshots of the user interface for our method. The slider is used to control mask (in green overlay) propagation. (The reader is referred to the electronic version of the paper for full interpretability of colors in the images.) (Color figure online)

There has been a surge of interest in this topic with the fast development of deep learning models [2,4,10,14,36]. State-of-the-art (SOTA) results show that such methods can achieve a 90% intersection over union (IoU) with the ground truth (GT) with under a dozen clicks by the user. Recent works target reduction of user effort by locally refining the segmentation mask in a cropped image around a user click [5,13] or simulating intermediate user clicks [15]. [38] uses a

Gaussian process classification model to explicitly utilise the contribution of the positive and negative user clicks. However, such approaches rely on extensively annotated datasets for training and evaluation, such as COCO [12], LVIS [8], ADE20K [37], GrabCut [23], etc. Commonly, the publicly available datasets contain low-accuracy annotations. As a result, deep learning (DL) interactive segmentation methods do not target IoU above the modest 90% [28]. The evaluation of new interactive segmentation algorithms is typically based on only 20 iterations and a target of 85% or 90% IoU similarity [14,27] with the ground truth masks. Interestingly, as our experiments reported below show, DL methods tend to converge in terms of accuracy after 20–100 iterations. There is virtually no further improvement in IoU scores afterward, even after 500 iterations. This makes highly accurate segmentation results very difficult to attain using solely current DL methods. On the other hand, real-life applications like photo editing, e-commerce, etc., require a much higher accuracy of segmentation. Thus, when segmentation algorithms target such domains, the ability of an algorithm to achieve higher segmentation accuracy should be prioritized.

To address this limitation we propose a traditional computer vision algorithm which can be considered either as a stand-alone tool or as a refinement on top of existing coarse segmentation methods. Such a tool can be used to generate high-quality training and evaluation data for DL approaches while being faster and easier to use than manual segmentation. To summarize, the main contributions of this paper are:

- A simple, yet highly accurate tool for segmentation;
- Full control for the user over segmentation results;
- Large-scale comparison showing our clear advantage over SOTA DL methods which have stagnating accuracy for high-accuracy annotations.

2   Interactive Segmentation Method

The proposed method allows users to get high-quality segmentation masks by clicking on the objects of interest in images and growing the mask region with the help of a slider as shown in Fig. 1[1] Although our method can be used to segment out all object instances separately, below, for simplicity, we consider the case of foreground/background separation.

2.1   Overview of the Method

The pipeline of our approach is presented in Fig. 2. This is an iterative process each iteration of which is composed of user interaction and backend operation steps.

First, the user zooms into a region of interest in the image. During the first few iterations, this may be a subimage enclosing the whole foreground object. Later

---

[1] Image   source: https://www.shutterstock.com/image-illustration/lineup-black-cosmetic-beauty-products-3d-1610305420.

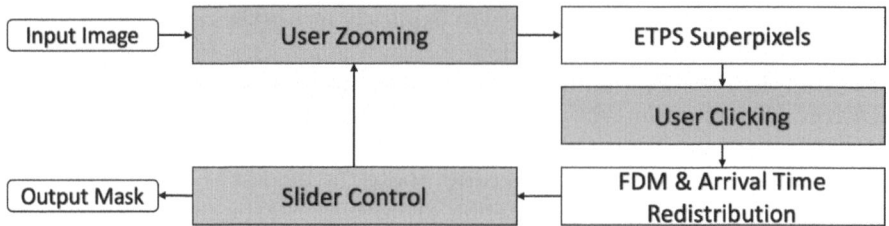

Fig. 2. The workflow of our method. The highlighted steps correspond to user interactions.

this is usually a small subimage with a detail of the foreground. The second step is to split the zoomed subimage into a constant number of superpixels (1000 chosen experimentally) by using Efficient Topology Preserving Segmentation [33]. The choice of the superpixel algorithm was made by taking the top six algorithms from the survey [30] and comparing them based on execution time and boundary adherence of superpixels.

Then the user initiates a mask growth process by clicking on several foreground pixels. The foreground mask growth is similar to a wave propagation process from seeds. The chosen pixels are treated as the seeds from which the wave starts its expansion. Thus each pixel in the zoomed subimage has a corresponding time for the wave to arrive. The arrival times are calculated by using the Fast Dashing Method (FDM) [34]. FDM is a generalization of the Fast Marching Method (FMM) [24,25]—for solving the Eikonal Eq. 1—to estimate the wave propagation arrival times for superpixels instead of pixels. This results in one arrival time for a whole superpixel. In the cases when the subimage contains fewer than 2000 pixels, we treat each pixel as a separate superpixel, thus switching to the FMM algorithm instead of FDM.

After the arrival times are obtained, a naive approach is to control the mask region propagation by a slider, the values of which are the arrival times of superpixels. However, in practice, the distribution of arrival times is far from uniform. Despite small changes in slider position, we may observe at times large, while at times small changes of the mask. To address this problem we introduce a redistribution technique and construct the slider in a way to avoid such discrepancies in the mask wave propagation process.

As the last step of an iteration, the user chooses a slider value to threshold the redistributed arrival times to contain as many superpixels inside the foreground as possible, but exclude all superpixels intersecting the background. As a result of the iteration, more pixels are labeled as foreground. At this point, the user can decide to export the segmentation mask or apply another iteration. It is worth noting that the workflow described is applicable to images of any resolution: the user can just zoom into the small details of the image and perfect the segmentation mask.

2.2   Fast Dashing Method and Arrival Time Redistribution

As mentioned above, FDM is used to obtain the mask wave propagation arrival times of superpixels. FDM is an extension of FMM, a numerical method for solving the Eikonal equation

$$|\nabla u(x)| = \frac{1}{f(x)}, \tag{1}$$

where $u(x)$ and $f(x)$ show the arrival time and the speed of the wave at the location x, respectively.

When FMM is applied for image segmentation, $f(x) = \exp\left(-|\nabla I(x)|\right)$ is a common choice for speed, where $I(x)$ is the image value at pixel x [17]. Instead, as a wave propagation barrier we use superpixel mean color dissimilarity. It is calculated using CIEDE2000 [26] color difference of CIELAB color space, i.e. $f(x) = \exp(-\Delta(c(x), c(y)))$, where $c(x)$ is mean color and Δ is CIEDE2000 distance between x and neighbouring superpixel y, from which the wave spreads to x. FDM pseudocode is presented in Alg. 1.

Algorithm 1. Fast dashing method

```
 1: for all s in superpixels do                            ▷ initialisation
 2:     if isSeed(s) then
 3:         time(s) ← 0
 4:         Queue.push(0, s)                                ▷ min priority queue
 5:     else
 6:         time(s) ← ∞
 7: while Queue ≠ ∅ do                                      ▷ arrival time computation
 8:     s ← Queue.pop()                                     ▷ superpixel with the lowest time
 9:     for all n in neighbours(s) do
10:         f ← exp(−Δ(color(n), color(s)))                ▷ use CIEDE2000
11:         time_upd ← SolveEikonal(f)                      ▷ see [34] for details
12:         if time(s) ≤ time_upd < time(n) then
13:             time(n) ← time_upd
14:             Queue.push(time_upd, n)
```

After calculating the arrival times for all superpixels inside the current subimage, we redistribute them to the slider values for mask region propagation. The main challenge is to obtain an even growth of the region every time the slider value is incremented. At first, we group all superpixels in one bin. In every iteration, we choose the bin with the highest number of pixels and calculate the lowest and the highest arrival times in it, denoted by a and b, respectively. Next, we split the range $[a, b]$ into ranges $[a, (a+b)/2]$ and $((a+b)/2, b]$, thus splitting superpixels of the current bin into two bins with corresponding arrival times. The process stops if we reach a sufficient number of bins or if all superpixels in each bin have equal arrival times. Finally, we distribute the bins sorted by arrival time into slider values.

3   Experiments and Results

For a high-quality segmentation DL approaches usually require data annotated in a very detailed manner (with an accuracy up to one pixel). Recent
learning-based interactive segmentation approaches sometimes struggle to provide detailed annotations on edges or for objects out of the scope of their training
set. In this section, we show that our method has an ability to overcome such
difficulties since it provides control over the foreground mask growth.

As our goal is to show the effectiveness of our method especially when one
tries to annotate the foreground with very high accuracy, most existing benchmarks, e.g. COCO [12], ADE20K [37] or GrabCut [23], are not suitable for
testing. While we include two existing benchmarks in our experiments below, we
have, in addition, collected a new dataset, composed of 100 logo images, carefully segmented by human annotators. An in-house survey of photo editor usage
patterns suggests that the category of illustrations, especially logo images, is a
common target for background removal.

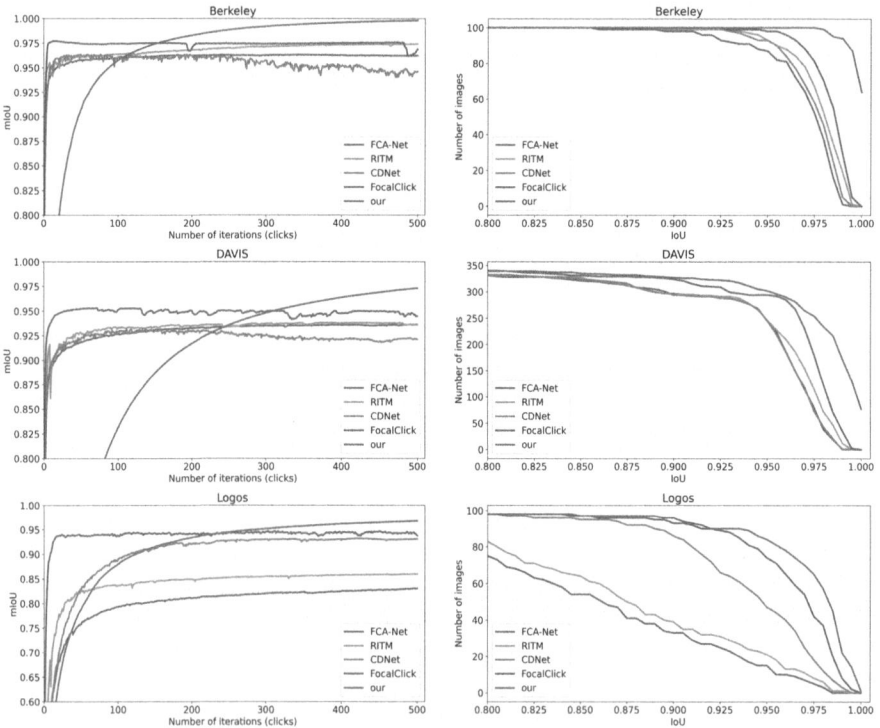

Fig. 3. Performance on Berkeley, DAVIS, and Logos datasets: mean IoU per number of
iterations, or clicks, (left), cumulative number of images to achieve IoU (right). Capping
the number of iterations at 500 affects the graphs.

(a) Image/GT   (b) 100 iterations   (c) 300 iterations   (d) 500 iterations

Fig. 4. Different steps of the algorithm on a sample image. (b)–(d) on odd rows illustrate the work of our algorithm; even rows correspond to [5]. (The reader is referred to the electronic version of the paper for full interpretability of colors in the images.)

To show the effectiveness of our method we compare it with recent SOTA approaches RITM [29], FCA-Net [14], CDNet [4], and FocalClick [5]. We chose those versions of the pretrained models that performed best, as reported in the corresponding papers. To compare interactive segmentation methods, one should take into account both the accuracy and the amount of time/effort spent by the user. Similar to previous work, we design an automatic process imitating the user actions and measuring the user effort by the number of iterations of the interactive segmentation algorithms. As shown in Fig. 2, in each iteration, the human interaction steps are zooming, seeding the wave propagation, and choosing the optimal slider value. In order to automate zooming, we crop the image with a slightly enlarged bounding box of the largest connected component of the difference mask between the current segmented region and the GT. As a seed, we use the point furthest from the boundaries of that connected component. The slider value is selected to stop the propagation process just before it overflows the GT mask. Unlike previous work, we allow the automated interaction simulation to execute for 500 iterations; most papers use only 20, and some consider 100 [29].

3.1 Datasets

In our experiments, we use three different datasets, described below. **Berkeley** is a set of 100 masks for 96 images [19] taken from the original Berkeley segmentation dataset of 300 images [18]. From **DAVIS** [20], a collection of 50

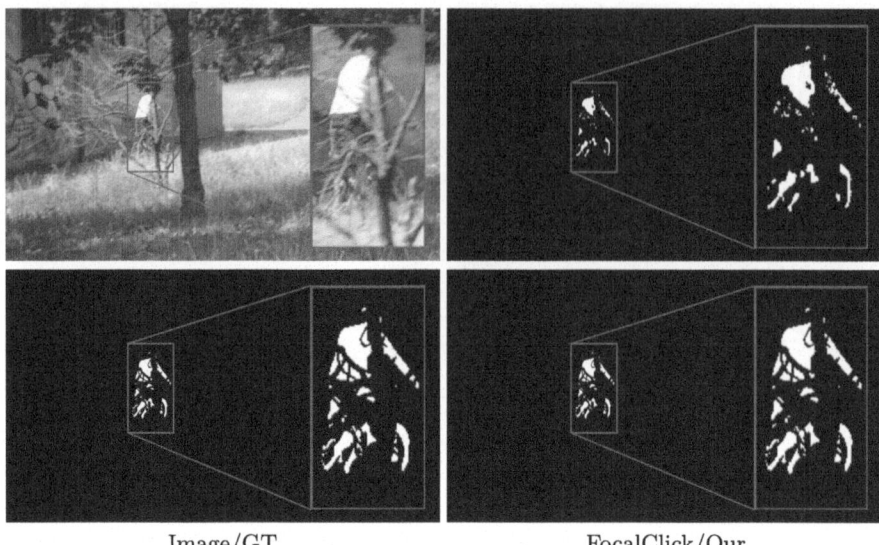

Image/GT   FocalClick/Our

Fig. 5. Comparison of segmentation results with FocalClick and our algorithm after 500 automated iterations on an image from the DAVIS dataset. Our algorithm achieves a high IoU score of 100%, indicating perfect segmentation alignment, while FocalClick achieves an IoU score of 64.3%.

full HD video sequences totalling 3455 frames, we use the same 10% of images as [14,29]. **Logos** is our in-house dataset of 100 logo illustrations, commonly on a plain background. The masked objects typically contain long boundaries and consist of multiple components. While in the first two datasets, only DAVIS annotations are of high accuracy, the Logos dataset has been constructed specifically to offer only pixel-accurate masks.

Following the common practice in interactive segmentation [13,29,38], we report the mean intersection over union (mIoU) after every iteration, i.e. IoU–NoC curves. Further insights into the algorithms' performance for high-accuracy segmentation could be acquired by using evaluation metrics that focus on boundary errors [6,35].

3.2 Quantitative and Qualitative Comparison

We report the results of the experimental comparison in Fig. 3. For all five approaches, we performed 500 iterations for each image. An iteration involves a single simulated click, and simulated zooming and slider value choice for our approach. The improvement of mean IoU scores over iterations are reported in the left column for all three datasets. The right column shows the cumulative number of images to achieve each IoU target between 80% and 100% after the 500 iterations.

On Berkeley, the SOTA methods achieve a mean IoU above 95% within a few iterations. However, their accuracy stagnates after that, staying below 97.8% IoU throughout all 500 iterations. While slow to catch up at the beginning, our approach eventually surpasses the SOTA at the 154th iteration, achieving a mean IoU of 99.8% and segmenting 87 instances out of 100 with IoU of at least 99.5%. The performance pattern is similar for DAVIS and Logos, where the best-performing SOTA approach is again FocalClick with mean IoU climbing up to 95.3% and 94.7%, respectively. Our algorithm surpasses FocalClick on the 311th and 224th iterations, respectively, reaching eventually a mean IoU of 97.3% on DAVIS and 96.8% on Logos. On DAVIS, our algorithm reaches IoU of 99% or above on half of the images (172/345); FocalClick does that for around 10% of images (35/345). RITM and FCA-Net achieve mean IoU scores of 85.9% and 83%, respectively, on Logos. We believe that the reason for the poorer performance of some DL approaches on this specific dataset is that image contents are so different from their training data. In contrast with learning methods, our approach has no such limitations in generalization.

Qualitative comparison of our method and FocalClick can be found in Fig. 4. In contrast with our approach, FocalClick fails to reach high-quality results around the object boundaries with the first 100 or even 300 automated iterations. Moreover, even with 500 automated iterations their masks are still not as accurate as the results provided by our method.

Figure 5 illustrates an example image from DAVIS with a complex GT mask with multiple components due to an obstructing object. After the 500 automated iterations, FocalClick achieves an IoU of only 64.3%. Our algorithm, on the other hand, produces a perfect segmentation with an IoU of 100%. This is due to the ability of our approach to incrementally and independently fill in small details of the foreground, preserving sharp edges and discontinuities without altering other parts of the mask, produced in preceding iterations.

All four DL approaches are implemented in Python, using the PyTorch library. Our approach is implemented in C++ with the use of the OpenCV [1] library, including the graphical user interface features in it. It is currently not optimized for faster execution, although certain parts of our algorithm, e.g. FDM, can be parallelized and executed on a graphics processing unit (GPU) [34].

In order to estimate relative execution times of all considered algorithms, we ran them for 100 auto-iterations on the first 50 images from the DAVIS dataset. We got the following results: FocalClick—2h 54mins; FCA-Net—6h 6mins; CDNet—6h 15mins; RITM—11h 51mins. Our algorithm, on the other hand, completed the same experiment in 21mins. The hardware used for this experiment was a MacBook Pro with an Intel Core i9 processor and 16GB RAM.

3.3   Comparison with Naive Approach

We compare the user effort required while using our tool versus the naive approach with varying brush sizes. Since this comparison process is expensive, we consider only the two images in Fig. 4. We asked 3 human annotators to segment these as accurately as possible by using brushes of different sizes. In addition, one annotator applied our tool for the same purpose. The annotation with our tool took 3 and 5 mins on the logo and real-life images, while the naive approach was more time-consuming (9 and 11 mins). We calculated the IoU of our mask against each of the 3 GT masks, and all the pairwise IoUs among the GT masks. The lowest IoUs of our mask (99.74% and 99.81%) are at the same level as the lowest pairwise IoUs between GT masks (99.61% and 99.87%).

Image/GT   FocalClick/Our

Fig. 6. An illustrative example of a challenging textured and blurry image. Even after 500 automated iterations, both FocalClick and our algorithms face difficulties in achieving satisfactory segmentation results. The IoU scores for our algorithm and FocalClick are 47.8% and 52.7%, respectively, indicating the difficulty of segmenting such images.

4   Conclusions

Recent SOTA methods for interactive segmentation (mostly DL approaches) can achieve acceptable initial segmentation with several iterations. However, it is much harder to reach IoU values higher than 95%. We presented a novel framework for interactive segmentation, which allowed users to get high-accuracy results using foreground clicks and a slider for mask growing. We showed that our tool outperformed the SOTA on certain classes of images (multi-component and long boundary images like logos). Furthermore, while SOTA methods easily stagnate without accuracy improvement through hundreds of iterations, our approach iteratively improves its accuracy and reaches extremely high similarity scores. As our experiments show, our method can easily replace naive approaches for fast and high-quality segmentation.

4.1   Limitations

Figure 6 considers a challenging textured and blurry image. As expected, even after 500 iterations, both DL and our algorithms face difficulties in achieving satisfactory segmentation results. The IoU score of our algorithm is only 47.8%. But FocalClick is only marginally better with 52.7% IoU.

While giving full control over segmentation results, our method struggles with textured images due to its current propagation speed function. It takes more user effort to segment such regions. Thus, there is space to refine the superpixel similarity measure. We also plan to optimize the implementation and add negative seed propagation for a smoother user experience. This will also enable the easy construction of hybrid approaches, combining DL methods for initial segmentation and mask refinement with our method.

Acknowledgements. Varduhi Yeghiazaryan's participation in this project was supported by Picsart AI Research. We also thank the annotators for their time and contribution to our research.

References

1. OpenCV library. https://opencv.org/. Accessed 28 Aug 2023
2. Benenson, R., Popov, S., Ferrari, V.: Large-scale interactive object segmentation with human annotators. In: CVPR, pp. 11700–11709 (2019)
3. Chen, L.C., Papandreou, G., Kokkinos, I., Murphy, K., Yuille, A.L.: Deeplab: semantic image segmentation with deep convolutional nets, atrous convolution, and fully connected crfs. IEEE Trans. Pattern Anal. Mach. Intell. **40**(4), 834–848 (2017)
4. Chen, X., Zhao, Z., Yu, F., Zhang, Y., Duan, M.: Conditional diffusion for interactive segmentation. In: ICCV, pp. 7345–7354 (2021)
5. Chen, X., Zhao, Z., Zhang, Y., Duan, M., Qi, D., Zhao, H.: Focalclick: towards practical interactive image segmentation. In: CVPR, pp. 1300–1309 (2022)
6. Cheng, B., Girshick, R., Dollár, P., Berg, A.C., Kirillov, A.: Boundary iou: Improving object-centric image segmentation evaluation. In: Proceedings of the IEEE/CVF Conference on Computer Vision and Pattern Recognition, pp. 15334–15342 (2021)
7. Galeev, D., Popenova, P., Vorontsova, A., Konushin, A.: Contour-based interactive segmentation. arXiv preprint arXiv:2302.06353 (2023)
8. Gupta, A., Dollar, P., Girshick, R.: LVIS: a dataset for large vocabulary instance segmentation. In: CVPR (2019)
9. He, K., Gkioxari, G., Dollár, P., Girshick, R.: Mask r-cnn. In: ICCV, pp. 2961–2969 (2017)
10. Jang, W.D., Kim, C.S.: Interactive image segmentation via backpropagating refinement scheme. In: CVPR, pp. 5297–5306 (2019)
11. Levin, A., Rav-Acha, A., Lischinski, D.: Spectral matting. IEEE Trans. Pattern Anal. Mach. Intell. **30**(10), 1699–1712 (2008)
12. Lin, T.Y., et al.: Microsoft COCO: Common objects in context. In: ECCV, pp. 740–755 (2014)
13. Lin, Z., Duan, Z.P., Zhang, Z., Guo, C.L., Cheng, M.M.: Focuscut: diving into a focus view in interactive segmentation. In: CVPR, pp. 2637–2646 (2022)

14. Lin, Z., Zhang, Z., Chen, L.Z., Cheng, M.M., Lu, S.P.: Interactive image segmentation with first click attention. In: CVPR, pp. 13339–13348 (2020)
15. Liu, Q., et al.: Pseudoclick: interactive image segmentation with click imitation. In: ECCV, pp. 728–745 (2022)
16. Long, J., Shelhamer, E., Darrell, T.: Fully convolutional networks for semantic segmentation. In: CVPR, pp. 3431–3440 (2015)
17. Malladi, R., Sethian, J.A.: An O(N log N) algorithm for shape modeling. Proc. National Acad. Sci. **93**(18), 9389–9392 (1996)
18. Martin, D., Fowlkes, C., Tal, D., Malik, J.: A database of human segmented natural images and its application to evaluating segmentation algorithms and measuring ecological statistics. In: ICCV, pp. 416–423 (2001)
19. McGuinness, K., O'Connor, N.E.: A comparative evaluation of interactive segmentation algorithms. Pattern Recogn. **43**(2), 434–444 (2010)
20. Perazzi, F., Pont-Tuset, J., McWilliams, B., Van Gool, L., Gross, M., Sorkine-Hornung, A.: A benchmark dataset and evaluation methodology for video object segmentation. In: CVPR, pp. 724–732 (2016)
21. Redmon, J., Divvala, S., Girshick, R., Farhadi, A.: You only look once: unified, real-time object detection. In: CVPR, pp. 779–788 (2016)
22. Ronneberger, O., Fischer, P., Brox, T.: U-net: convolutional networks for biomedical image segmentation. In: MICCAI, pp. 234–241 (2015)
23. Rother, C., Kolmogorov, V., Blake, A.: "grabcut": Interactive foreground extraction using iterated graph cuts. ACM Trans. Graph. **23**(3), 309–314 (2004)
24. Sethian, J.A.: Fast marching methods. SIAM Rev. **41**(2), 199–235 (1999)
25. Sethian, J.A.: Evolution, implementation, and application of level set and fast marching methods for advancing fronts. J. Comput. Phys. **169**(2), 503–555 (2001)
26. Sharma, G., Wu, W., Dalal, E.N.: The ciede2000 color-difference formula: Implementation notes, supplementary test data, and mathematical observations. Color. Res. Appl. **30**(1), 21–30 (2005)
27. Sofiiuk, K., Petrov, I., Barinova, O., Konushin, A.: F-brs: rethinking backpropagating refinement for interactive segmentation. In: CVPR, pp. 8623–8632 (2020)
28. Sofiiuk, K., Petrov, I.A., Konushin, A.: Reviving iterative training with mask guidance for interactive segmentation. arXiv preprint arXiv:2102.06583 (2021)
29. Sofiiuk, K., Petrov, I.A., Konushin, A.: Reviving iterative training with mask guidance for interactive segmentation. In: ICIP, pp. 3141–3145 (2022)
30. Stutz, D., Hermans, A., Leibe, B.: Superpixels: an evaluation of the state-of-the-art. Comput. Vis. Image Underst. **166**, 1–27 (2018)
31. Wang, X., Kong, T., Shen, C., Jiang, Y., Li, L.: Solo: Segmenting objects by locations. In: ECCV, pp. 649–665 (2020)
32. Yang, M., Yu, K., Zhang, C., Li, Z., Yang, K.: Denseaspp for semantic segmentation in street scenes. In: CVPR, pp. 3684–3692 (2018)
33. Yao, J., Boben, M., Fidler, S., Urtasun, R.: Real-time coarse-to-fine topologically preserving segmentation. In: CVPR, pp. 2947–2955 (2015)
34. Yeghiazaryan, V.: Parallel Front Propagation in Medical Image Segmentation. DPhil thesis, University of Oxford (2018)
35. Yeghiazaryan, V., Voiculescu, I.: Family of boundary overlap metrics for the evaluation of medical image segmentation. J. Med. Imag. **5**(1), 015006–015006 (2018)
36. Zhang, S., Liew, J.H., Wei, Y., Wei, S., Zhao, Y.: Interactive object segmentation with inside-outside guidance. In: CVPR, pp. 12234–12244 (2020)

37. Zhou, B., Zhao, H., Puig, X., Fidler, S., Barriuso, A., Torralba, A.: Scene parsing through ADE20K dataset. In: CVPR, pp. 633–641 (2017)
38. Zhou, M., et al.: Interactive segmentation as gaussion process classification. In: Proceedings of the IEEE/CVF Conference on Computer Vision and Pattern Recognition, pp. 19488–19497 (2023)

Gesture Recognition on Video Data

Georgy Gunkin[1] and Ilya Makarov[2,3]($\boxtimes$)

[1] MIPT, Moscow, Russia
[2] AIRI, Moscow, Russia
makarov@airi.net
[3] HSE University, Moscow, Russia

Abstract. Recognizing dynamic hand gestures presents a significant challenge for artificial intelligence systems, as it requires integrating visual information across a sequence of frames.

Current approaches to hand gesture recognition face limitations. Some methods rely solely on visual inputs, processing them without considering occlusion and illumination issues, leading to performance degradation. Others employ wearable sensors, which can be bulky and prone to noise. Alternatively, recurrent neural network-based models have been proposed to classify gestures using transformed hand keypoints in a sequential manner.

To address these challenges and identify existing problems and bottlenecks, we conduct experiments using different modalities, datasets, and existing solutions. We analyze various datasets, evaluate their quality [1,65], and select the most suitable ones for each modality, including text and images.

Ultimately, we propose an artificial intelligence system inspired by Contrastive Language-Image Pretraining, leveraging the notion that hand gestures represent their own language. Our system aims to link textual representations of gestures with their visual inputs. The system comprises two main components: a pre-trained text encoder that transforms textual labels of dataset examples into meaningful embeddings, and a vision-transformer-based image sequence encoder that generates the desired embeddings from the visual modality. We pretrain the text encoder and subsequently use it to train the image encoder with text supervision.

Keywords: Gesture recognition · dynamic hand gestures · metric learning · computer vision

1 Introduction

One of the challenging tasks in video-based video classification [59,61,74] is hand gestures recognition [13]. As hand gestures provide a natural way of interacting

---

The work on Section 2 was supported by RSF under grant 22-11-00323 and performed at HSE University, Moscow, Russia, in 2023; improved camera-ready paper was submitted for publication in 2024.

© The Author(s), under exclusive license to Springer Nature Switzerland AG 2024
D. I. Ignatov et al. (Eds.): AIST 2023, CCIS 1905, pp. 180–193, 2024.
https://doi.org/10.1007/978-3-031-67008-4_14

with computers, and accurately interpreting moving hand gestures in real-time has numerous applications. These include virtual reality systems [32,36,38,71], interactive gaming platforms [20,35,49], sign language recognition [22,28,50,78], enabling young children to interact with computers [76], robot control [45], music conducting practice [63], television control [10,29], automotive interfaces [43,44], emotion analysis [55,56,72], learning and teaching assistance [52,53,57], hand gesture generation [67], audio-visual speech recognition [51,60,62] and general human-computer interaction [15,19,23,48,58,66]. In the past, specialized hardware equipment like gloves [21], smartwatches [25], head-mounted gear [41], joysticks [77], and other external tools were commonly used. However, these methods are considered unnatural and burdensome. In this paper, we focus on using a single camera to detect and recognize known gestures, assuming no additional hardware. Hand gestures can be categorized as static or dynamic. Static gestures refer to fixed hand positions that don't change over time, while dynamic gestures involve varying hand keypoints across subsequent frames. The number of frames required for a dynamic gesture isn't predetermined, as it depends on the gesture speed and camera frame rate.

Numerous studies have concentrated on the recognition of static hand gestures, leading to several successful methodologies. These range from Template Matching, a straightforward technique that compares the hand gesture image to predefined templates but can be sensitive to variations in lighting, orientation, and scale [16], to machine learning algorithms such as Support Vector Machines (SVMs) that are often employed with a variety of feature descriptors extracted from input images [14]. Other approaches utilize machine learning algorithms like random forests and gradient boosting. Additionally, Convolutional Neural Networks (CNNs) have been used for classification-based approaches [27,42]. Histogram of Oriented Gradients (HOG) descriptors have also been applied to analyze input images [45].

However, this paper focuses solely on recognizing dynamic hand gestures, which pose greater challenges due to the diversity in how people perform them [11].

Gesture classification isn't the only problem tackled in this paper. Other challenges addressed include:

1. Defining a suitable benchmark for both classification and metric learning scenarios.
2. Selecting an appropriate time window for running the model and obtaining results. Determining the perfect window size is crucial, as it should capture gestures of varying speeds without overlapping.
3. Evaluating the effectiveness of working on pure images versus detecting hand keypoints before processing. Both approaches are investigated to determine the most suitable one.

The paper proceeds by discussing relevant datasets, reviewing state-of-the-art (SOTA) methods reported by other researchers, presenting original experiments from different perspectives, and comparing the obtained results with SOTA solutions. Finally, the conclusion highlights the need for further research in this area.

Fortunately, numerous high-quality datasets have been compiled specifically for hand gesture recognition tasks. While many of these datasets focus on static gestures, this section highlights the most suitable datasets for dynamic gesture recognition.

2   Related Work

2.1   Datasets

Fortunately, there are many quality datasets available specifically for the task of recognizing hand gestures. While several of them focus on static gestures, this section provides information on the most appropriate datasets for dynamic gestures.

The Cambridge Hand Gesture Dataset consists of 900 sequences, each containing pure images, representing 9 gesture classes. The dataset offers diversity by including multiple subjects performing the gestures under 5 different illumination conditions [24]. This dataset is suitable for models that work solely on image data, but it can also be preprocessed using hand keypoints extractors to enable time series solutions.

The DHG Dataset 14/28 comprises 2800 sequences of varying lengths, with each sequence containing a depth image for every frame, along with 22 2D and 3D hand keypoint coordinates that describe the hand movements. The dataset consists of 14 gesture classes performed by the right hand of 20 participants [6]. It is particularly useful for benchmarking approaches that work with time series data. However, this dataset lacks colored images, limiting its suitability for inter-domain solution comparison.

The SHREC 2017 Dataset is similar to the DHG Dataset in terms of setup and gesture classes. It provides an online scenario where each person executes all gestures in succession [7]. This dataset serves as a powerful benchmark for gesture splitting, as discussed earlier.

The EgoGesture Dataset contains 83 classes of gestures, predominantly dynamic with a small portion of static gestures. The dataset captures gestures using a camera attached to a strap belt on subjects' heads. It includes 2081 RGB-D videos, 24161 gesture samples, and 2953224 frames from 50 distinct subjects [4,79]. Unfortunately, this dataset does not provide hand keypoint data. Notably, the gestures are captured against various backgrounds, enhancing its diversity.

The Dataset for Dynamic Hand Gesture Recognition Systems consists of 27 dynamic hand gesture classes, totaling 1701 videos and 204120 video frames. This dataset solely includes RGB-D images and does not provide hand keypoint information. However, all participants underwent special training and performed the gestures under strict control to ensure correctness. The dataset's substantial size makes it a valuable resource [11].

The Nvidia Dynamic Hand Gesture Dataset comprises 25 dynamic gesture classes, recorded by multiple sensors including color, depth, and infrared streams. The dataset contains a total of 1532 dynamic hand gestures, captured under various lighting conditions [42].

The Jester Dataset consists of 148092 short video clips, each 3 s in length, performed by 1376 actors, resulting in more than 5 million frames. The dataset provides only RGB frames and does not include hand keypoint information. It contains 27 dynamic gesture classes [40].

The IPN Hand Dataset comprises more than 4000 gesture instances and 800000 frames from 50 subjects. The dataset includes RGB frames, optical flow data, and hand segmentation for each frame. However, it lacks hand keypoint information. It consists of 13 static and dynamic hand gestures designed for interaction with touchless screens [2].

The Chalearn LAP ConGD Database consists of 47933 RGB-D gestures in 22535 RGB-D gesture videos. Each video may contain one or more gestures, with a total of 249 gesture labels performed by 21 different individuals [70]. This dataset does not provide hand keypoint data.

The Sheffield Kinect Gesture Dataset includes 1080 RGB-D video sequences recorded by 6 subjects. It encompasses 10 dynamic gesture classes representing figures and signs [30]. This dataset does not provide additional hand keypoint information.

The American Sign Language (ASL) Dataset contains 25000 annotated samples comprising 1000 signs, recorded in challenging and unconstrained real-life conditions. The dataset includes RGB data and does not provide hand keypoint information. It covers sign classes such as numbers, letters, words, and sentences.

The Slovo dataset is a multimodal dataset that contains RGB data and hand keypoint sequences of sign language [22]. Hand keypoint information was annotated using MediaPipe Hands [31], which is a state-of-the-art method for detecting hand keypoints. However, it is important to note that, like any machine learning algorithm, MediaPipe Hands may have errors and inaccuracies that could impact subsequent models relying on the annotated data.

2.2 Current SOTA Methods

In this section, we will briefly discuss solutions to dynamic hand gesture recognition that perform the best on the given datasets. As discussed above, we will review both directions - using hand keypoints only and using images only.

We start from Hand keypoints SOTA. For benchmarking models that work with keypoints only, the 14 class DHG dataset has been chosen due to its quality and suitability.

Several approaches that report SOTA do not use only hand keypoints as the input to their models. For example DD-NET [75] model feed in 3D keypoints as well as some handcrafted features of dimension 231. First of all, feeding in the Z-coordinate is not very stable as it is a prediction of the depth taken relative to the wrist point. This is common with the use of MediaPipe Hands framework [31] and hence provides a source of inaccuracy, especially during inference stage in changing conditions. Secondly, modifying the dataset with own data will not be considered in this work, and therefore we will no take such models into account for fairness.

Another example is FPPR-PCB [3]. In this work they use the same dataset but treat depth image as a point cloud and work with it. Depth frames are also a source of severe inaccuracies for many cameras used in the industry (unless they are extremely expensive, which does not make sense for this project), as they contain substantial amount of noise, especially with occluded objects. This model will also not be considered due to the different modality used.

Building a pipeline of several neural networks that work on top of each other will likely accumulate the error. For example, if the hand keypoints detector produces too much noise, this can severely affect the classifier. However, it is useful to know to which extent these models can perform and hence report their scores.

One of the SOTA methods that uses pure hand keypoints is Parallel-Conv [8]. It is implemented via 1D convolutions applied to each keypoint channel in the sequence, treated as channel-wise feature extractors, followed by a classification Multi Layer Perceptron (MLP).

The best, however, verified method that works on hand keypoints data is DG-STA [5]. Their idea lies in the implementation of the so called dynamic graph-based spatial and temporal attention heads to model the behaviour of hand keypoints relative to each other on a single frame as well as sequence-wise. They also implemented Positional Encoding as suggested by [69].

Let's proceed to Image-based SOTA. For benchmarking models that work with image sequences, the 11 class IPN Hand dataset has been chosen due to its suitability and applicability of classes to real world task. However, its quality is far from the best, the participants did not execute the gestures with care and introduce large noise rate.

Here and later, the 3D prefix of the models mean that they work on a sequence of images, that we shall, from now on, call a clip with 32 images in the sequence.

There are many networks that can work with image sequences or that can be adapted from single image to multi image input, such as CNN to C3D transitions [68]. However, we will only consider a single SOTA model that beats the others by a high margin.

The current SOTA image sequence based dynamic gesture recognition network is ResNeXt-101 with cardinality 32.

3   Experiments and Results

Since there are no publicly available datasets containing both 2D images and accurate 2D hand keypoints of dynamic gestures, it was decided that benchmarking will be done separately for keypoint approaches and for image-based approaches. Unfortunately, given that setup, we can not compare the two approaches. However, in future works, it is possible to annotate some given dataset, for example EgoGesture dataset with the help of frameworks like MediaPipe Hands [31] and manual corrections. For this project manual annotation and dataset collection is considered to be out of scope.

During the proposed experiments, firstly, baselies that do not consider the order of the sequence have been evaluated, such as MLP and gradient boosting, then an LSTM is verified to have smaller metrics than the models before. These results show a nontrivial connection between the hand keypoints in the sequence and the fact that the order introduces high noise by itself, since the keypoints are detected with a small (but still significant) error. Therefore, it has been proposed to exploit the other sequence based hand keypoints models as well as start working on image-based approaches, as they introduce less errors in the input data.

3.1   Hand Keypoints Approaches

We will start by discussing different approaches to the problem and defining baselines. The methods below will be briefly described, and the final metrics will be summarized in a Table 1. As mentioned earlier, we will be using the DHG dataset for benchmarking.

Through experimentation, it has been verified that a sequence of size 32 is optimal. Most gestures consist of approximately that many frames on average, and we aim to ensure that a single sequence contains the majority of each gesture.

One approach involves utilizing a simple linear fully connected model with 5 hidden layers and a ReLU nonlinearity [12]. This model already demonstrates better metrics than the aforementioned LSTM model, despite not considering the order in the sequence and treating it as a single chunk.

To further improve the model, Catboost [46] has been selected. It exhibits significantly better performance, as shown in the results Table 1, even with the default hyperparameters. However, it still fails to consider the order of the keypoints sequence.

Additionally, a simple baseline has been established using an LSTM model [18] with 8 hidden layers and a hidden vector size of 128. Surprisingly, this model performs the worst out of all the tested models, despite being designed to capture long-term relationships within the given sequence of keypoints.

Another time-series approach that has been proposed is the Transformer model [69]. It employs 22 attention heads alongside 12 encoder layers, with a hidden vector size of 512. Theoretically, this model should outperform Catboost since it incorporates positional encoding and is aware of the order. However, in practice, for this dataset, it performs worse and only surpasses the performance of LSTM and MLP models.

3.2   Image-Based Approaches

During the inference stage, the pure 2D image has demonstrated the least amount of noise. It contains sufficient information to infer the gesture class and allows for the construction of a shallow pipeline with just a single model. We argue that this approach is the best due to its flexibility and simplicity.

Table 1. Results of experiments on DHG-14 dataset

Keypoint-based methods (lower is better)				
Method	Accuracy	Precision	Recall	F1
Ours				
LSTM	0.534	0.377	0.419	0.397
MLP	0.552	0.511	0.460	0.484
Vanilla Transformer	0.601	0.481	0.496	0.488
Catboost	0.687	0.558	0.544	0.551
Current SOTA methods				
Parallel-Conv [8]	0.669	0.477	0.480	0.478
DG-STA [5]	**0.703**	**0.693**	**0.630**	**0.66**

Initially, 3D convolutional networks were proposed to address video classification tasks [54,68]. These networks consist of 3D convolutions followed by fully connected layers for classification.

This family of models incorporates residual connections on top of C3D networks, which facilitate gradient flow to the earlier layers in deep neural networks [17,73]. ResNext is a modification of the original ResNet, introducing a different internal block architecture and performance enhancements.

More recently, vision transformers have demonstrated outstanding results in the field of computer vision, surpassing classical convolutional-based approaches on public benchmarks [9]. Results are summarized in the Table 2.

Table 2. Results of experiments on IPN Hand dataset

Image-based methods (lower is better)		
Method	Accuracy	Inference Time (ms/clip)
C3D	0.778	76.2
3D ResNet-50	0.731	**18.2**
3D ResNeXt-101	0.836	27.7
MViTv2_small	0.839	29.1
Ours		
CLIP [47] based MViTv2_small	-	-

4   Proposed Approach

Our proposed approach is based on the understanding that hand gestures can be considered as a distinct language, whether it is sign language or a command language. We aim to combine the text and image modalities into a unified vector space, ensuring a strong connection between what we see and what we read. To accomplish this, we leverage a CLIP-like architecture [47].

The list of key differences from the CLIP approach:

1. Textual descriptions should represent dynamic actions. Instead of using descriptions like "A photo of a object," we utilize descriptions like "A video of a person's arm moving action." These dynamic action descriptions are then fed into the original pre-trained text encoder, generating a single vector representation for the image sequence.
2. The image encoder in CLIP is replaced with an image sequence (visual) encoder based on a 3D vision transformer. This visual encoder processes the image sequence in the given order and produces separate embeddings for each chunk.

To train our models, we optimize the visual encoder to produce similar embeddings to the ones generated by the text encoder using cosine similarity distance. We employ a metric learning setting for model training.

The concept of using a metric learning scenario enables us to adapt to various gestures by potentially clustering them and finding the most similar ones. Additionally, we can extend this idea to create a setup where, during inference, a new unknown gesture is presented, memorized, and subsequently recognized.

By combining the strengths of text and image modalities through a unified vector space, our proposed approach aims to enhance gesture understanding and enable effective recognition and classification of hand gestures.

4.1 Training

The training process is divided into two parts:

1. Initially, we employ a frozen pre-trained vision encoder to generate embeddings. These embeddings are then utilized to fine-tune the text encoder.
2. Subsequently, we freeze the text encoder and utilize it for supervision while training the vision encoder. This involves generating embeddings for each batch and minimizing the distance between the resulting embeddings.

4.2 Inference

During the inference stage, we propose a fixed time window approach with a size of 32 frames and an overlap of 16 frames. This means that we process a sequence of 32 frames at a time, while sliding only 16 frames between each run. This strategy ensures that crucial parts of the gestures are not missed and facilitates combining predictions between neighboring runs.

Moreover, we suggest employing a pre-trained hand detector to filter out irrelevant visual information and focus solely on the hands. Additionally, hand tracking techniques such as SORT or DeepSORT with embeddings can be utilized to stabilize detections over time.

By combining the hand detections into sequences of frames, we directly input them into the visual encoder to retrieve embeddings. These embeddings can then be used to identify similar known gestures.

5   Conclusion

We have successfully developed a novel system that relies solely on visual information to recognize dynamic hand gestures and exhibits strong generalization capabilities even for unknown gestures. Our approach leverages multimodal data, incorporating both visual appearance and textual descriptions to learn gesture representations.

Through experimental evaluation, we have determined that using a sliding window of 32 sequential frames provides the best results for the studied sets of gestures, as this window size aligns with the average number of frames required for most gestures.

During the evaluation of various methods based on hand keypoints, we have observed that the keypoint data is inherently noisy and challenging to annotate precisely. Constructing a pipeline that propagates this noise downstream to other networks is likely to yield inferior results compared to methods that solely rely on visual information. This aligns with how humans comprehend gestures – we do not track hand keypoints but rather visually understand the sequence of hand views that lead to the desired gesture. Consequently, we propose to focus on visual-only approaches for further research in this domain.

5.1   Future Work

In our current approach, since we utilize prebuilt textual representations for each class in the dataset, there is no variation in the textual embeddings between examples from the same class. However, each example differs due to factors such as the performer of the gesture, the surroundings, and even the execution speed of the gesture. Manual annotation for each gesture is a costly task, especially considering the original difficulty of collecting data for dynamic gestures.

To address this challenge in future research, we propose the use of a pre-trained video-to-text encoder. This encoder would generate textual annotations for each example in the dataset, allowing us to produce embeddings for textual supervision on a per-example basis. This approach would ensure that there are as many embeddings as there are examples in the training set, in contrast to our current methodology, where only one textual description of the action and one embedding per class are available.

The practical implementation of the proposed research is left for further work, followed by careful comparison with the existing methods for recognition of dynamic hand gestures and an ablation study, discussing the extent to which every proposed idea increases the overall accuracy. Moreover, the sliding window size and the overlap size have been chosen based on several experiments, however it depends largely on the actual dataset that is used. So, a complete study that determines the optimal window size, based on the distribution of sequence lengths in the dataset, would well supplement the invented solution. Apart from that we are planing to test the performance of depth-aware models [26,33,34, 37,39,64] on the task of hand gesture recognition.

References

1. Antsiferova, A., Lavrushkin, S., Smirnov, M., Gushchin, A., Vatolin, D., Kulikov, D.: Video compression dataset and benchmark of learning-based video-quality metrics. In: Koyejo, S., Mohamed, S., Agarwal, A., Belgrave, D., Cho, K., Oh, A. (eds.) Advances in Neural Information Processing Systems, vol. 35, pp. 13814–13825. Curran Associates, Inc. (2022). https://proceedings.neurips.cc/paper_files/paper/2022/file/59ac9f01ea2f701310f3d42037546e4a-Paper-Datasets_and_Benchmarks.pdf
2. Benitez-Garcia, G., Olivares-Mercado, J., Sanchez-Perez, G., Yanai, K.: IPN hand: a video dataset and benchmark for real-time continuous hand gesture recognition. In: 2020 25th International Conference on Pattern Recognition (ICPR), pp. 4340–4347. IEEE (2021)
3. Bigalke, A., Heinrich, M.P.: Fusing posture and position representations for point cloud-based hand gesture recognition. In: 2021 International Conference on 3D Vision (3DV), pp. 617–626. IEEE (2021)
4. Cao, C., Zhang, Y., Wu, Y., Lu, H., Cheng, J.: Egocentric gesture recognition using recurrent 3D convolutional neural networks with spatiotemporal transformer modules. In: Proceedings of the IEEE International Conference on Computer Vision, pp. 3763–3771 (2017)
5. Chen, Y., Zhao, L., Peng, X., Yuan, J., Metaxas, D.N.: Construct dynamic graphs for hand gesture recognition via spatial-temporal attention. arXiv preprint arXiv:1907.08871 (2019)
6. De Smedt, Q., Wannous, H., Vandeborre, J.P.: Skeleton-based dynamic hand gesture recognition. In: Proceedings of the IEEE Conference on Computer Vision and Pattern Recognition Workshops, pp. 1–9 (2016)
7. De Smedt, Q., Wannous, H., Vandeborre, J.P., Guerry, J., Le Saux, B., Filliat, D.: Shrec'17 track: 3D hand gesture recognition using a depth and skeletal dataset. In: 3DOR-10th Eurographics Workshop on 3D Object Retrieval, pp. 1–6 (2017)
8. Devineau, G., Moutarde, F., Xi, W., Yang, J.: Deep learning for hand gesture recognition on skeletal data. In: 2018 13th IEEE International Conference on Automatic Face & Gesture Recognition (FG 2018), pp. 106–113. IEEE (2018)
9. Dosovitskiy, A., et al.: An image is worth 16x16 words: transformers for image recognition at scale. arXiv preprint arXiv:2010.11929 (2020)
10. Freeman, W.T., Weissman, C.D.: Television control by hand gestures. In: Proceedings of International Workshop on Automatic Face and Gesture Recognition, pp. 179–183 (1995)
11. Fronteddu, G., Porcu, S., Floris, A., Atzori, L.: A dynamic hand gesture recognition dataset for human-computer interfaces. Comput. Netw. 205, 108781 (2022)
12. Fukushima, K.: Cognitron: a self-organizing multilayered neural network. Biol. Cybern. 20(3–4), 121–136 (1975)
13. Garg, M., Ghosh, D., Pradhan, P.M.: Multiscaled multi-head attention-based video transformer network for hand gesture recognition. IEEE Signal Process. Lett. 30, 80–84 (2023)

14. Ghosh, D.K., Ari, S.: Static hand gesture recognition using mixture of features and SVM classifier. In: 2015 Fifth International Conference on Communication Systems and Network Technologies, pp. 1094–1099. IEEE (2015)
15. Grechikhin, I., Savchenko, A.V.: User modeling on mobile device based on facial clustering and object detection in photos and videos. In: Morales, A., Fierrez, J., Sánchez, J.S., Ribeiro, B. (eds.) IbPRIA 2019. LNCS, vol. 11868, pp. 429–440. Springer, Cham (2019). https://doi.org/10.1007/978-3-030-31321-0_37
16. Hassink, N., Schopman, M.: Gesture recognition in a meeting environment. Master's thesis, University of Twente (2006)
17. He, K., Zhang, X., Ren, S., Sun, J.: Deep residual learning for image recognition. In: Proceedings of the IEEE Conference on Computer Vision and Pattern Recognition, pp. 770–778 (2016)
18. Hochreiter, S., Schmidhuber, J.: Long short-term memory. Neural Comput. 9(8), 1735–1780 (1997)
19. Andreeva, E., Ignatov, D.I., Grachev, A., Savchenko, A.V.: Extraction of visual features for recommendation of products via deep learning. In: van der Aalst, W.M.P., et al. (eds.) AIST 2018. LNCS, vol. 11179, pp. 201–210. Springer, Cham (2018). https://doi.org/10.1007/978-3-030-11027-7_20
20. Ilya, M., Mikhail, T., Lada, T.: Imitation of human behavior in 3D-shooter game. In: AIST'2015 Analysis of Images, Social Networks and Texts, p. 64 (2015)
21. Iwai, Y., Watanabe, K., Yagi, Y., Yachida, M.: Gesture recognition by using colored gloves. In: 1996 IEEE International Conference on Systems, Man and Cybernetics. Information Intelligence and Systems (Cat. No. 96CH35929), vol. 1, pp. 76–81. IEEE (1996)
22. Kapitanov, A., Kvanchiani, K., Nagaev, A., Petrova, E.: Slovo: Russian sign language dataset. arXiv preprint arXiv:2305.14527 (2023)
23. Kharchevnikova, A., Savchenko, A.: Neural networks in video-based age and gender recognition on mobile platforms. Opt. Mem. Neural Netw. 27, 246–259 (2018)
24. Kim, T.K., Wong, S.F., Cipolla, R.: Tensor canonical correlation analysis for action classification. In: 2007 IEEE Conference on Computer Vision and Pattern Recognition, pp. 1–8. IEEE (2007)
25. Kimura, N.: Self-supervised approach for few-shot hand gesture recognition. In: Adjunct Proceedings of the 35th Annual ACM Symposium on User Interface Software and Technology, pp. 1–4 (2022)
26. Korinevskaya, A., Makarov, I.: Fast depth map super-resolution using deep neural network. In: 2018 IEEE International Symposium on Mixed and Augmented Reality Adjunct (ISMAR-Adjunct), pp. 117–122. IEEE (2018)
27. Kumar, N.D., Suresh, K., Dinesh, R.: CNN based static hand gesture recognition using RGB-D data. In: 2022 2nd International Conference on Artificial Intelligence and Signal Processing (AISP), pp. 1–6. IEEE (2022)
28. Kuznetsova, A., Leal-Taixé, L., Rosenhahn, B.: Real-time sign language recognition using a consumer depth camera. In: Proceedings of the IEEE International Conference on Computer Vision Workshops, pp. 83–90 (2013)
29. Lian, S., Hu, W., Wang, K.: Automatic user state recognition for hand gesture based low-cost television control system. IEEE Trans. Consum. Electron. 60(1), 107–115 (2014)
30. Liu, L., Shao, L.: Learning discriminative representations from RGB-D video data. In: Twenty-Third International Joint Conference on Artificial Intelligence (2013)
31. Lugaresi, C., et al.: Mediapipe: a framework for building perception pipelines. arXiv preprint arXiv:1906.08172 (2019)

32. Lv, Z., Halawani, A., Feng, S., Ur Réhman, S., Li, H.: Touch-less interactive augmented reality game on vision-based wearable device. Pers. Ubiquit. Comput. **19**, 551–567 (2015)
33. Makarov, I., Bakhanova, M., Nikolenko, S., Gerasimova, O.: Self-supervised recurrent depth estimation with attention mechanisms. PeerJ Comput. Sci. **8**, e865 (2022)
34. Makarov, I., Borisenko, G.: Depth inpainting via vision transformer. In: 2021 IEEE International Symposium on Mixed and Augmented Reality Adjunct (ISMAR-Adjunct), pp. 286–291. IEEE (2021)
35. Makarov, I., Kashin, A., Korinevskaya, A.: Learning to play pong video game via deep reinforcement learning. In: AIST (Supplement), pp. 236–241 (2017)
36. Makarov, I., et al.: Adapting first-person shooter video game for playing with virtual reality headsets. In: The Thirtieth International Flairs Conference (2017)
37. Makarov, I., Korinevskaya, A., Aliev, V.: Sparse depth map interpolation using deep convolutional neural networks. In: 2018 41st International Conference on Telecommunications and Signal Processing (TSP), pp. 1–5. IEEE (2018)
38. Makarov, I., et al.: First-person shooter game for virtual reality headset with advanced multi-agent intelligent system. In: Proceedings of the 24th ACM International Conference on Multimedia, pp. 735–736 (2016)
39. Maslov, D., Makarov, I.: Online supervised attention-based recurrent depth estimation from monocular video. PeerJ Comput. Sci. **6**, e317 (2020)
40. Materzynska, J., Berger, G., Bax, I., Memisevic, R.: The jester dataset: a large-scale video dataset of human gestures. In: Proceedings of the IEEE/CVF International Conference on Computer Vision Workshops (2019)
41. Mo, G.B., Dudley, J.J., Kristensson, P.O.: Gesture knitter: a hand gesture design tool for head-mounted mixed reality applications. In: Proceedings of the 2021 CHI Conference on Human Factors in Computing Systems, pp. 1–13 (2021)
42. Molchanov, P., Yang, X., Gupta, S., Kim, K., Tyree, S., Kautz, J.: Online detection and classification of dynamic hand gestures with recurrent 3D convolutional neural network. In: Proceedings of the IEEE Conference on Computer Vision and Pattern Recognition, pp. 4207–4215 (2016)
43. Ohn-Bar, E., Trivedi, M.: The power is in your hands: 3D analysis of hand gestures in naturalistic video. In: Proceedings of the IEEE Conference on Computer Vision and Pattern Recognition Workshops, pp. 912–917 (2013)
44. Ohn-Bar, E., Trivedi, M.M.: Hand gesture recognition in real time for automotive interfaces: a multimodal vision-based approach and evaluations. IEEE Trans. Intell. Transp. Syst. **15**(6), 2368–2377 (2014)
45. Prasuhn, L., Oyamada, Y., Mochizuki, Y., Ishikawa, H.: A hog-based hand gesture recognition system on a mobile device. In: 2014 IEEE International Conference on Image Processing (ICIP), pp. 3973–3977. IEEE (2014)
46. Prokhorenkova, L., Gusev, G., Vorobev, A., Dorogush, A.V., Gulin, A.: Catboost: unbiased boosting with categorical features. In: Advances in Neural Information Processing Systems, vol. 31 (2018)
47. Radford, A., et al.: Learning transferable visual models from natural language supervision. In: International Conference on Machine Learning, pp. 8748–8763. PMLR (2021)
48. Rautaray, S.S., Agrawal, A.: Vision based hand gesture recognition for human computer interaction: a survey. Artif. Intell. Rev. **43**, 1–54 (2015)
49. Ren, Z., Yuan, J., Meng, J., Zhang, Z.: Robust part-based hand gesture recognition using kinect sensor. IEEE Trans. Multimedia **15**(5), 1110–1120 (2013)

50. Ren, Z., Yuan, J., Zhang, Z.: Robust hand gesture recognition based on finger-earth mover's distance with a commodity depth camera. In: Proceedings of the 19th ACM International Conference on Multimedia, pp. 1093–1096 (2011)

51. Ryumin, D., Ivanko, D., Ryumina, E.: Audio-visual speech and gesture recognition by sensors of mobile devices. Sensors **23**(4), 2284 (2023)

52. Sathayanarayana, S., et al.: Towards automated understanding of student-tutor interactions using visual deictic gestures. In: Proceedings of the IEEE Conference on Computer Vision and Pattern Recognition Workshops, pp. 474–481 (2014)

53. Sathyanarayana, S., Littlewort, G., Bartlett, M.: Hand gestures for intelligent tutoring systems: dataset, techniques & evaluation. In: Proceedings of the IEEE International Conference on Computer Vision Workshops, pp. 769–776 (2013)

54. Savchenko, A.: Facial expression recognition with adaptive frame rate based on multiple testing correction. In: Krause, A., Brunskill, E., Cho, K., Engelhardt, B., Sabato, S., Scarlett, J. (eds.) Proceedings of the 40th International Conference on Machine Learning (ICML). Proceedings of Machine Learning Research, vol. 202, pp. 30119–30129. PMLR (2023)

55. Savchenko, A.V.: MT-EmotiEffNet for multi-task human affective behavior analysis and learning from synthetic data. In: Proceedings of European Conference on Computer Vision (ECCV) Workshops, Part VI, pp. 45–59 (2022)

56. Savchenko, A.V.: EmotiEffNets for facial processing in video-based valence-arousal prediction, expression classification and action unit detection. In: Proceedings of the IEEE/CVF Conference on Computer Vision and Pattern Recognition, pp. 5715–5723 (2023)

57. Savchenko, A.V., Makarov, I.: Neural network model for video-based analysis of student's emotions in e-learning. Opt. Mem. Neural Netw. **31**(3), 237–244 (2022)

58. Savchenko, A.V., Savchenko, L.V., Makarov, I.: Fast search of face recognition model for a mobile device based on neural architecture comparator. IEEE Access **11**, 65977–65990 (2023)

59. Savchenko, A.: Deep neural networks and maximum likelihood search for approximate nearest neighbor in video-based image recognition. Opt. Mem. Neural Netw. **26**, 129–136 (2017)

60. Savchenko, A., Khokhlova, Y.I.: About neural-network algorithms application in viseme classification problem with face video in audiovisual speech recognition systems. Opt. Mem. Neural Netw. **23**, 34–42 (2014)

61. Savchenko, A., Savchenko, L.: Three-way classification for sequences of observations. Inf. Sci. 119540 (2023)

62. Savchenko, V.V., Savchenko, A.V.: Criterion of significance level for selection of order of spectral estimation of entropy maximum. Radioelectron. Commun. Syst. **62**(5), 223–231 (2019)

63. Schramm, R., Jung, C.R., Miranda, E.R.: Dynamic time warping for music conducting gestures evaluation. IEEE Trans. Multimedia **17**(2), 243–255 (2014)

64. Semenkov, I., Karpov, A., Savchenko, A.V., Makarov, I.: Inpainting semantic and depth features to improve visual place recognition in the wild. IEEE Access **12**, 5163–5176 (2024). https://doi.org/10.1109/ACCESS.2024.3350038

65. Shumitskaya, E., Antsiferova, A., Vatolin, D.: Towards adversarial robustness verification of no-reference image- and video-quality metrics. Comput. Vision Image Underst. **240**, 103913 (2024). https://doi.org/10.1016/j.cviu.2023.103913. https://www.sciencedirect.com/science/article/pii/S107731422300293X

66. Tang, H., Liu, H., Xiao, W., Sebe, N.: Fast and robust dynamic hand gesture recognition via key frames extraction and feature fusion. Neurocomputing **331**, 424–433 (2019)

67. Tang, H., Wang, W., Xu, D., Yan, Y., Sebe, N.: Gesturegan for hand gesture-to-gesture translation in the wild. In: Proceedings of the 26th ACM International Conference on Multimedia, pp. 774–782 (2018)
68. Tran, D., Bourdev, L., Fergus, R., Torresani, L., Paluri, M.: Learning spatiotemporal features with 3D convolutional networks. In: Proceedings of the IEEE International Conference on Computer Vision, pp. 4489–4497 (2015)
69. Vaswani, A., et al.: Attention is all you need. In: Advances in Neural Information Processing Systems, vol. 30 (2017)
70. Wan, J., Zhao, Y., Zhou, S., Guyon, I., Escalera, S., Li, S.Z.: Chalearn looking at people RGB-D isolated and continuous datasets for gesture recognition. In: Proceedings of the IEEE Conference on Computer Vision and Pattern Recognition Workshops, pp. 56–64 (2016)
71. Wang, C., Liu, Z., Chan, S.C.: Superpixel-based hand gesture recognition with kinect depth camera. IEEE Trans. Multimedia **17**(1), 29–39 (2014)
72. Wu, J., Zhang, Y., Sun, S., Li, Q., Zhao, X.: Generalized zero-shot emotion recognition from body gestures. Appl. Intell. 1–19 (2022)
73. Xie, S., Girshick, R., Dollár, P., Tu, Z., He, K.: Aggregated residual transformations for deep neural networks. In: Proceedings of the IEEE Conference on Computer Vision and Pattern Recognition, pp. 1492–1500 (2017)
74. Yan, S., et al.: Multiview transformers for video recognition. In: Proceedings of the IEEE/CVF Conference on Computer Vision and Pattern Recognition (CVPR), pp. 3333–3343 (2022)
75. Yang, F., Wu, Y., Sakti, S., Nakamura, S.: Make skeleton-based action recognition model smaller, faster and better. In: Proceedings of the ACM Multimedia Asia, pp. 1–6. Association for Computing Machinery (2019)
76. Yao, Y., Fu, Y.: Contour model-based hand-gesture recognition using the kinect sensor. IEEE Trans. Circuits Syst. Video Technol. **24**(11), 1935–1944 (2014)
77. Yeh, S.C., Wu, E.H.K., Lee, Y.R., Vaitheeshwari, R., Chang, C.W.: User experience of virtual-reality interactive interfaces: a comparison between hand gesture recognition and joystick control for xrspace manova. Appl. Sci. **12**(23), 12230 (2022)
78. Yu, J., Qin, M., Zhou, S.: Dynamic gesture recognition based on 2D convolutional neural network and feature fusion. Sci. Rep. **12**(1), 4345 (2022)
79. Zhang, Y., Cao, C., Cheng, J., Lu, H.: Egogesture: a new dataset and benchmark for egocentric hand gesture recognition. IEEE Trans. Multimedia **20**(5), 1038–1050 (2018)

Semantic-Aware GAN Manipulations for Human Face Editing

Pavel Khlusov[1] and Ilya Makarov[1,2(✉)]

[1] HSE University, Moscow, Russia
pvkhlusov@edu.hse.ru
[2] AIRI, Moscow, Russia
makarov@airi.net

Abstract. Generative Adversarial Networks (GANs) have greatly advanced image generation, producing high-quality images, close to real. However, manipulation of the semantic properties of generated images remains a challenge. This study focuses on the study of manipulation methods in the intermediate latent space StyleGAN2. A comparison was made of several methods for identifying semantic meaningful directions in the latent space of StyleGAN2, which do not require labeled data. The quality of the received images and the received metrics are analyzed. The results obtained provide an understanding of the capabilities and limitations of existing methods for editing faces using the manipulation of GAN latent spaces.

A method based on mapping the latent code z into the extended intermediate latent space $W+$ and using a pretrained discriminator to control the quality of the edited image was also proposed. The results of experiments and the values of metrics are presented, confirming the improvement in image quality at large shifts in the latent space of Style-GAN2.

Keywords: Computer Vision · Generative Adversarial Networks · Human Face Editing

1 Introduction

Generative adversarial networks (GANs) [5] have made remarkable progress in computer vision over the past decade, enabling the generation of high-quality images that closely resemble real ones. As a result, GANs have found applications in various domains, including video games, animation, and more [4,20].

Despite their advantages, the traditional GAN architecture lacks control over the semantic properties of generated images. While some models have been proposed to generate images with specific attributes [39], change object semantics

---

The work on Section 2 was supported by RSF under grant 22-11-00323 and performed at HSE University, Moscow, Russia.

The original version of the chapter has been revised. Author name not been displayed correctly. This has been corrected. A correction to this chapter can be found at https://doi.org/10.1007/978-3-031-67008-4_23

© The Author(s), under exclusive license to Springer Nature Switzerland AG 2024, corrected publication 2024
D. I. Ignatov et al. (Eds.): AIST 2023, CCIS 1905, pp. 194–208, 2024.
https://doi.org/10.1007/978-3-031-67008-4_15

while preserving characteristics [18], or perform style mixing [12], these capabilities fall short for real image editing tasks. However, it is especially important to change various facial attributes in order to train better models for face recognition and other similar tasks [2,14,21,22,28,29,32], age and gender prediction [9,15,25] and emotion analysis [10,17,24,27,30]. This research aims to explore modern approaches in detecting semantically meaningful directions in the latent space of GANs, particularly those that do not rely on labeled data. These approaches offer extensive possibilities for manipulating images, such as altering gender, background, or hair volume in face editing tasks [35].

A significant milestone in GAN development was the introduction of style-based architectures [11]. These architectures not only improved image quality but also introduced properties of the intermediate latent space useful for tasks like image inversion and editing. In particular, [1] proposed inverting real images into the extended latent space $W+$, resulting in improved inversion quality.

Several studies [6,37,38] have investigated the properties of the latent space and proposed methods for finding semantically meaningful directions using unsupervised learning. Additionally, [3] highlighted the possibility of identifying semantic features and visual effects in the GAN parameter space that cannot be replicated by manipulating the latent code.

While unsupervised methods can uncover directions in the latent space enabling a wide range of changes in original images, experiments have shown that significant modifications to global semantics (e.g., gender, race, age) require large shifts in the latent space, often resulting in unexpected artifacts and decreased image quality.

This issue is particularly relevant for methods based on finding principal components in the latent space or singular values in the parameter space, as the lack of a specific loss function during direction discovery prevents direct control over the identified directions' properties and quality.

Building on the observation that inverting into the extended latent space $W+$ enhances inversion quality and provides more flexibility to the generator, this research explores the transformation of latent codes z into the extended intermediate latent space $W+$ instead of the regular intermediate latent space W to enhance the quality of images with large shifts in the latent space.

Thus, the objective of this research is to analyze existing methods for detecting semantically significant directions in GAN latent spaces for human face editing tasks.

The research will focus on studying methods for manipulating the intermediate latent [6,34,38] and parameter [3] spaces of StyleGAN2 [13,26] without the need for labeled data. It will evaluate the extent to which these manipulations allow for changing facial properties while preserving the quality and identity of faces before and after editing. The application of this method for manipulating real images inverted in the extended latent space [1] will also be considered.

In addition to the comparative analysis of methods for discovering directions in the latent and parameter spaces of GANs, this work proposes a method based on the use of the extended latent space to improve the image quality when applying large shifts to StyleGAN2.

2   Related Work

Generative Adversarial Networks (GANs). GANs [5] have gained popularity for photorealistic image synthesis. They consist of a generator G and a discriminator D, which are trained simultaneously to capture the distribution of real data. The GAN architecture incorporates features like ReLU, dropout regularization, and convolution layers from discriminative models.

Style-Based Generator Architecture. The Style-based generator architecture [12] introduces a non-linear mapping $f : \mathcal{Z} \to \mathcal{W}$ that transforms the latent code z into an intermediate latent vector $w \in \mathcal{W}$. Adaptive instance normalization (AdaIn) [8] is applied to each feature map x_i using the style vector w. The architecture also includes explicit noise input, providing more variation in details while styles control global effects. Path length regularization was introduced to make the generator smoother [1].

Inversion to the Latent Space. Inverting images into the latent space W is more suitable for style-based GANs [1,13]. Various methods have been proposed, including optimization-based approaches and encoder-based approaches. Direct embedding into W can result in poor quality, but using different w for layers related to different scales significantly improves reconstruction quality [1]. Linear transformations in the latent space have been used for meaningful editing [33,40], while a Neural ODE-based method controls semantically independent attributes [16].

Unsupervised GAN Latent Space Directions Discovery. Unsupervised approaches for latent space editing have been proposed. PCA-based editing can be directly applied due to the intermediate latent space W in StyleGAN [6]. Layer-wise editing techniques enable editing related to specific components within a range of layers [6]. Methods like [38] discover linear directions using trainable matrices, while [34] suggests using singular vectors of weight matrices as directions.

These methods have advanced the field of GAN-based image synthesis and manipulation, offering various techniques for exploring and controlling latent spaces in GANs.

3   Methods

3.1   Comparable Methods

This work compares three methods for discovering and manipulating directions in StyleGAN2 latent space and parameter space. The first method, proposed by Voynov et al. [38], utilizes an optimize-based approach to find linear directions in the latent space. It involves optimizing a trainable matrix, that contains the learnable directions and a reconstruction, that predicts chosen direction and shift size. In the training procedure, these two parts are optimized jointly. The objective includes cross-entropy and mean squared error functions.

The second method, introduced by Harkonen et al. [6], employs PCA to determine principal components as directions in the latent space. It computes the basis using random vectors and their corresponding synthesized vectors, and allows editing by varying PCA coordinates.

The third approach, called Closed-Form Factorization of Latent Semantics (SeFa) [34], treats the latent code transformation as an affine transformation. It aims to discover meaningful full latent directions by decomposing the transformation matrix through singular value decomposition.

We can consider a shift in some direction obtained from previously described methods as following model (Formula 1):

$$edit(G(w)) = G(w + \alpha n) \tag{1}$$

where n is shift directions, α is shift size, w is intermediate latent code, $G(w)$ and $edit(G(w))$ initially generated and edited images respectively.

3.2 Latent Space Manipulation Improvements

To generate images from the intermediate latent code $w+$ instead of w in Style-GAN2, a modification was made to the generator architecture. Instead of using a single mapping network whose output is fed to all layers of the synthesis network, separate mapping networks are used for each layer Fig. 1.

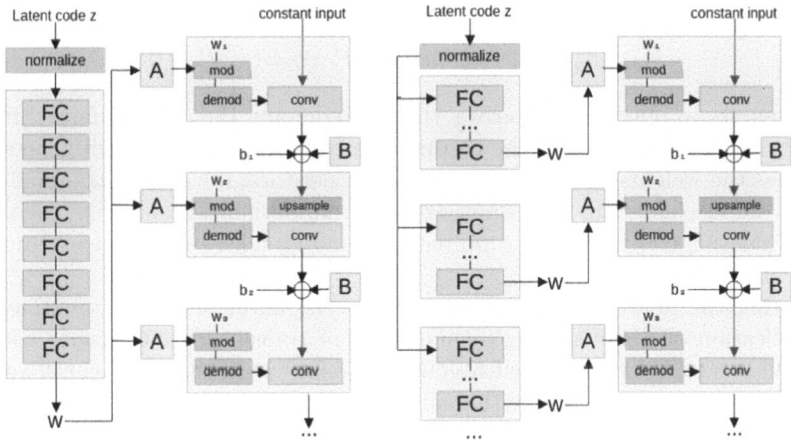

Fig. 1. Schemes of the original (left side) and modified (right side) generator architecture.

Initially, all the mapping networks are initialized with the same weights as the pre-trained mapping network of the original StyleGAN2, which was trained on the FFHQ dataset. These mapping networks take the latent code z obtained from normal distribution as input and generate the corresponding style vectors for each layer of the synthesis network.

During training, the weights of the last k layers of each mapping network are adjusted. The value of k is a hyperparameter that determines the depth of the mapping network to be trained.

We can assume that the edited image, obtained by Formula 1 in an ideal case should belong to the same distribution. Also, we aim that images obtained from latent codes from $W+$ also belong to this distribution. One way to evaluate how likely an image belongs to the space of images that produce a generator is to use a discriminator that was trained together with a generator during adversarial training. So we can formulate the first term of training objective:

$$
\begin{aligned}
L_D(P, z, n) =& |D(G(f(P, z))) - D(G(f'(P, z)))| \\
& + |D(G(f(P, z))) - D(G(f'(P, z) + n))|
\end{aligned}
\tag{2}
$$

where P are weights of trainable layers of mapping networks $z \in \mathcal{N}$ is latent code obtained from normal distribution, n is shift direction, $D : \mathcal{I} \rightarrow \mathcal{R}$ is pre-trained discriminator, $G : \mathcal{W} \rightarrow \mathcal{I}$ is original pre-trained synthesis network of StyleGAN2, $f : \mathcal{Z} \rightarrow \mathcal{W}$ is original pre-trained mapping network of StyleGAN2, and $f' : \mathcal{Z} \rightarrow \mathcal{W}+$ is modified mapping network described above.

Also we want that images obtained from the same z using mapping to W and $W+$ preserve identity, for this goal, LPIPS is used, so the second term of the objective is

$$
L_{identity}(P, z) = lpips(G(f(P, z)), G(f'(P, z)))
\tag{3}
$$

Attempts to optimize objectives that include only these terms failed. If the weights of trainable layers of the mapping network was initialized with the same values as the weights of the original mapping network, during the training they almost didn't change. In case when weights were initialized randomly, the value LPIPS doesn't converge to acceptable values, and generated images were very different.

During the analysis was mentioned that the norm of vectors obtained as multiplication of weights using in first steps of transformation of style vectors and style vectors, for images with good editability significantly larger than the norm of vectors for images with bad editability. So to encourage optimizing weights of mapping networks to be changed, the third term was added to the objective.

$$
L_n(P, z) = -\sum_{i=1}^{m} ||A f_i'(P_i, z)||_2
\tag{4}
$$

where P_i is matrix of weights of i-th mapping network $f_i(\cdot)$, m is the number of mapping networks, A_i is the weight used in the first steps of transformation of style vectors in i-th layer, $f_i'(\cdot)$ is the mapping network that produce style vector that fed to i-th layer.

So learning is performed by minimizing the following loss function:

$$\min_{P} \mathbb{E}_{z \sim \mathcal{N}(0,1)} [L_D(P, z) + L_{identity}(P, z) + L_n(P, z)] \tag{5}$$

3.3 Experiment Settings

This work compares unsupervised methods of manipulation in GAN latent space and GAN parameter space for human face editing. All methods are applied to the StyleGAN2 model trained on the FFHQ dataset [12].

Directions are obtained as follows: For the optimize-based method, an orthogonal configuration of the directions matrix A is used, and the model is trained for $1e5$ epochs with a batch size of 32. Principal components are obtained by sampling $1e5$ random codes from a standard Gaussian distribution, which are then mapped to the latent space W of StyleGAN2 using a pre-trained generator. The closed-form decomposition method obtains 512 singular vectors of the weight matrix of the first transformation for each layer, and their mean forms the direction matrix.

The model described in Sect. 3.2 is trained as follows: Optimization of the term related to the norm 4 is performed consistently over 5,000 iterations for layers 1–13, with a batch size of 8. The last 4 layers of each mapping network are adjusted during the training procedure. For shifts in the latent space, directions obtained with SeFa are chosen uniformly in each iteration, and shift sizes are also chosen uniformly from two values: -4 and 4. The ADAM optimizer is used for optimization, with a learning rate of 0.0005 for layers 1–6 and 0.0001 for layers 7–13.

The following metrics are used for evaluation: Ability to find interpretable directions (human evaluation), Frechet inception distance (FID) [7], cosine similarity of images before and after manipulation using the FaceNet model [31], target attribute changes evaluated using attributes from the CelebA dataset [19], independent attribute changes calculated for all attributes excluding the target attribute, LPIPS for measuring perceptual similarity between two images, and predictions of the pretrained discriminator for images obtained by mapping latent code z to $W+$. Weights of pretrained discriminator are taken from official repository of StyleGAN2 and converted to PyTorch format.

The inversion of real images into the $W+$ space of StyleGAN2 uses the encoder presented in [36].

4 Results

4.1 Manipulations in GAN Latent Space

Overall, all the methods discussed for discovering directions in the latent space are capable of finding directions that change global semantics such as gender Fig. 2, age, race, and so on. When shifting in the directions discovered by all the methods Fig. 2, the facial features and age are relatively well-preserved when shifting in the direction of gender inversion, while accessories typically associated with females may appear or disappear. However, when shifting in the opposite direction, the gender remains unchanged but the age changes. It is also worth noting that when shifting in the opposite direction, both the optimize-based and SeFa methods add baldness. This can be explained by the distribution of the data on which StyleGAN2 is trained, as men in general have less hair than women as they age. It should be noted that when editing in the direction revealed by the optimize-based method, the face color changes, highlights are added, and the eye shape changes, which does not allow for preserving identity and adversely affects the image quality.

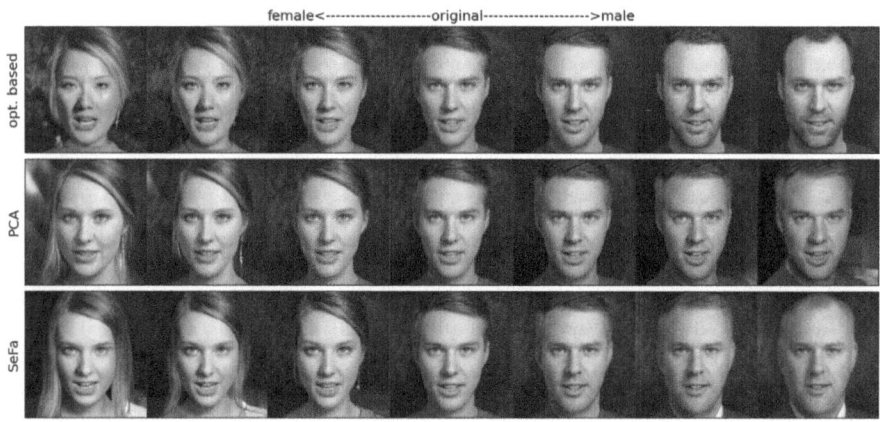

Fig. 2. Gender change. The first row - change in direction received an optimize-based approach, the second row - PCA-based approach, third-row SeFa.

Regarding gender editing, PCA and SeFa demonstrate the best performance. This can be observed visually, as the photorealism and identity are preserved in both directions, as well as through metric values (Fig. 3). The direction discovered by PCA preserves facial identity to the greatest extent, as it has the lowest cosine distance. The direction discovered by SeFa performs the best in changing the target attribute, but it has the highest FID value, which can be attributed to the decrease in the quality of the edited images and a significant shift in the data distribution, as age undergoes significant changes when shifting in one direction.

The optimize-based method reveals the best direction for age manipulation. Shifting in this direction results in the lowest FID value and the highest change in the target attribute. The worst results are obtained when shifting in the direction discovered by the PCA method. It is worth noting that when shifting in directions identified by all methods, the best performance is achieved when changing the age for individuals of middle age and older. In the case of editing images of young people, the age change is not as pronounced. When shifting

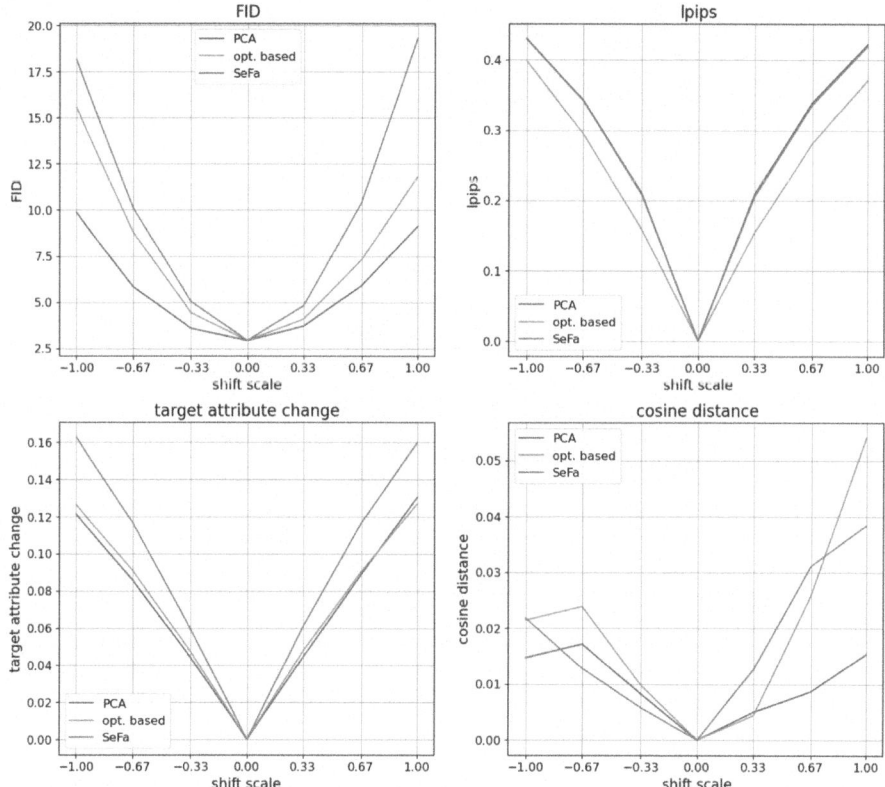

Fig. 3. Metrics for gender change. The first plot is FID, the second plot is lpips. Shift magnitude: for optimize-based approach [−6, 6], for PCA-based approach [−3, 3], for SeFa magnitude [−3, 3].

towards decreasing age, image quality often suffers (artifacts appear, skin color changes significantly, head size changes), and overall identity is not preserved.

It is also worth noting that by applying shifts only to certain layers, you can achieve changes in local attributes (adding glasses, lipstick, etc.) with minimal changes in other attributes while maintaining image quality.

4.2   Manipulations with Real Images

The real image is inverted into the $W+$ space using the E4E framework [23]. In general, there is some loss of identity already during the inversion process. The metrics LPIPS and cosine distance show that there are some differences between the real and inverted images, but overall, the similarity is well-preserved, allowing the face to remain recognizable.

In general, the considered methods allow for performing semantic manipulations by changing both global and local semantics (Fig. 4). It is difficult to single

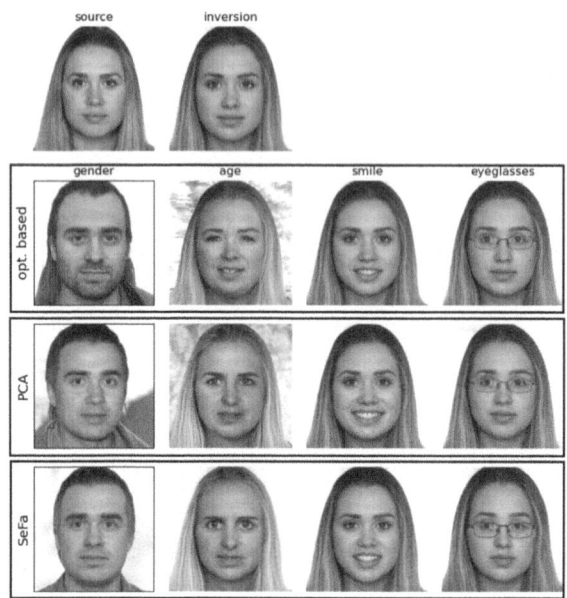

Fig. 4. Shift applied for real image inverted in $W+$ space by e4e encoder [36]. The first row - change in direction received an optimize-based approach, the second row - PCA-based approach, third-row SeFa.

out any one method that would be superior to others, according to different semantics, different methods show the best performance.

4.3 Improving Quality Manipulation with Translation in Extended Latent Space

Although the model described in Sect. 3.2 was trained only on directions obtained from SeFa, an improvement in the quality of images was observed when applying large shifts in directions discovered by other methods. This improvement can be noticed both visually and in terms of the metric values.

The original images obtained by mapping the latent code z to W and $W+$ are nearly identical, with only subtle differences. For example, images may differ slightly in the color of the face and the shape of the accessories. The identity of the images is further supported by the values of the LPIPS metric and cosine distance for zero shift.

It can be hypothesized that the transformation of the latent code z to $W+$ using the proposed model prevents the emergence of details that could lead to significant quality degradation during shifts. This hypothesis is supported by the metric values for zero shift. We observe slightly higher FID values, but the discriminator still recognizes the images as real with the same level of confidence as for images obtained without mapping to $W+$. This can be interpreted as a trade-off between the variability of generated images and maintaining quality.

This behavior can be attributed to the loss function used during training, specifically, its first term (Formula 2).

As shifts were introduced, notable differences in image quality became more apparent. Regarding the images obtained through the optimize-based method Fig. 5, one image appeared more realistic without any noticeable artifacts, while another exhibited unnatural skin color and flaws in the lower part of the face. Additionally, some images showed pronounced artifacts such as pink patches, light spots, wrinkles, and dark spots. However, there were instances where the artifacts were minimal or absent.

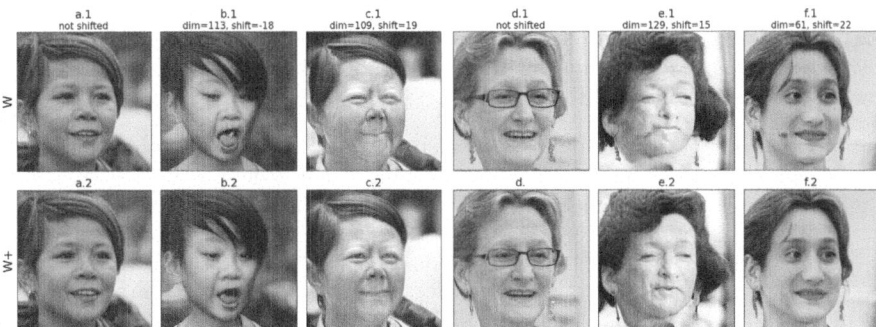

Fig. 5. Improving quality manipulation with direction obtained an optimize-based method. The first row contains images obtained with the translation of latent code to W, the second row latent code is translated to $W+$

Similar observations were made when shifts were applied using the PCA-based Fig. 6 and SeFa Fig. 7 methods. In some cases, images exhibited preserved facial features with minimal artifacts, while others demonstrated significant visual defects and lower overall quality. Notably, certain images lacked fully

Fig. 6. Improving quality manipulation with direction obtained a PCA method. The first row contains images obtained with the translation of latent code to W, the second row latent code is translated to $W+$

Fig. 7. Improving quality manipulation with direction obtained a SeFa method. The first row contains images obtained with the translation of latent code to W, the second row latent code is translated to $W+$

Table 1. Metrics for transformations in directions obtained an optimize-based method. A shift is applied to images obtained by mapping latent code z in W and $W+$. The same direction as for experiments with age transformation described above is used.

Shift	FID		LPIPS		discriminator prediction		cosine distance	
	W	$W+$	W	$W+$	W	$W+$	W	$W+$
-15	89.57	53.26	0.642	0.576	-0.625	-0.278	0.071	0.066
-10	30.64	22.1	0.48	0.424	0.204	0.442	0.05	0.045
-6	11.15	9.04	0.312	0.278	0.738	0.841	0.028	0.025
0	2.92	3.47	0.0	0.094	1.088	1.091	0	0.004
6	13.38	12.0	0.256	0.26	0.92	0.95	0.022	0.022
10	26.4	22.76	0.36	0.35	0.684	0.754	0.036	0.034
15	43.9	38.63	0.447	0.43	0.31	0.414	0.049	0.047

formed facial shapes and displayed artifacts resembling hair or patterns on the skin.

The metrics presented in the Tables 1, 2, 3 confirm the improvement in quality. With almost all manipulations, we get better values for images obtained by mapping the latent code z to $W+$

5   Conclusion

In this research, we tackled the challenge of controlling semantic properties in generated images by manipulating GAN latent spaces, specifically focusing on human face editing using StyleGAN2 [13]. We conducted a comparative analysis of existing methods for discovering semantically significant directions in the latent space and examined their applicability to face manipulation tasks.

Our analysis revealed that layer-wise shifts in the intermediate latent space can reduce entanglement in image editing and improve manipulation quality.

Table 2. Metrics for transformations in directions obtained a PCA-based method. A shift is applied to images obtained by mapping latent code z in W and $W+$. The same direction as for experiments with age transformation described above is used.

Shift	FID		LPIPS		discriminator prediction		cosine distance	
	W	$W+$	W	$W+$	W	$W+$	W	$W+$
−8	64.99	55.29	0.613	0.575	−0.019	0.137	0.089	0.087
−6	45.38	39.88	0.536	0.502	0.33	0.469	0.076	0.075
−4	25.94	22.79	0.427	0.401	0.694	0.782	0.057	0.055
0	2.92	3.47	0.0	0.095	1.088	1.091	0	0.004
4	15.71	15.99	0.405	0.391	0.551	0.544	0.051	0.049
6	28.4	27.21	0.501	0.485	0.071	0.095	0.066	0.063
8	43.34	40.68	0.564	0.549	−0.398	−0.338	0.074	0.07

Table 3. Metrics for transformations in directions obtained the SeFa method. A shift is applied to images obtained by mapping latent code z in W and $W+$. The same direction as for experiments with age transformation described above is used.

Shift	FID		LPIPS		discriminator prediction		cosine distance	
	W	$W+$	W	$W+$	W	$W+$	W	$W+$
−8	21.54	17.28	0.403	0.373	0.327	0.544	0.038	0.035
−6	14.25	11.72	0.324	0.303	0.629	0.781	0.029	0.026
−4	8.51	7.57	0.227	0.223	0.881	0.96	0.018	0.017
0	2.92	3.47	0.0	0.094	1.088	1.091	0	0.004
4	10.02	8.98	0.238	0.235	0.805	0.858	0.019	0.018
6	20.4	16.99	0.34	0.32	0.527	0.626	0.031	0.028
8	33.55	27.84	0.424	0.395	0.21	0.351	0.043	0.038

However, significant shifts often lead to unexpected artifacts and image quality degradation. This observation motivated us to propose a novel method that transforms latent codes into the extended latent space $(W+)$ to mitigate these issues and enhance the quality of edited images, particularly for large shifts in the intermediate latent space. It was also proposed to use a pre-trained discriminator, the weights of which were obtained during the training of StyleGan2 to control the quality of edited images.

Moving forward, our future research will focus on further refining our proposed method. We aim to explore techniques that facilitate target attribute editing, enabling more precise control over specific facial properties during manipulation. Additionally, we plan to investigate the effects of introducing Gaussian noise to the intermediate latent code and analyze its impact on manipulation quality.

By advancing the field of GAN-based image editing and offering insights into the capabilities and limitations of existing methods, our research contributes to the ongoing development of more powerful and effective tools for manipulating GAN latent spaces and enhancing the quality of generated images.

Overall, this study highlights the potential for improving image manipulation quality in the intermediate latent space of StyleGAN2 and opens up avenues for future advancements in GAN-based image editing [13].

References

1. Abdal, R., Qin, Y., Wonka, P.: Image2stylegan: how to embed images into the stylegan latent space? In: Proceedings of the IEEE/CVF International Conference on Computer Vision (ICCV) (2019)
2. Boutros, F., Struc, V., Fierrez, J., Damer, N.: Synthetic data for face recognition: current state and future prospects. Image Vis. Comput. 104688 (2023)
3. Cherepkov, A., Voynov, A., Babenko, A.: Navigating the GAN parameter space for semantic image editing. In: Proceedings of the IEEE/CVF Conference on Computer Vision and Pattern Recognition, pp. 3671–3680 (2021)
4. Golyadkin, M., Makarov, I.: Semi-automatic manga colorization using conditional adversarial networks. In: van der Aalst, W.M.P., et al. (eds.) AIST 2020. LNCS, vol. 12602, pp. 230–242. Springer, Cham (2021). https://doi.org/10.1007/978-3-030-72610-2_17
5. Goodfellow, I., et al.: Generative adversarial networks. Commun. ACM **63**(11), 139–144 (2020)
6. Härkönen, E., Hertzmann, A., Lehtinen, J., Paris, S.: Ganspace: discovering interpretable GAN controls. In: Larochelle, H., Ranzato, M., Hadsell, R., Balcan, M., Lin, H. (eds.) Advances in Neural Information Processing Systems, vol. 33, pp. 9841–9850. Curran Associates, Inc. (2020)
7. Heusel, M., Ramsauer, H., Unterthiner, T., Nessler, B., Hochreiter, S.: GANs trained by a two time-scale update rule converge to a local nash equilibrium. In: Advances in Neural Information Processing Systems, vol. 30 (2017)
8. Huang, X., Belongie, S.: Arbitrary style transfer in real-time with adaptive instance normalization. In: Proceedings of the IEEE International Conference on Computer Vision, pp. 1501–1510 (2017)
9. Huang, Z., Zhang, J., Shan, H.: When age-invariant face recognition meets face age synthesis: a multi-task learning framework. In: Proceedings of the IEEE/CVF Conference on Computer Vision and Pattern Recognition, pp. 7282–7291 (2021)
10. Ilya, M., Mikhail, T., Lada, T.: Imitation of human behavior in 3D-shooter game. In: AIST'2015 Analysis of Images, Social Networks and Texts, p. 64 (2015)
11. Karras, T., Aila, T., Laine, S., Lehtinen, J.: Progressive growing of GANs for improved quality, stability, and variation. arXiv preprint arXiv:1710.10196 (2017)
12. Karras, T., Laine, S., Aila, T.: A style-based generator architecture for generative adversarial networks. In: Proceedings of the IEEE/CVF Conference on Computer Vision and Pattern Recognition (CVPR) (2019)
13. Karras, T., Laine, S., Aittala, M., Hellsten, J., Lehtinen, J., Aila, T.: Analyzing and improving the image quality of stylegan. In: Proceedings of the IEEE/CVF Conference on Computer Vision and Pattern Recognition (CVPR) (2020)
14. Khanzhina, N., Kashirin, M., Filchenkov, A.: New Bayesian focal loss targeting aleatoric uncertainty estimate: pollen image recognition. In: Proceedings of the IEEE/CVF Conference on Computer Vision and Pattern Recognition (CVPR) Workshops, pp. 4253–4262 (2023)

15. Kharchevnikova, A., Savchenko, A.: Neural networks in video-based age and gender recognition on mobile platforms. Opt. Mem. Neural Netw. **27**, 246–259 (2018)
16. Khrulkov, V., Mirvakhabova, L., Oseledets, I., Babenko, A.: Latent transformations via neuralodes for GAN-based image editing. In: Proceedings of the IEEE/CVF International Conference on Computer Vision, pp. 14428–14437 (2021)
17. Kollias, D.: ABAW: learning from synthetic data & multi-task learning challenges. In: Karlinsky, L., Michaeli, T., Nishino, K. (eds.) ECCV 2022. LNCS, vol. 13806, pp. 157–172. Springer, Cham (2023). https://doi.org/10.1007/978-3-031-25075-0_12
18. Liang, X., Zhang, H., Xing, E.P.: Generative semantic manipulation with contrasting GAN. arXiv preprint arXiv:1708.00315 (2017)
19. Liu, Z., Luo, P., Wang, X., Tang, X.: Deep learning face attributes in the wild. In: Proceedings of the IEEE International Conference on Computer Vision, pp. 3730–3738 (2015)
20. Lomov, I., Makarov, I.: Generative models for fashion industry using deep neural networks. In: 2019 2nd International Conference on Computer Applications & Information Security (ICCAIS), pp. 1–6. IEEE (2019)
21. Makarov, I., Veldyaykin, N., Chertkov, M., Pokoev, A.: American and Russian sign language dactyl recognition. In: Proceedings of the 12th ACM International Conference on PErvasive Technologies Related to Assistive Environments, pp. 204–210 (2019)
22. Makarov, I., Veldyaykin, N., Chertkov, M., Pokoev, A.: Russian sign language dactyl recognition. In: 2019 42nd International Conference on Telecommunications and Signal Processing (TSP), pp. 726–729. IEEE (2019)
23. Richardson, E., et al.: Encoding in style: a stylegan encoder for image-to-image translation. In: Proceedings of the IEEE/CVF Conference on Computer Vision and Pattern Recognition, pp. 2287–2296 (2021)
24. Savchenko, A.: Facial expression recognition with adaptive frame rate based on multiple testing correction. In: Krause, A., Brunskill, E., Cho, K., Engelhardt, B., Sabato, S., Scarlett, J. (eds.) Proceedings of the 40th International Conference on Machine Learning (ICML). Proceedings of Machine Learning Research, vol. 202, pp. 30119–30129. PMLR (2023)
25. Savchenko, A.V.: Facial expression and attributes recognition based on multi-task learning of lightweight neural networks. In: Proceedings of the 19th International Symposium on Intelligent Systems and Informatics (SISY), pp. 119–124. IEEE (2021)
26. Savchenko, A.V.: MT-EmotiEffNet for multi-task human affective behavior analysis and learning from synthetic data. In: Karlinsky, L., Michaeli, T., Nishino, K. (eds.) ECCV 2022, Part VI. LNCS, vol. 13806, pp. 45–59. Springer, Cham (2022). https://doi.org/10.1007/978-3-031-25075-0_4
27. Savchenko, A.V.: EmotiEffNets for facial processing in video-based valence-arousal prediction, expression classification and action unit detection. In: Proceedings of the IEEE/CVF Conference on Computer Vision and Pattern Recognition, pp. 5715–5723 (2023)
28. Savchenko, A.V., Savchenko, L.V., Makarov, I.: Fast search of face recognition model for a mobile device based on neural architecture comparator. IEEE Access **11**, 65977–65990 (2023)
29. Savchenko, A.: Deep neural networks and maximum likelihood search for approximate nearest neighbor in video-based image recognition. Opt. Mem. Neural Netw. **26**, 129–136 (2017)

30. Savchenko, A., Savchenko, L.: Three-way classification for sequences of observations. Inf. Sci. 119540 (2023)
31. Schroff, F., Kalenichenko, D., Philbin, J.: Facenet: a unified embedding for face recognition and clustering. In: Proceedings of the IEEE Conference on Computer Vision and Pattern Recognition, pp. 815–823 (2015)
32. Semenkov, I., Karpov, A., Savchenko, A.V., Makarov, I.: Inpainting semantic and depth features to improve visual place recognition in the wild. IEEE Access **12**, 5163–5176 (2024). https://doi.org/10.1109/ACCESS.2024.3350038
33. Shen, Y., Yang, C., Tang, X., Zhou, B.: Interfacegan: interpreting the disentangled face representation learned by GANs. IEEE Trans. Pattern Anal. Mach. Intell. **44**(4), 2004–2018 (2020)
34. Shen, Y., Zhou, B.: Closed-form factorization of latent semantics in GANs. In: Proceedings of the IEEE/CVF Conference on Computer Vision and Pattern Recognition, pp. 1532–1540 (2021)
35. Sokolova, A., Savchenko, A.: Open-set face identification with sequential analysis and out-of-distribution data detection. In: Proceedings of International Joint Conference on Neural Networks (IJCNN), pp. 1–8. IEEE (2022)
36. Tov, O., Alaluf, Y., Nitzan, Y., Patashnik, O., Cohen-Or, D.: Designing an encoder for stylegan image manipulation. ACM Trans. Graph. (TOG) **40**(4), 1–14 (2021)
37. Voynov, A., Babenko, A.: Rpgan: GANs interpretability via random routing. arXiv preprint arXiv:1912.10920 (2019)
38. Voynov, A., Babenko, A.: Unsupervised discovery of interpretable directions in the GAN latent space. In: III, H.D., Singh, A. (eds.) Proceedings of the 37th International Conference on Machine Learning. Proceedings of Machine Learning Research, vol. 119, pp. 9786–9796. PMLR (2020)
39. Yan, X., Yang, J., Sohn, K., Lee, H.: Attribute2Image: conditional image generation from visual attributes. In: Leibe, B., Matas, J., Sebe, N., Welling, M. (eds.) ECCV 2016. LNCS, vol. 9908, pp. 776–791. Springer, Cham (2016). https://doi.org/10.1007/978-3-319-46493-0_47
40. Zhuang, P., Koyejo, O., Schwing, A.G.: Enjoy your editing: controllable GANs for image editing via latent space navigation. arXiv preprint arXiv:2102.01187 (2021)

Learning Facial Expression Recognition In-the-Wild from Synthetic Data Based on an Ensemble of Lightweight Neural Networks

Long Nguyen[1] and Andrey V. Savchenko[1,2(✉)] (iD)

[1] HSE University, Laboratory of Algorithms and Technologies for Network Analysis,
Nizhny Novgorod, Russia
avsavchenko@hse.ru
[2] Sber AI Lab, Moscow, Russia

Abstract. This paper deals with one of the problems of recognizing the emotion from a photo gathered from in-the-wild settings, namely, facial expression recognition. We study various ensemble approaches that combine the lightweight MT-EmotiEffNet, attention-based transformer, and the graph-based model. The models have been implemented on an edge device (Jetson Nano) to be integrated into a demo application for video analytics to evaluate their performance in practical conditions. The experimental results demonstrated that the proposed approach outperformed competitive solutions using the Learning from Synthetic Data challenge datasets of the fourth Affective Behavior Analysis in-the-Wild competition.

Keywords: Affective Behavior Analysis in-the-Wild (ABAW) · facial expression recognition (FER) · synthetic dataset · MT-EmotiEffNet · edge device

1 Introduction

Automatic emotion understanding is a significant problem of many information systems with human-computer interfaces [1]. Human emotions are represented in such modalities as voice [2–5], physiological signals [6,7], dept data [8,9] and faces [10,11]. Facial expression recognition (FER) is a classification problem that associates an input photo or video with one or multiple emotions from such categories as anger, fear, surprise, sadness, and happiness [12]. This task is solved in various applications, including enhancement of human-computer interfaces, health monitoring, robotics, and advertising, and sometimes goes in hand with sign language dactyl recognition [13,14]. By analyzing facial expressions, computers can gain insight into how a person feels, improving communication and quality of decisions about their behavior.

One challenge of facial expression recognition tasks is their performance degradation when applied to in-the-wild conditions, which is caused of lacking real-world data for training [15]. Acquiring sufficient amounts of labeled

© The Author(s), under exclusive license to Springer Nature Switzerland AG 2024
D. I. Ignatov et al. (Eds.): AIST 2023, CCIS 1905, pp. 209–221, 2024.
https://doi.org/10.1007/978-3-031-67008-4_16

emotional data for training these models is a complex and costly process, often hindered by privacy concerns and difficulties in obtaining individual consent. An emerging solution is using synthetic data [16], which allows researchers to construct a large and diverse dataset without concerns about ensuring compliance with legal restrictions and safeguarding personal privacy. The usage of synthetic data for FER has been widely studied for several years [17,18]. Some solutions apply Graph Neural Networks [19,20]. The 4th Affective Behavior Analysis in-the-Wild (ABAW) [21] workshop in 2022 addressed this issue by organizing the Learning from Synthetic Dataset (LSD) competition, requested participants to develop their FER model on a synthetic dataset [22], generated from the Aff-Wild-2 dataset [11]. In this competition, the winner utilized the MTEmotiEffNet model [23], based on a lightweight architecture, EfficientNet-B0. It achieves the current state-of-the-art results on several facial datasets, including AffectNet [24]. The second-ranked solution employed a graph convolutional-based model named Graph embedded Uncertainty Suppressing (GUS) [25]. Another prominent method used by the third-ranked participants was the transformer-based model, known as Distract your Attention Network (DAN) [26]. In 2021, DAN held the position of being the state-of-the-art model on the AffectNet dataset. Each of the mentioned solutions has its own set of advantages and disadvantages. MT-EmotiEffNet is well-suit for many applications from on-device computing to large servers due to its small size and high quality. The DAN is acclaimed for its strong generalization capabilities. The GUS [25] was designed to possess superior generalization capabilities for handling diverse expression patterns and enhanced robustness in capturing expression representations.

Thus, in this paper, we proposed to combine the best solutions into a single ensemble model, leveraging their strengths to achieve higher scores. In addition, we deal with another critical challenge for FER in the wild, namely, personal data privacy [27]. The capture and manipulation of human faces for FER tasks are subject to legal restrictions in many countries. Furthermore, transferring face images to external servers poses potential risks of misuse and deepfake [28] attacks. One cost-effective and efficient solution is deploying the application on an internal edge device to perform computations and store output data locally. However, running deep learning models on edge devices encounters limitations due to their limited computing resources and software stack [29–31]. To address this issue, we deployed and evaluated the performance and feasibility of our models on Jetson Nano [32], a well-known edge device for running deep learning applications, then proposed an end-to-end video analytics system that could utilize the benefits of using FER in-the-wild while ensuring compliance with stringent data privacy requirements.

In summary, the primary objective of this paper is to propose an efficient ensemble approach that combines the top performers' solutions from the LSD competition in the 4th ABAW workshop. The secondary goal is to assess the practical performance of those ensemble models on Jetson Nano. In the following sections, we will delve into the details of our proposed solution.

Fig. 1. Sample images from the 4th ABAW competition [4]: (a) LSD; (b) MTL.

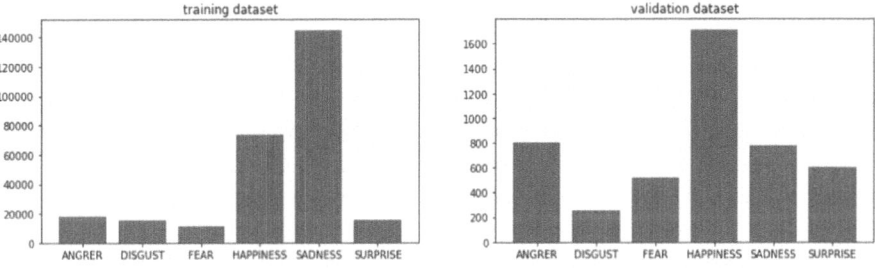

Fig. 2. Number of images per class in LSD dataset.

2   Research Methods

2.1   Dataset

We used the dataset from the 4th ABAW competition [21]. It is based on the Aff-Wild2 database [11], which is an extension of the Aff-wild database [33] and consists of the Multi-Task-Learning (MTL) and Learning from Synthetic Data (LSD) challenges. For the LSD competition, some frames from the Aff-Wild2 database [11] were selected by the organizers of this challenge and then used to generate synthetic face images with various facial expressions [34,35].

The synthetic training set consists of approximately 280K images (Fig. 1a) and their corresponding annotations for six basic facial expressions (anger, disgust, fear, happiness, sadness, surprise) used in model training/methodology development. A set of original facial images of the subjects who also appeared in the training set was provided for validation. The dataset is very imbalanced (Fig. 2), which can influence the overall score of the trained model.

BEFORE   AFTER

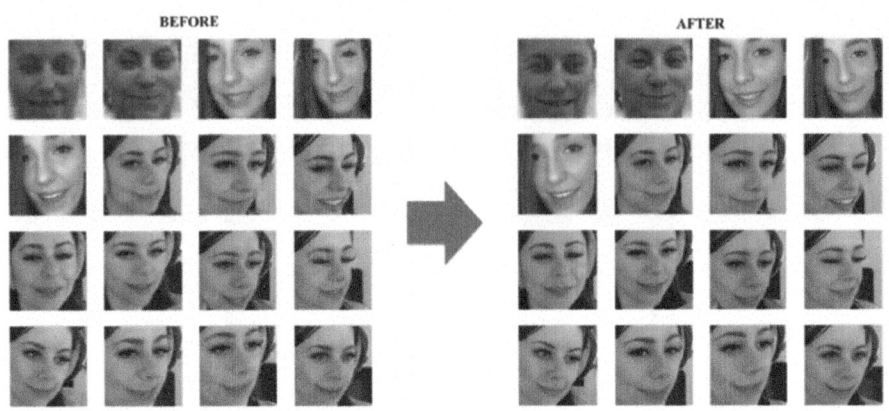

Fig. 3. Sample images before and after enhancing quality by super-resolution Real-ESRGAN [36].

Finally, following the testing procedure of the winner [23], a subset of real facial images from MTL datasets (Fig. 1b) was sampled to evaluate the model performance. Because the MTL dataset has multiple labels, including valence-arousal, six basic expressions, and two additional labels (neutral, others), 12 action units, and those images were not labeled completely. Hence, all labeled images belonging to 6 types of expressions in the LSD competition were sampled. Similar to the LSD competition, the mean F1 score across all six categories was used as the metric to evaluate the quality of the output model.

Finally, following the testing procedure of the winner [23], a subset of real facial images from MTL datasets (Fig. 1b) was sampled to evaluate the model performance. Similar to the LSD competition, the mean F1 score across all six categories was used as the metric to assess the quality of the output model.

2.2 Data Preprocessing

The original resolution of images from LSD and MTL datasets is 128×128, which is relatively low. Furthermore, some photos include noise, blur, and degradation, affecting the model performance. Therefore, to enhance the training robustness, the original datasets were upscale and improved quality using the super-resolution Real-ESRGAN [36] to receive another variant of the dataset with a larger size of 256×256 and better quality.

The quality change before and after applying the super-resolution Real-ESRGAN is shown in Fig. 3. Due to the high computational requirements, all datasets were used to the super-resolution and saved to a new directory instead of applying the super-resolution enhancement during the training process.

Despite the unbalanced nature of the LSD and MTL datasets (Fig. 2), no data augmentation was performed to ensure a fair comparison across all experiments. The input images were resized to a uniform size of 224×224 and converted to tensor format, the expected input for many deep neural networks.

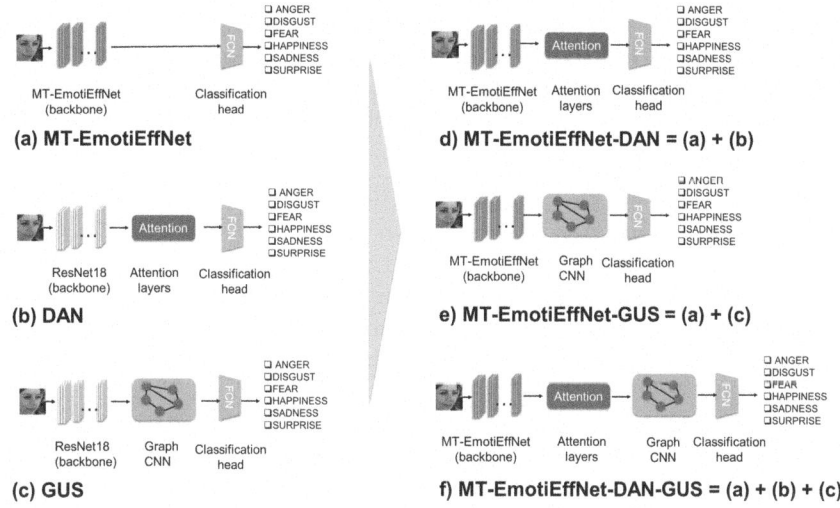

Fig. 4. Our ensemble model architectures (right-hand side) which combined solutions of top performers (left-hand side) in LSD challenge.

2.3 Ensemble Approach for Facial Expressions Recognition

Model Architecture. In this paper, we study various ensemble (blending) techniques [37] that combine different solutions from top-performing models into a single unified model. This approach aimed to enhance overall performance, as measured similarly by the original metrics from the LSD competition, the F1-score. The original model architectures and our proposed ensemble model architectures are illustrated in Fig. 4. To be more precise, the original models in our study consist of the best participants of the LSD competition, namely:

(a) MT-EmotiEffNet [23,38], the recently state-of-the-art EfficientNetB0 architecture pre-trained on the AffectNet dataset;
(b) an attention-based transformer architecture, DAN [26], and
(c) GUS [25].

In this paper, we introduced three ensemble approaches (Fig. 4):

(d) "MT-EmotiEffNet-DAN", that combines MT-EmotiEffNet and DAN by replacing the original backbone of DAN with the SoTA EfficientNetB0;
(e) "MT-EmotiEffNet-GUS", merges MT-EmotiEffNet and GUS by substituting the original backbone of GUS with the SoTA EfficientNetB0 model.
(f) "MT-EmotiEffNet-DAN-GUS", leverages MT-EmotiEffNet, DAN, and GUS together, utilizing the state-of-the-art EfficientNetB0 as the backbone, DAN as the intermediary layer, and GUS as the classification head.

Training. We applied a similar and simple training procedure to ensure a fair comparison across all experiments. All models were trained for five epochs using Cross Entropy as the loss function, Adam as the optimizer, and weight decay as the regularization strategy. An exponential learning rate scheduler was also employed.

Evaluation. To evaluate the accuracy of our proposed solutions, we use the F1-score as the same evaluation metric as the LSD competition. We employed the validation dataset from the LSD competition as our primary validation dataset. Additionally, to measure the performance of our solutions in a real-world condition, we used a sample dataset selected from the training dataset and a validation dataset from the MTL competition.

We took the following steps to evaluate our models' inference speed. First, we converted the models to the ONNX format, enabling us to utilize the TensorRT accelerator on Jetson Nano [32,39]. Then, we performed inference on our models 100 times, using a randomly generated input tensor with a shape of (1, 3, 224, 224). Finally, the average frames per second (FPS) across these inference runs is calculated.

2.4   End-to-End Video Analytics Application

Our end-to-end video analysis application aims to evaluate the performance of our FER models, showcase the benefit of using FER in the wild, and effectively address concerns regarding personal data privacy. Figure 5 illustrates the proposed solution architecture for our video analytics application, comprising a video analytics module and a backend module. The video analytics module, running on Jetson Nano, encompasses functionalities such as frame capture from the environment, face detection, and subsequent FER. On the other hand, the backend module can operate either on-premises or in the cloud, facilitating data storage and providing an API for retrieving emotion results. Figure 6 illustrated the application running in practice. The source code of our application is publicly available [40].

Video Analytics Module. To capture frames from the environment and enable continuous streaming, we employed the Logitech 720 webcam in conjunction with OpenCV. Face detection was performed using the YuNet face detector [41], while facial tracking utilized DeepSORT [42] with a customized facial ReID. We leveraged the YuNet face recognition model for facial feature extraction within our facial ReID framework. The emotions were extracted using the most optimal FER model derived from our previous experiments, and the results were stored in a predefined database as part of the backend module. This design ensures that the necessary FER computations can be conducted locally, guaranteeing that personal data remains securely within the system and eliminating the need to transmit it externally.

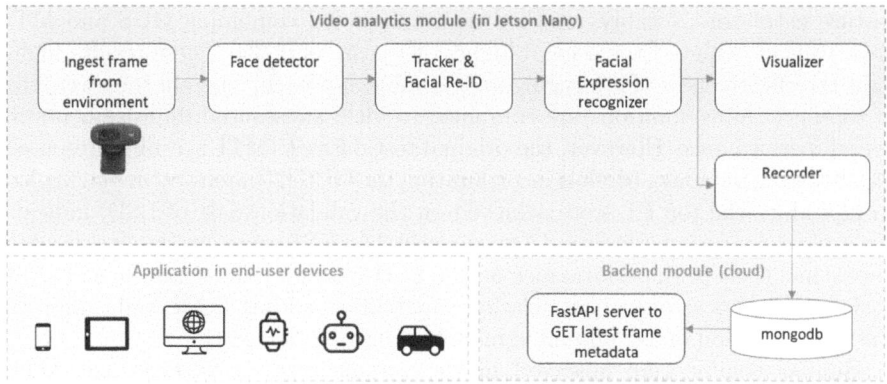

Fig. 5. Our end-to-end video analytics system.

Fig. 6. Running facial expression recognition in-the-wild on Jetson Nano and streaming emotion data to MongoDB server.

Backend Module. We advocate for using MongoDB, a widely recognized cloud database solution, to store the output results generated by the video analytics module. This design ensures that the data is readily accessible. It enables a straightforward implementation of a FastAPI server that can retrieve and deliver this information to any client requesting it for further application, all while maintaining strict adherence to personal privacy requirements.

3   Experimental Results

The F1-score of all models computed on the validation sets from the LSD and MTL challenges [21] are summarized in Table 1. Among the models evaluated, the ensemble model stood out with the highest F1-score of 0.771 on the original validation dataset (LSD). This ensemble model utilized a combination of DAN [26], GUS [25], and the state-of-the-art MT-EmotiEffNet [23] as its feature extractor. Notably, another ensemble model combining GUS and MT-EmotiEffNet achieved the second-highest F1 score of 0.734. These results highlight the effectiveness of leveraging an ensemble approach, wherein the strengths of top-performing solutions are combined to yield substantial improvements in overall performance. However, the original test dataset (MTL) results presented a contrasting outcome. Models incorporating the GUS [25] architecture struggled to reproduce the top F1 score achieved on the validation part of LSD, indicating a clear case of overfitting. Consequently, their F1 scores were considerably lower than their peak performance on the LSD validation dataset. Since DAN is acclaimed for its strong generalization capabilities, adding DAN could improve the generalization of the ensemble model on unseen datasets.

Similar results were observed in the variant versions of LSD and MTL datasets, which enhanced quality using the super-resolution method Real-ESRGAN [36]. The Real-ESRGAN super-resolution technique has the potential to significantly improve image quality while preserving intricate details and patterns of the original information, thereby minimizing any negative impact on overall quality. While it has proven effective in specific experiments, improving the F1-score, there are instances where it has decreased the F1-score.

Table 1. F1-score of facial expression models on the validation dataset from LSD and MTL competition in 4th ABAW Workshop.

Model	Original dataset		Real-ESRGAN enhancement	
	LSD	MTL	LSD	MTL
Original models				
GUS [25]	0.648	0.312	0.627 (−0.021)	0.323 (+0.011)
DAN [26]	0.628	0.383	0.643 (+0.015)	0.354 (−0.029)
MT-EmotiEffNet [23]	0.663	0.409	0.640 (−0.023)	0.386 (−0.023)
Our models				
MT-EmotiEffNet-DAN	0.663	**0.419**	0.643 (−0.020)	**0.401 (−0.018)**
MT-EmotiEffNet-GUS	0.734	0.369	0.728 (−0.006)	0.369 (0.000)
MT-EmotiEffNet-DAN-GUS	**0.771**	0.352	**0.749 (−0.022)**	0.343 (−0.009)

The inference speed is presented in Table 2. The original GUS model demonstrated the highest inference speed at 47.62 fps. Following closely behind were the MT-EmotiEffNet, DAN [26], and the ensemble model of MT-EmotiEffNet and GUS, which achieved a second-highest inference speed of 34.48 FPS. The remaining models also reached near real-time inference speeds at 26.32 and 27.03 FPS. One possible explanation is that ResNet18, the original backbone of GUS, was initially optimized by the TensorRT framework, contributing to their superior performance. In contrast, other models, such as MT-EmotiEffNet [23], include several unsupported layers in Jetson Nano TensorRT, such as padding, which could account for its comparatively slower inference speed.

Table 2. Inference speed, measured by frame per second (FPS) on Jetson Nano.

Model	Number of parameters, M	FPS, onnx-trt
Original models		
GUS [25]	11.71	47.62
DAN [26]	19.72	34.48
MT-EmotiEffNet [23]	4.02	34.48
Our models		
MT-EmotiEffNet-DAN	13.21	27.03
MT-EmotiEffNet-GUS	5.19	34.48
MT-EmotiEffNet-DAN-GUS	13.73	26.32

Based on the results from Table 1 and Table 2, it is evident that the MT-EmotiEffNet, which achieved an optimal balance between F1-score and inference speed, emerged as the primary choice for the end-to-end video analytics application on the Jetson Nano. We ran our application and evaluated its frame rate by fps with different numbers of faces appearing on the frame (Fig. 7).

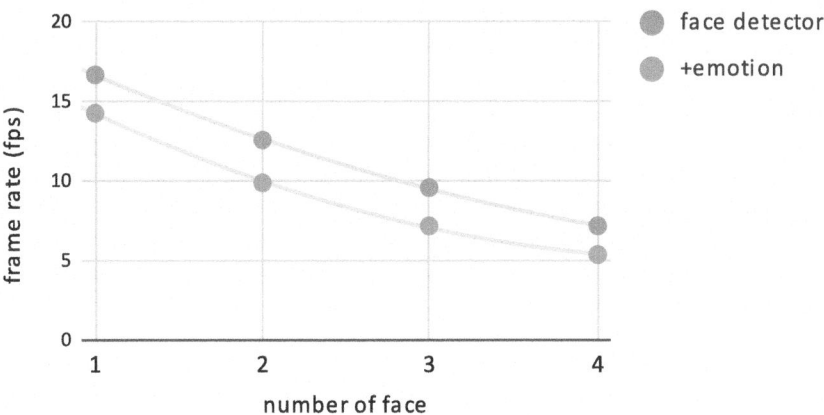

Fig. 7. Frame rate (FPS) of end-to-end video analytics application on Jetson Nano. Face tracking is DeepSORT [42] with a modified facial ReID. Face detector and facial ReID are using YuNet models [41]. FER model is MT-EmotiEffNet [23].

While performing the inference task on a single input tensor can achieve real-time speed, the frame rate experienced a significant drop in an end-to-end video analytics application. Even with only a single face appearing in the frame, the frame rate decreased notably to 12.7 fps and further declined as the number of faces increased. This speed may not be optimal for applications that demand near real-time affective behavior analysis. These findings indicate the need for additional efforts to enhance the frame rate, making the models more viable and production-ready for a broader range of applications.

4   Conclusion

In this paper, we proposed an efficient ensemble approach for the FER problem using synthetic training data. We have trained our models using the training dataset from the LSD challenge of the 4th ABAW competition [21]. Notably, our ensemble model achieved an impressive validation F1-score of 0.773, surpassing the performance of any individual model, all of which attained F1-scores lower than 0.70 (Table 1). These outcomes demonstrate the feasibility and effectiveness of utilizing synthetic datasets for facial expression recognition and other tasks involving affective behavior analysis. We sampled a subset dataset from the MTL task from the ABAW competition to evaluate the model performance on a real-life dataset. Among the models tested, the MT-EmotiEffNet [23] stood out with the highest F1-score of 0.419. These results emphasize the gap between synthetic and real-life datasets, underscoring the need for further advancements in bridging this disparity. While the data sets used in this paper are not widespread, we are going to apply our approach to other popular datasets such as Aff-Wild2 [11], AffectNet [24].

We also implemented our models to run on Jetson Nano to assess their practical performance (Table 2). Although the real-time speed can be achieved when performing the inference task on a single input tensor, the frame rate substantially decreases in an end-to-end video analytics application. Even with just one face present in the frame, the frame rate drops significantly to 12.7 fps and further declines with an increasing number of faces. This speed may not meet the requirements of applications that necessitate near real-time affective behavior analysis. Consequently, there is a clear need for additional endeavors to improve the frame rate [43–45], thereby enhancing these models' feasibility and production readiness for a broader range of applications.

Acknowledgements. The article was implemented in the framework of the Basic Research Program at the National Research University Higher School of Economics (HSE University).

References

1. Maithri, M., et al.: Automated emotion recognition: current trends and future perspectives. Comput. Methods Programs Biomed. **215**, 106646 (2022)
2. Savchenko, A., Savchenko, V.: Method for measurement the intensity of speech vowel sounds flow for audiovisual dialogue information systems. Meas. Tech. **65**(3), 219–226 (2022)
3. Ryumina, E., Dresvyanskiy, D., Karpov, A.: In search of a robust facial expressions recognition model: a large-scale visual cross-corpus study. Neurocomputing **514**, 435–450 (2022)

4. Kondratenko, V., Karpov, N., Sokolov, A., Savushkin, N., Kutuzov, O., Minkin, F.: Hybrid dataset for speech emotion recognition in Russian language. In: Proceedings of INTERSPEECH, pp. 4548–4552 (2023)

5. Savchenko, V.V., Savchenko, A.V.: Criterion of significance level for selection of order of spectral estimation of entropy maximum. Radioelectron. Commun. Syst. **62**(5), 223–231 (2019)

6. Dhall, A., Sharma, G., Goecke, R., Gedeon, T.: EmotiW 2020: driver gaze, group emotion, student engagement and physiological signal based challenges. In: Proceedings of the 2020 International Conference on Multimodal Interaction (ICMI), pp. 784–789 (2020)

7. Li, X., et al.: EEG based emotion recognition: a tutorial and review. ACM Comput. Surv. **55**(4), 1–57 (2022)

8. Oyedotun, O.K., Demisse, G., El Rahman Shabayek, A., Aouada, D., Ottersten, B.: Facial expression recognition via joint deep learning of RGB-depth map latent representations. In: Proceedings of the IEEE International Conference on Computer Vision (ICCV) Workshops, pp. 3161–3168 (2017)

9. Savchenko, A., Khokhlova, Y.I.: About neural-network algorithms application in viseme classification problem with face video in audiovisual speech recognition systems. Opt. Mem. Neural Netw. **23**, 34–42 (2014)

10. Demochkina, P., Savchenko, A.V.: MobileEmotiFace: efficient facial image representations in video-based emotion recognition on mobile devices. In: Del Bimbo, A., et al. (eds.) ICPR 2021. LNCS, vol. 12665, pp. 266–274. Springer, Cham (2021). https://doi.org/10.1007/978-3-030-68821-9_25

11. Kollias, D., Zafeiriou, S.: Expression, affect, action unit recognition: Aff-wild2, multi-task learning and ArcFace. arXiv preprint arXiv:1910.04855 (2019)

12. Ekman, P., Friesen, W.: Constants across cultures in the face and emotion. J. Pers. Soc. Psychol. **17**(2), 124 (1971)

13. Makarov, I., Veldyaykin, N., Chertkov, M., Pokoev, A.: Russian sign language dactyl recognition. In: Proceedings of the 42nd International Conference on Telecommunications and Signal Processing (TSP), pp. 726–729. IEEE (2019)

14. Makarov, I., Veldyaykin, N., Chertkov, M., Pokoev, A.: American and Russian sign language dactyl recognition and Text2Sign translation. In: van der Aalst, W.M.P., et al. (eds.) AIST 2019. LNCS, vol. 11832, pp. 309–320. Springer, Cham (2019). https://doi.org/10.1007/978-3-030-37334-4_28

15. Samadiani, N., et al.: A review on automatic facial expression recognition systems assisted by multimodal sensor data. Sensors **19**(8), 1863 (2019)

16. Nikolenko, S.I.: Synthetic Data for Deep Learning, vol. 174. Springer, Cham (2021). https://doi.org/10.1007/978-3-030-75178-4

17. Abbasnejad, I., Sridharan, S., Nguyen, D., Denman, S., Fookes, C., Lucey, S.: Using synthetic data to improve facial expression analysis with 3D convolutional networks. In: Proceedings of the IEEE International Conference on Computer Vision Workshops, pp. 1609–1618 (2017)

18. Zeng, J., Shan, S., Chen, X.: Facial expression recognition with inconsistently annotated datasets. In: Proceedings of the European Conference on Computer Vision (ECCV), pp. 222–237 (2018)

19. Makarov, I., Korovina, K., Kiselev, D.: Jonnee: joint network nodes and edges embedding. IEEE Access **9**, 144646–144659 (2021)

20. Shi, H., Peng, W., Chen, H., Liu, X., Zhao, G.: Multiscale 3D-shift graph convolution network for emotion recognition from human actions. IEEE Intell. Syst. **37**(4), 103–110 (2022)

21. Kollias, D.: ABAW: learning from synthetic data & multi-task learning challenges. In: Karlinsky, L., Michaeli, T., Nishino, K. (eds.) ECCV 2022, Part VI. LNCS, vol. 13806, pp. 157–172. Springer, Cham (2022). https://doi.org/10.1007/978-3-031-25075-0_12

22. Mao, S., Li, X., Chen, J., Peng, X.: AU-supervised convolutional vision transformers for synthetic facial expression recognition. arXiv preprint arXiv:2207.09777 (2022)

23. Savchenko, A.V.: MT-EmotiEffNet for multi-task human affective behavior analysis and learning from synthetic data. In: Karlinsky, L., Michaeli, T., Nishino, K. (eds.) ECCV 2022, Part VI. LNCS, vol. 13806, pp. 45–59. Springer, Cham (2022). https://doi.org/10.1007/978-3-031-25075-0_4

24. Mollahosseini, A., Hasani, B., Mahoor, M.H.: AffectNet: a database for facial expression, valence, and arousal computing in the wild, vol. 10, pp. 18–31. IEEE (2017)

25. Lei, J., et al.: Facial expression recognition with mid-level representation enhancement and graph embedded uncertainty suppressing. In: Karlinsky, L., Michaeli, T., Nishino, K. (eds.) ECCV 2022, Part VI. LNCS, vol. 13806, pp. 93–103. Springer, Cham (2022). https://doi.org/10.1007/978-3-031-25075-0_7

26. Wen, Z., Lin, W., Wang, T., Xu, G.: Distract your attention: multi-head cross attention network for facial expression recognition. Biomimetics 8(2), 199 (2023)

27. Mireshghallah, F., Taram, M., Vepakomma, P., Singh, A., Raskar, R., Esmaeilzadeh, H.: Privacy in deep learning: a survey. arXiv preprint arXiv:2004.12254 (2020)

28. Masood, M., Nawaz, M., Malik, K.M., Javed, A., Irtaza, A., Malik, H.: Deepfakes generation and detection: state-of-the-art, open challenges, countermeasures, and way forward. Appl. Intell. 53(4), 3974–4026 (2023)

29. Kharchevnikova, A., Savchenko, A.: Neural networks in video-based age and gender recognition on mobile platforms. Opt. Mem. Neural Netw. 27, 246–259 (2018)

30. Wu, Y., Zhang, L., Gu, Z., Lu, H., Wan, S.: Edge-AI-driven framework with efficient mobile network design for facial expression recognition. ACM Trans. Embed. Comput. Syst. 22(3), 1–17 (2023)

31. Savchenko, A.V., Savchenko, L.V., Makarov, I.: Fast search of face recognition model for a mobile device based on neural architecture comparator. IEEE Access 11, 65977–65990 (2023)

32. Sati, V., Sánchez, S.M., Shoeibi, N., Arora, A., Corchado, J.M.: Face detection and recognition, face emotion recognition through NVIDIA Jetson Nano. In: Novais, P., Vercelli, G., Larriba-Pey, J.L., Herrera, F., Chamoso, P. (eds.) ISAmI 2020. AISC, vol. 1239, pp. 177–185. Springer, Cham (2021). https://doi.org/10.1007/978-3-030-58356-9_18

33. Kollias, D., et al.: Deep affect prediction in-the-wild: Aff-wild database and challenge, deep architectures, and beyond. Int. J. Comput. Vision 127(6–7), 907–929 (2019)

34. Kollias, D., Cheng, S., Ververas, E., Kotsia, I., Zafeiriou, S.: Deep neural network augmentation: generating faces for affect analysis. Int. J. Comput. Vision 128, 1455–1484 (2020)

35. Kollias, D., Zafeiriou, S.: VA-StarGAN: continuous affect generation. In: Blanc-Talon, J., Delmas, P., Philips, W., Popescu, D., Scheunders, P. (eds.) ACIVS 2020. LNCS, vol. 12002, pp. 227–238. Springer, Cham (2020). https://doi.org/10.1007/978-3-030-40605-9_20

36. Wang, X., Xie, L., Dong, C., Shan, Y.: Real-ESRGAN: training real-world blind super-resolution with pure synthetic data. In: Proceedings of the IEEE/CVF International Conference on Computer Vision (ICCV), pp. 1905–1914 (2021)

37. Jeong, J.Y., et al.: Ensemble of multi-task learning networks for facial expression recognition in-the-wild with learning from synthetic data. In: Karlinsky, L., Michaeli, T., Nishino, K. (eds.) ECCV 2022. LNCS, vol. 13806, pp. 60–75. Springer, Cham (2022). https://doi.org/10.1007/978-3-031-25075-0_5

38. Savchenko, A.V.: EmotiEffNets for facial processing in video-based valence-arousal prediction, expression classification and action unit detection. In: Proceedings of the IEEE/CVF Conference on Computer Vision and Pattern Recognition, pp. 5715–5723 (2023)

39. Pascual, A.M., et al.: Light-FER: a lightweight facial emotion recognition system on edge devices. Sensors **22**(23), 9524 (2022)

40. Long, N.B.: Facial affective behavior analysis in-the-wild (2023). https://github.com/billynguyenlss/jetson-nano-face-abaw

41. Wu, W., Peng, H., Yu, S.: YuNet: a tiny millisecond-level face detector. Mach. Intell. Res. 1–10 (2023)

42. Wojke, N., Bewley, A., Paulus, D.: Simple online and realtime tracking with a deep association metric. In: Proceedings of International Conference on Image Processing (ICIP), pp. 3645–3649. IEEE (2017)

43. Savchenko, A.: Facial expression recognition with adaptive frame rate based on multiple testing correction. In: Krause, A., Brunskill, E., Cho, K., Engelhardt, B., Sabato, S., Scarlett, J. (eds.) Proceedings of the 40th International Conference on Machine Learning (ICML). Proceedings of Machine Learning Research, vol. 202, pp. 30119–30129. PMLR (2023)

44. Park, S.J., Kim, B.G., Chilamkurti, N.: A robust facial expression recognition algorithm based on multi-rate feature fusion scheme. Sensors **21**(21), 6954 (2021)

45. Savchenko, A., Savchenko, L.: Three-way classification for sequences of observations. Inf. Sci. 119540 (2023)

Acne Recognition: Training Models with Experts

Nikolic Stefan[1]([✉]) [iD], Dmitry I. Ignatov[1] [iD], and Peter Fedorov[2]

[1] National Research University Higher School of Economics, Moscow, Russia
futurelifestefan@gmail.com
[2] Skin Research Institute, Santa Monica, CA, USA

Abstract. Acne vulgaris is a skin condition which occurs frequently within the population. An important step in diagnosing the condition is grading the severity of the case. For this purpose dermatologists often use different grading scales and criteria. To aid this grading process, the usage of deep learning algorithms has been proposed by multiple authors. This paper explores the usage of deep learning algorithms on two separate datasets, one of which is reinforced with bounding boxes for acne.

Keywords: skin conditions · deep learning · acne grading

1 Introduction

Acne is a skin condition that can greatly impact quality of life for an individual in a negative way. Indeed, research shows that it can affect one's mental health greatly, especially among young people [20].

Nowadays the diagnosis of acne is done by the dermatologists who carefully observe skin of the patient to come up with conclusion. A lot of dermatologists are using so-called acne severity grading systems such as Global Acne Grading System (GAGS) [2] or Investigator Global Assessment of Acne (IGA) [3]. Severity grade is a helpful tool for dermatologists to select an appropriate treatment. It could also play a major role in designing clinical trials.

In this work we explore deep learning (DL) based approaches for automated acne grading. For this purpose, a dataset has been collected and labelled by a professional dermatologist with the grading criteria outlined. Along with the main dataset we use an additional one with annotated acne lesions, which allows applying semantic segmentation as well as object detection techniques.

The paper is organized as follows. In Sect. 2, we briefly summarise relevant research papers and projects. Section 3 describes our data and problem statement. Section 4 describes our experiments and obtained results. Finally, Sect. 5 concludes the paper and outline future prospects.

© The Author(s), under exclusive license to Springer Nature Switzerland AG 2024
D. I. Ignatov et al. (Eds.): AIST 2023, CCIS 1905, pp. 222–231, 2024.
https://doi.org/10.1007/978-3-031-67008-4_17

2   Related Work

DL methods have been recently successfully applied to different computer vision problems. Convolutional neural networks (CNNs) in particular have been showing great performances for such tasks as image classification [13], object detection [14] and semantic segmentation [9] among others.

This development has also brought advances to the medical domain. Work [7] explores different CNN architectures to build the model for detecting the presence of Lyme disease. In [4] authors apply Inception-v3 [18] architecture to detect diabetic retinopathy in retinal fundus photographs. As a result they achieve sensitivity and specificity scores of 97.5% and 93.4% correspondingly.

There are also several works that deal with acne specifically. Zhao et al. [23] claim to have developed a deep learning model which is capable of assessing acne severity from selfie images as accurately as dermatologists. They used transfer learning paradigm by extracting image features using a ResNet architecture pretrained model, then adding and training a fully connected layer to learn the target severity level from labeled images. Notably, authors also consider only four face areas of the original images (forehead, both cheeks, and chin) thus restricting the access of the model to the rest of the face. Work done by Zhang et al. [22] uses the ensemble approach solving classification and detection problems simultaneously. Classification block of the ensemble model uses ResNet architecture while the detection block is a you only look once (YOLO) [15] model. Finally, [19] train object detector models such as YOLOv4, faster region-based convolutional neural network (faster R-CNN) [16] with different backbones and single shot multibox detector (SSD) [12]. The work also uses predicted bounding boxes to evaluate severity by counting them.

3   Data and Problem Statement

Two datasets are used in this work. The first one, consists of 668 images of the so-called selfies (self-portraits of individuals usually taken with the help of a smartphone). The average resolution across the whole dataset is 931 by 674 ($H \times W$). The labels for this dataset are real numbers ranging from 0 to 1. They indicate the severity of acne, where the higher number indicates more severe case. More precisely, criteria used to obtain labels are shown in Table 1. To come to consensus regarding the criteria three dermatologists labeled image sets independently first, then analysis of their scoring differences was performed. To perform the analysis for each pair of dermatologists we built three figures. First one is constructed by sorting grades for both dermatologists according to one of them and plotting them in that order. Example is shown in Fig. 1. The second Fig. 3 demonstrates differences in form of the scatter plot, Pearson correlation score is 0.89. Finally, we compare distributions of their plots in Fig. 2.

According to this criteria, the set of labels was obtained with the help of professional dermatologist. Distribution for the labels is shown in Fig. 4. In order to fine-tune and test our models we split this dataset into training, test and

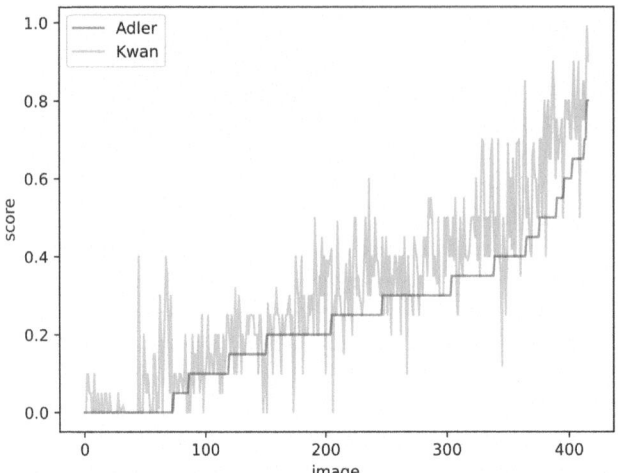

Fig. 1. Score differences: Dr. Adler vs. Dr. Kwan

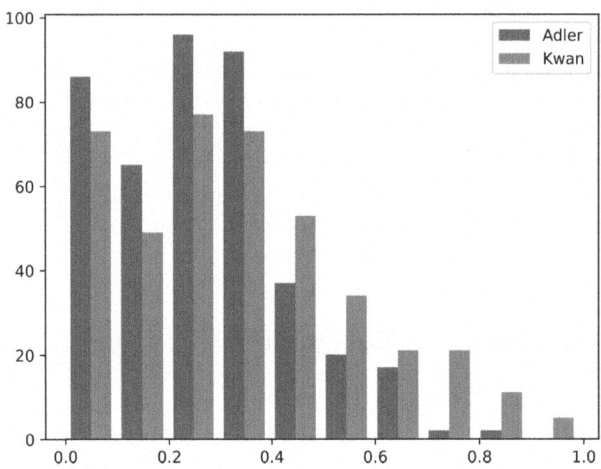

Fig. 2. Distribution differences: Dr. Adler vs. Dr. Kwan

validation sets with 8:1:1 ratio. To measure the quality of the models, we evaluate their performances on the validation set with the mean absolute error (MAE) and symmetric mean absolute percentage error (sMAPE) metrics. sMAPE is defined as follows:

$$sMAPE(p,t) = \frac{100}{n} \sum_{i=1}^{n} \frac{|p_i - t_i|}{(|p_i| + |t_i|)/2},$$

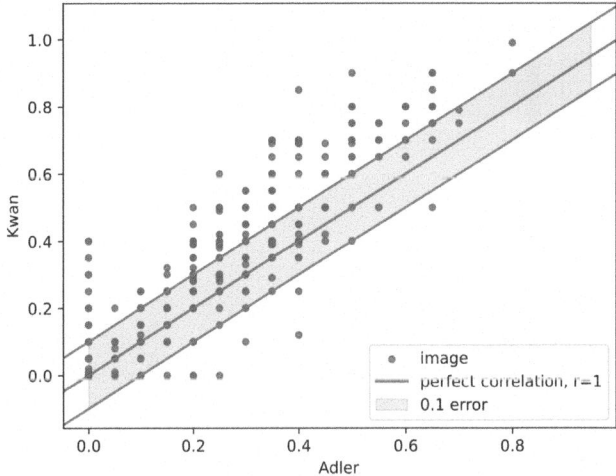

Fig. 3. Scatter plot: Dr. Adler vs. Dr. Kwan

Table 1. Grading criteria

Scores	Verdict	Criteria
0	No acne	cNo acne Present
[0.01, 0.39]	Mild	–Lesion number is less in count –Distribution is localized
[0.4, 0.69]	Moderate	–Lesion number greater in count –Lesion size greater in size –May have a few nodules –Distribution is in more areas of the face than mild
[0.7, 1]	Severe	–Lesions, all over the face: pustules, nodules –Presence of 1 or more cystic acne lesions

where p is prediction vector, t truth vector and n is the size of both vectors. This quality metric is a symmetric version of MAPE and accounts for average relative error while MAE focuses on the average absolute error.

The second dataset, ACNE04, was proposed in [21]. Like the first one, it consists of face images. The difference is that it follows Hayashi's [5] requirements, so all images are taken at an approximately 70-degree angle from the front of patients. In total, there are 1457 images with the average resolution of 3027 by 2918 ($H \times W$). This dataset has both acne severity labels as well as bounding boxes of lesions annotated by professional dermatologists. There are 18 983 bounding boxes in total and their distribution is shown in Fig. 5. Severity grade is obtained from the lesion count. The criteria is provided in Table 2. For training purposes this dataset is already split with 8:2 ratio. The quality of the

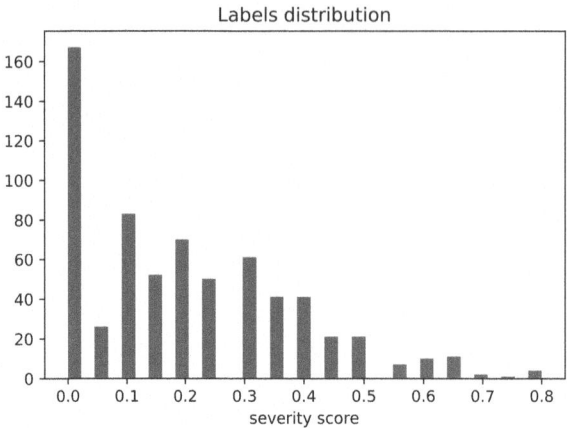

Fig. 4. Label distribution

models trained on this dataset is evaluated via Intersection over Union (IoU) and Dice Coefficient for semantic segmentation and mAP@0.5 for object detection.

Table 2. Grading criteria 2

Class	Num. of lesions, l
Mild	$l \in [1, 5]$
Moderate	$l \in [6, 20]$
Severe	$l \in [21, 50]$
Very severe	$l > 50$

4 Experiments and Results

All experiments are performed on the machine with Intel Core i5-8250U CPU @ 1.60GHz and 16 GB RAM memory capacity.

Initially, we use only the main dataset to solve the regression problem. For this, transfer learning approach was adopted. We use backbone CNN with pre-trained weights with a modified last layer to yield one real number from [0,1] interval. ResNet-18 [6] as well as MobileNet-v3 [8] were used as a backbone. For the loss function we choose MAE while the optimizer is Adam [10]. For the experiments an image processing pipeline was constructed. It includes resizing the images to 224 by 224 resolution, normalisation with mean and standard deviation of ImageNet [1] dataset. In order to increase the size of training data and robustness of the models, we also use augmentation techniques during the training procedures. It includes horizontal flips, Gaussian noise, small rotations and

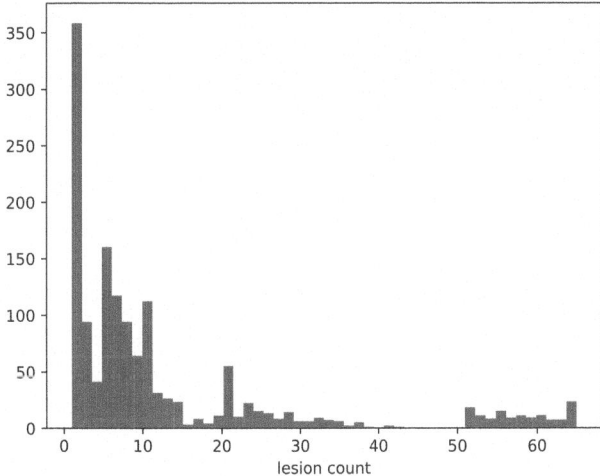

Fig. 5. Lesion count distribution

shifts. Pre-trained weights for both architectures were obtained on ImageNet. The resulting models however had underperforming metric scores with respect to acceptable quality level (see Table 4).

Next, we explore the approaches to localize acne before grading severity. We use ACNE04 data for this purpose. The starting approach was to use semantic segmentation to highlight acne lesions. To achieve this, we firstly transform our bounding boxes by making all pixels inside of them equal to 1 indicating target class zones. In this way, for each image in the additional dataset, we get a corresponding binary mask. We employ the U-Net model [17] next. For the backbone, we again test ResNet-18 and MobileNet-v3 pre-trained on ImageNet dataset. We preserve the same data processing pipeline as for the regression case with the only change being performed, i.e. resizing to higher resolution (480 by 480). For the loss function, the pixel-wise cross-entropy was used and the training phase was done with Adam optimizer. We observe that MAE scores on the validation are positive, while the SMAPE scores indicate that individually some predicted mask were far off. After the error analysis step, we note that the model performs better for severe cases, while the predictions for mild cases contains many false positives. Examples are provided in Fig. 6. The results are presented in Table 4.

An alternative approach to localize acne was to use object detection techniques. We choose YOLO [15] model, namely, YOLOv8s implementation by ultralytics with the parameters pre-trained on the common objects in context (COCO) [11] dataset. The resizing was applied again to suit 640 by 640 COCO format. Resulting scores are presented in Table 3 and the prediction example is shown in Fig. 7. We make use of the built model to solve the original regression problem on the main dataset. The approach is count based. Number of detected acne lesions by YOLO model were used as the only feature for linear regression.

To count lesions we experimented with different confidence threshold for detected bounding boxes from 0 to 1 with step size 0.05 and choose the optimal one at 0.25. This resulted in an improvement (Table 4) compared to the initial transfer learning approach. To improve further upon this we introduce the other factors such as the coverage and positioning. Coverage (C) is defined as a normalised total area taken by bounding boxes if treated as continuous rectangles:

$$C[i] = \frac{1}{H \times W} \sum_{j=1}^{M[i]} (b_i[j][x_{max}] - b_i[j][x_{min}]) \times (b_i[j][y_{max}] - b_i[j][y_{min}]),$$

where i is the current image, H and W are height and width of the image i, $M[i]$ is the amount of bounding boxes detected, $b_i[j]$ is the jth bounding box of image i and indices $x_{max}, x_{min}, y_{max}, y_{min}$ indicate corresponding coordinates of the bounding boxes.

Positioning is defined as follows: we split the image into $n \times n$ grid and for each of the n^2 cells we count how many of detected bounding boxes fall into that cell. Bounding box b_i falls into the cell c_{rc} if their center of b_i is closer (Eucledian distance) to the center of c_{rc} compared to other cells. This way obtain n^2 new features related to relative positioning of bounding boxes. Using both Coverage and Positioning ($n = 2$) we obtain the improvement in performance (Table 4).

Table 3. Results for ACNE04 dataset

Dataset	Backbone	Model	Metric	Score
ACNE04	ResNet18	U-net	IoU@0.2	0.16
ACNE04	ResNet18	U-net	Dice@0.2	0.28
ACNE04		YOLOv8s	mAP@50	0.33

Table 4. Results for original dataset

Dataset	Backbone	Model	MAE	SMAPE
Original	ResNet18	Backbone + FC linear layer	0.13	130
Original	MobileNetV3	Backbone + FC linear layer	0.09	84
Original		YOLOv8s + LR	0.08	64
Original		YOLOv8s + features + LR	0.078	63

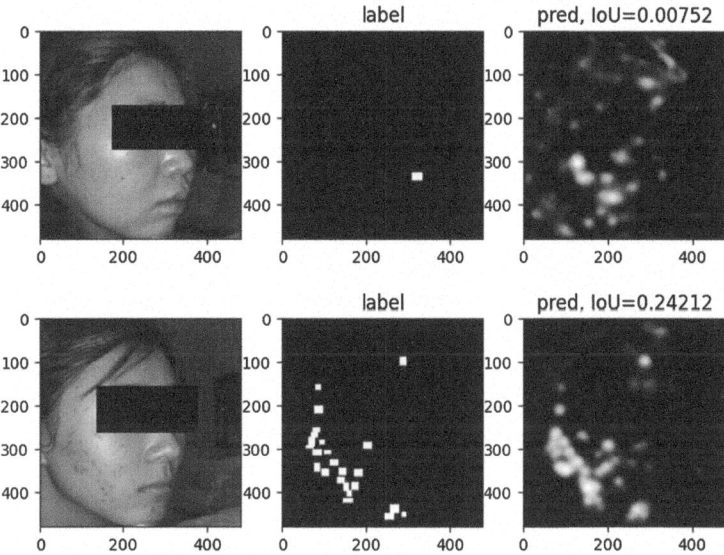

Fig. 6. U-net prediction examples. First column shows original images, second ground truth masks and the last one depicts predictions. For each prediction there is an IoU score shown above.

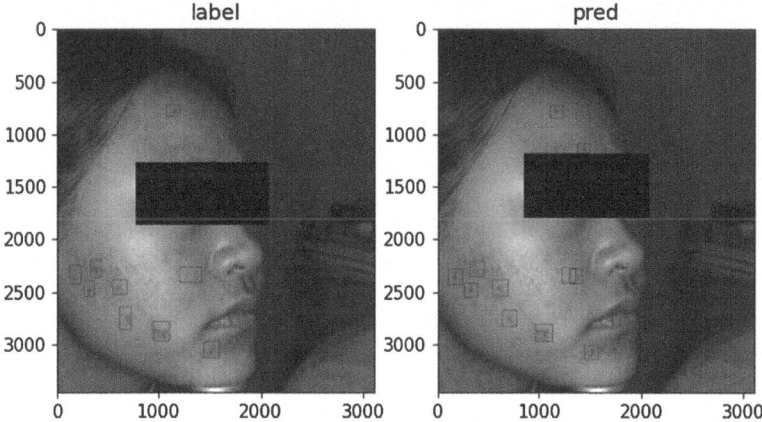

Fig. 7. YOLOv8 prediction examples

5   Conclusion and Future Work

As acne vulgaris continues to be a widespread skin condition worldwide, the need for automatic severity grading grows.

In this study, we explored approaches to achieve automatic grading that is as precise as possible with respect to criteria from a professional dermatologist. The initial approach was to construct baseline grading models using CNN backbones. To improve upon this, and considering the absence of annotations for each acne lesion in the main dataset, we use an additional dataset, ACNE04, in order to train an acne detector. With the use of YOLOv8s, we achieved an mAP@0.5 score of 0.33 on the additional dataset. After training the detector, we use it to build the new grader. We use a combination of a detector and simple regression to improve the baseline results. Then, we propose heuristic features based on the coverage and positioning of detected boxes. These features are well aligned with the grading criteria at hand. With the additional features at hand, we again train simple regression to achieve a final improvement with an MAE of 0.078 and SMAPE of 63.

Acknowledgements. The authors would like to thank Dr. Adler and Dr. Kwan for their expertise. The contribution of the first two authors is also an output of a research project implemented as part of the Basic Research Program at the National Research University Higher School of Economics (HSE University). This research was also supported in part through computational resources of HPC facilities at HSE University.

References

1. Deng, J., Dong, W., Socher, R., Li, L.J., Li, K., Fei-Fei, L.: ImageNet: a large-scale hierarchical image database. In: 2009 IEEE Conference on Computer Vision and Pattern Recognition, pp. 248–255. IEEE (2009)
2. Doshi, A., Zaheer, A., Stiller, M.J.: A comparison of current acne grading systems and proposal of a novel system. Int. J. Dermatol. **36**(6), 416–418 (1997)
3. Food, Administration, D., et al.: Acne vulgaris: Establishing effectiveness of drugs intended for treatment. Guidance for industry
4. Gulshan, V., et al.: Development and validation of a deep learning algorithm for detection of diabetic retinopathy in retinal fundus photographs. JAMA **316**(22), 2402–2410 (2016). https://doi.org/10.1001/jama.2016.17216
5. Hayashi, N., Akamatsu, H., Kawashima, M.: Establishment of grading criteria for acne severity. J. Dermatol. **35**, 255–60 (2008). https://doi.org/10.1111/j.1346-8138.2008.00462.x
6. He, K., Zhang, X., Ren, S., Sun, J.: Deep residual learning for image recognition. CoRR **abs/1512.03385** (2015). http://arxiv.org/abs/1512.03385
7. Hossain, S.I., et al.: Exploring convolutional neural networks with transfer learning for diagnosing lyme disease from skin lesion images. Comput. Methods Programs Biomed. **215**, 106624 (2022). https://doi.org/10.1016/j.cmpb.2022.106624
8. Howard, A., et al.: Searching for mobilenetv3. In: 2019 IEEE/CVF International Conference on Computer Vision, ICCV 2019, Seoul, Korea (South), October 27 - November 2, 2019, pp. 1314–1324. IEEE (2019). https://doi.org/10.1109/ICCV.2019.00140

9. Hu, X., Jing, L., Sehar, U.: Joint pyramid attention network for real-time semantic segmentation of urban scenes. Appl. Intell. **52**(1), 580–594 (2022). https://doi.org/10.1007/s10489-021-02446-8
10. Kingma, D.P., Ba, J.: Adam: a method for stochastic optimization. In: Bengio, Y., LeCun, Y. (eds.) 3rd International Conference on Learning Representations, ICLR 2015, San Diego, CA, USA, 7–9 May 2015, Conference Track Proceedings (2015). http://arxiv.org/abs/1412.6980
11. Lin, T., et al.: Microsoft COCO: common objects in context. CoRR **abs/1405.0312** (2014). http://arxiv.org/abs/1405.0312
12. Liu, W., et al.: SSD: single shot multibox detector. In: Leibe, B., Matas, J., Sebe, N., Welling, M. (eds.) ECCV 2016. LNCS, vol. 9905, pp. 21–37. Springer, Cham (2016). https://doi.org/10.1007/978-3-319-46448-0_2
13. Liu, Z., Mao, H., Wu, C., Feichtenhofer, C., Darrell, T., Xie, S.: A convnet for the 2020s. In: IEEE/CVF Conference on Computer Vision and Pattern Recognition, CVPR 2022, New Orleans, LA, USA, 18–24 June 2022, pp. 11966–11976. IEEE (2022). https://doi.org/10.1109/CVPR52688.2022.01167
14. Pham, V., Nguyen, D., Donan, C.: Road damage detection and classification with YOLOv7. In: Tsumoto, S., et al. (eds.) IEEE International Conference on Big Data, Big Data 2022, Osaka, Japan, 17–20 December 2022, pp. 6416–6423. IEEE (2022). https://doi.org/10.1109/BigData55660.2022.10020856
15. Redmon, J., Farhadi, A.: YOLOv3: an incremental improvement. arXiv (2018)
16. Ren, S., He, K., Girshick, R., Sun, J.: Faster R-CNN: towards real-time object detection with region proposal networks. In: Proceedings of the 28th International Conference on Neural Information Processing Systems, NIPS 2015, vol. 1, pp. 91–99. MIT Press, Cambridge (2015)
17. Ronneberger, O., Fischer, P., Brox, T.: U-Net: convolutional networks for biomedical image segmentation. CoRR **abs/1505.04597** (2015). http://arxiv.org/abs/1505.04597
18. Szegedy, C., Vanhoucke, V., Ioffe, S., Shlens, J., Wojna, Z.: Rethinking the inception architecture for computer vision. In: 2016 IEEE Conference on Computer Vision and Pattern Recognition, CVPR 2016, Las Vegas, NV, USA, 27–30 June 2016, pp. 2818–2826. IEEE Computer Society (2016). https://doi.org/10.1109/CVPR.2016.308
19. Wen, H., et al.: Acne detection and severity evaluation with interpretable convolutional neural network models. Technol. Health Care **30**(S1), 143–153 (2022)
20. Williams, H.C., Dellavalle, R.P., Garner, S.: Acne vulgaris. Lancet **379**(9813), 361–372 (2012)
21. Wu, X., et al.: Joint acne image grading and counting via label distribution learning. In: IEEE International Conference on Computer Vision (2019)
22. Zhang, H., Ma, T.: Acne detection by ensemble neural networks. Sensors **22**(18), 6828 (2022). https://doi.org/10.3390/s22186828
23. Zhao, T., Zhang, H., Spoelstra, J.: A computer vision application for assessing facial acne severity from selfie images. CoRR **abs/1907.07901** (2019). http://arxiv.org/abs/1907.07901

Data Analysis and Machine Learning

Application of Multimodal Machine Learning for Image Recommendation Systems

Mikhail Foniakov[1], Anatoly Bardukov[1], and Ilya Makarov[1,2(✉)]

[1] HSE University, Moscow, Russia
abardukov@hse.ru, makarov@airi.net
[2] AIRI, Moscow, Russia

Abstract. In the era of information overload, making decisions about various aspects of our lives has become increasingly challenging. The rise of the internet and online platforms has provided access to vast amounts of information, including recommendations based on user preferences. This paper focuses on the development of a unique multimodal recommender system for images, leveraging machine learning and deep learning techniques.

The results demonstrate the effectiveness of a complex recommender system, with an average accuracy of 0.814. The system outperforms algorithms based solely on CLIP embeddings or BERT vectors, showcasing the advantages of incorporating multiple modalities. The strengths of the model lie in its robust recognition of CLIP vectors and efficient processing of image features. However, there is room for improvement in text features and embeddings, suggesting the need for more detailed textual information and additional data sources.

The innovation of this system is to utilize both image data and text descriptions to provide personalized recommendations. Images are vectorized and combined with relevant metrics, while text descriptions are obtained through object recognition or dataset text. The model incorporates various data, such as location, creation date, and event information, to enhance the recommendation process.

Overall, this study highlights the potential of multimodal recommender systems for enhancing recommendation quality and providing users with a personalized experience. Ongoing efforts to refine the model with diverse data and parameter optimization can further improve its performance.

Keywords: Multimodal Models · Recommendation Systems · Image Processing

---

The work on Section 2 was supported by RSF under grant 22-11-00323 and performed at HSE University, Moscow, Russia.

© The Author(s), under exclusive license to Springer Nature Switzerland AG 2024
D. I. Ignatov et al. (Eds.): AIST 2023, CCIS 1905, pp. 235–249, 2024.
https://doi.org/10.1007/978-3-031-67008-4_18

1   Introduction

In today's information-rich era the relevance of a complex recommendendation system is very high [15, 19, 39]. Decision-making is a constant challenge due to the overwhelming amount of available information. Whether it's choosing a captivating book, an appealing music album, or a worthwhile movie, we face numerous questions daily. Traditionally, seeking advice from friends or experts helped us make these decisions. However, human input has limitations, and expert consultations can be expensive and inaccessible to many. The emergence of Internet technologies and the growth of online platforms have provided new sources of information, and neural networks have significantly contributed to the development of recommendations.

With the increasing number of images uploaded by users, the need arose to systematize and analyze them. Detection, segmentation, and checking for prohibited content became essential tasks. Furthermore, making images searchable required sophisticated algorithms, giving rise to the recommendation system.

1.1   Recommender System

A recommender system (RS) [12, 15, 33] is an AI-based class of machine learning algorithms designed to help people navigate an exponentially expanding array of options. These algorithms, including machine learning and neural networks, use Big Data to suggest additional products to consumers based on various criteria. Personalized image-based RS is employed in various fields, from online content platforms to medicine, aiming to estimate users' preferences and simplify the search for goods.

The foundation of any RS is a model that transforms users' preferences and other data into predictions of their potential interests. By training this model using vectorized images and corresponding target variables, the system can deliver high-quality recommendations. There are two main model types: one based on machine learning algorithms like DecisionTreeClassifier, and the other on deep learning concepts with pre-trained or non-pre-trained models.

However, these traditional recommendation algorithms have limitations [47] in fully considering users' interests. Different tasks necessitate diverse data for effective recommendations, leading to the development of multimodal RS [34].

1.2   Multimodality

Multimodality harnesses multiple approaches within one medium, offering an excellent opportunity to enhance recommendation quality. By extracting additional data from various input features, this approach adapts to the present digital age's diverse data sources. Multimodal data [36, 42, 46, 53] includes different embeddings [40, 43, 44], text descriptions, and various metrics, and multimodal deep learning [32] trains AI models to process and identify relationships between these features.

The model proposed in this thesis comprises two primary components: the image itself and its textual description. Both components are vectorized and concatenated, allowing the model to incorporate additional data such as location, date of creation, and event details. Utilizing multiple image data enhances the training process, resulting in superior recommendations.

Multimodality involves various communication practices, combining textual, spatial, visual, and other resources to compose image features. It optimizes recommendation systems by suggesting items with similar themes, styles, or ambiances, ultimately saving users valuable time.

In conclusion, multimodal RS stands out as a popular and captivating model, capable of incorporating diverse features to streamline the training process and deliver high-quality recommendations.

2   Related Work

The origins of the modern Recommender Systems (RS) go back 40 years [13]. The initial idea of using a system to recommend the best product for the user has been around since the dawn of computers. However, these were only the simplest systems without multimodality. These RSs were experimental: they only solved local problems and predicted users' preferences using only one feature.

Along with the rise of e-commerce, the industry realized the business value of recommendations [38]. A real breakthrough was with the implementation of Amazon RS [48]. It required using images, descriptions, and titles to categorize products to the maximum. Each product is assigned multiple labels, and the hierarchy in the labels is flattened and filtered. Then, it uses the modified Convolutional Neural Network (CNN) [22] architecture to classify descriptions and titles and modified Keras ResNet-50 [17] to classify the images. Amazon was the first to merge these three classifiers to categorize products [18]. The success story of Amazon gave rise to the development of many RS algorithms, including multimodal systems. From that time, RSs became very popular. They are implemented by many online systems and e-commerce.

The group of scientists from Asia wrote about the basis of multimodal RS. Generally, any RS is divided into two major groups of models: matrix factorization and neural networks. In the first group, the preference data are represented by a user-by-item matrix. The second group is based on non-linearity through activation functions and multiple compositional layers. The authors suggest using the Python framework Cornac [34], which is tailored for multimodal RS. They also represent a large number of models for texts, images, and graphs. This is the first article that tells about multimodal RS [52].

Multimodal RS is widely used for multimedia data analysis where the data are characterized by high dimensionality, large storage space required for the data, and rich data content. The following algorithm is usually used: A single-mode feature extraction is initially performed and followed by an application of cross-modal convergence [50].

Another option is to obtain recommendations using CNN-extracted features [12,27] in multimodal semantic representation models. This is achieved by concatenating a skip-gram linguistic [23] representation vector computed using the feature extraction layers of neural networks trained on a large labeled dataset. The multimodal representation consists of three parts: perceptual representations, using CNN, linguistic representation with the skip-gram model, and the multimodal representation itself [51].

Multimodal RS is an interesting research field that has grown rapidly and become popular. The increase in interest in this research topic has also been driven by major improvements in e-commerce. As mentioned, multimodality is primarily widely used in large e-commerce platforms [45], serving hundreds of millions of users with billions of items/products. Users visit the platforms to browse some items through general search or personalized recommendation systems. Each item is displayed by the item image with some descriptive texts. E-commerce platform is also a large target advertising system, which helps millions of advertisers to connect with potential customers. By identifying user interests, target ads are presented in various locations and efficiently deliver marketing messages to the right customers.

Click-Through-Rate (CTR) prediction is one of the most important tasks [21] and one of the key metrics for a building multimodal RS in e-commerce. It is the ratio between the number of clicks on a link to the number of users who visited the page. Taobao, the largest e-commerce platform in China, responds to billions of page views (PV) every day. For each view/request, the system selects the most suitable ad to display to each particular potential customer, using the advanced CTR Deep Learning model [54]. This model exploits abundant user behavior to identify whether they are interested in a candidate advertisement. It unites the users' preferences, users' behavioral features, and behavioral images and continuously trains with them. To systematize the results, the platform uses a novel and efficient distributed machine learning paradigm called Advanced Model Server (AMS). Each server node handles a separate part of the features and updates them independently. Then, the pre-ranking model shrinks the number of candidates, and finally, the ranking model predicts the CTR of ads and ranks them to provide the best choice. This happens in just tens of milliseconds [24].

Nowadays, a large number of people use social networks, mobile devices [20, 28,37,49] and various types of online services [25]. In doing so, they leave a lot of information about themselves, such as photos [19], videos [26,41], comments, texts [29], alerts, and purchase history, profile information. This information is valuable for constructing multimodal RS. The author identifies users' interests based on their actions in a social network and it's RS consists of four data types: images, texts, geolocation and ratings. He suggests to recognize object on the image using Inception-v3, to analyse text using Latent Dirichlet Allocation, and to analyse text sentiment using Word2Vec model with Keras library [31].

Nowadays the industry offered generous funding to implement RS's research. Researchers have developed a lot of image RS for almost every sphere like social networks, content-based sites (books, articles, e-learning), tourism, and a lot more, all dealing with the real world.

3 Model

3.1 Dataset Preparation

To build the recommender system, we require a dataset to train our model. We utilized a dataset from Yandex, comprising of 124695 images covering various random topics. Here the dataset contains of images pairs and follows the given structure:

- Related image: It may be a completely random image for absolutely different topics.
- Candidate image: The picture, which we recommended or not recommended together with the initial image.
- Target of the pair of images, with a binary value: 1 - indicating similar images and we recommend them together, 0 - indicating dissimilar images and we don't them together.

The dataset contains 100276 pairs of images: 72% of them with a target 1 (similar images, suitable for recommendation), and 28% with a target of 0 (dissimilar images, not recommended). As a result, the dataset is imbalanced and requires preparation. For building the multimodal system, we used different features for each image.

- The main part of our Recommender System is the description text of the picture. We obtained and parsed this information from Yandex Pictures [10], using the Beautiful Soup Python library [1]. This approach is simple and accessible, making it ideal for parsing. We collected texts in various languages for each picture.
- We processed all images using CLIP, a neural network for images trained on 400 million images and texts. This technique provides valuable image embeddings with a variety of models. Specifically, we employed the clip-vit-base-patch32 model [2], developed by researchers from OpenAI. This model consists of 512 dimensions.
- For text processing, we adopted another technique called BERT (Bidirectional Encoder Representations from Transformers). BERT, developed by Google, is a large language neural network with a diverse set of models. We used the model distilbert-base-multilingual-cased [35], which is trained on the titles of Wikipedia in 104 different languages and has 768 dimensions. Thus, for each image and text, we obtained CLIP and BERT vectors accordingly.
- The next preprocessing step involves calculating metrics for the data. For each pair of images and texts, we used two metrics to compute the paired distances between two CLIP-vectors for images and two BERT-vectors for texts: cosine similarity and Squared Euclidean distance.

- Cosine similarity represents the cosine of the angle between two non-zero vectors. It is widely used in Recommender System algorithms, where higher similarity indicates stronger recommendations between objects. Mathematically, cosine similarity is the division of the dot product of two vectors by the product of their Euclidean norms.
- The Squared Euclidean distance (L2 squared) measures the distance between two vectors by summing the squared values of their components. Larger distances imply greater dissimilarity between vectors and objects in the Recommender System.

3.2   Model

For our dataset, we computed these two metrics (Cosine similarity and L2 squared) for CLIP and BERT-vectors. The 95% confidence intervals for cosine similarity of CLIP vectors are [0.7435, 0.7451], and for BERT vectors are [0.9727, 0.9729]. The relatively high cosine similarity of BERT embeddings can be attributed to the specific construction of BERT [30]. The confidence intervals for the L2 distance of CLIP and BERT-vectors are [7.1531, 7.176] and [6.12, 6.1447], respectively (See Fig. 1). According to the Shapiro-Wilk test, all four metrics do not follow a normal distribution, with p-values extremely close to zero.

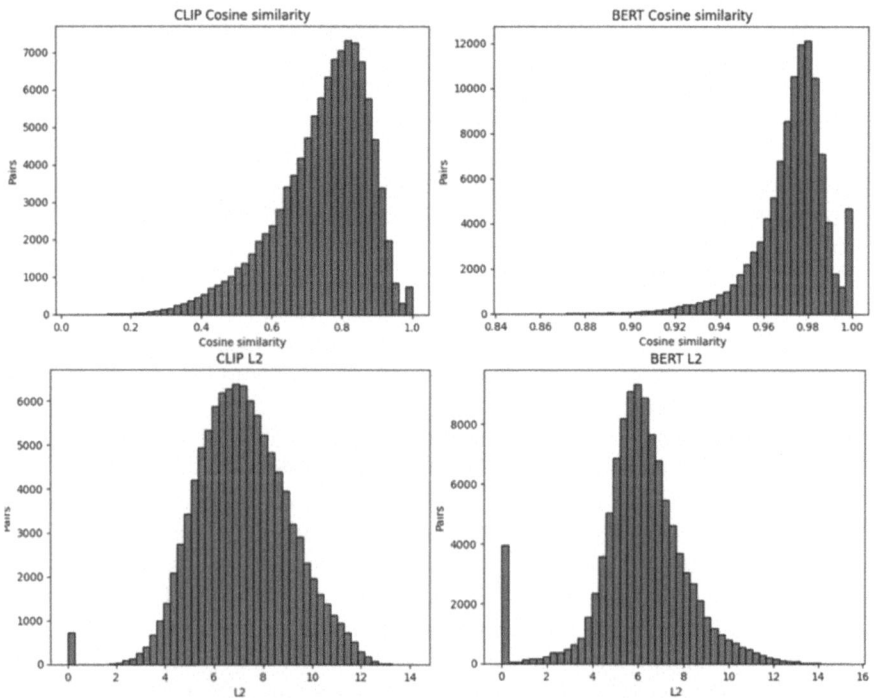

Fig. 1. Metrics distribution of the initial dataset

This dataset contains 28,023 related images, which we divided into this quantity of groups for our further model development. Thus, our dataset has the following structure:

- 124695 images, 100276 pairs of images
- 28,023 related images. 96672 candidate images. For one related image there are 3 candidate images on average.
- 10 features: 2 embedding features for CLIP dimensions, 2 embedding features for BERT dimensions, 4 numerical features for metrics, 1 feature for the binomial target variable, and 1 categorical feature for the group of related images.
- The metrics do not follow a normal distribution.
- The data is imbalanced.

We tested 5 different models for our recommender system: DecisionTreeClassifier [7], RandomForestClassifier [6], XGBoostClassifier [9] and two kinds of CatBoostClassifier [4] (with queries and without queries). Queries are groups, which quantity is equal to number of related images of our dataset. In the first version of CatBoostClassifier model we add one new categorical feature to our dataset - query. In the second version of this model we use dataset without this feature. The first two models (DecisionTreeClassifier and RandomForestClassifier) we use with SMOTE technology [8]. SMOTE is Synthetic Minority Oversampling TEchnique and it allows work with imbalanced data. It uses oversampling method: a method by balancing the data by adding new observation of minority class. The testing models showed that the best results there were in the CatBoostClassifier algorithm with adding queries. And we used it for building our recommender system [16] (See Table 1).

Table 1. The models comparisons

Model	AUC	Accuracy	PFound
DecisionTreeClassifier with SMOTE	0.66	0.75	–
RandomForestClassifier with SMOTE	0.81	0.85	–
XGBoostClassifier	0.83	0.87	–
CatBoostClassifier without queries	0.8	0.87	–
CatBoostClassifier with queries	0.83	0.89	0.77

Catboost is an algorithm for gradient boosting on decision trees, developed by Yandex engineers, and widely used in various recommender and search systems, personal assistants, weather forecasting, and other tasks. It is known to work effectively with data from different sources.

In this algorithm, we employ the CatboostClassifier package for classification, as our dataset consists of only two target variables (0 and 1), and unfortunately, we are unable to modify it.

Before constructing the model, our data requires some preprocessing:

– Metric features were scaled using the StandardScaler method, which standardizes them by removing the mean and scaling them to unit variance [14].
– We split the dataset into training, testing, and validation sets, allocating 60%, 20%, and 20% of the entire dataset, respectively.
– We create the necessary data structures using the Pool constructor [16], taking note of the embeddings features.

The main part of model constructing is selection of hyperparameters. The basic hyperparameters of our model are iterations, learning rate and max depth. We can adjust iterations using parameter "early stop rounds". If the parameters dictate, algorithm stops training. The only parameter that needs to be tested is learning rate.

We tested for different learning rates from 0.001 to 0.1 and found that the optimal learning rate is the smallest (See Table 2).

Table 2. Learning rate comparisons for CatBoostClassifier model

Learning rate	AUC	Accuracy	PFound
0.1	0.81	0.88	0.77
0.05	0.81	0.88	0.77
0.01	0.82	0.89	0.77
0.001	0.83	0.89	0.77

So, we trained our model using the following hyperparameters:

– Iterations: 2000. This is determined as the optimal number of iterations for learning the model with our data.
– Early stop rounds: 20. The algorithm stops training if the parameters dictate.
– Learning rate: 0.001. We opted for a small gradient step size to enhance the precision of our model's learning process.
– Max depth: 6. This depth is found to be optimal for our classification model.
– Scale pos weight: 2.64. As our data is imbalanced, we use this parameter with a value equal to the ratio of the majority class to the minority class.

For the loss function, we utilized the logarithmic loss function (LogLoss). It serves as a metric for evaluating the performance of binary classification models and indicates how close the prediction probability is to the corresponding true value.

As a custom metric, we used the Area Under the Curve (AUC). AUC is an effective method for visualizing the performance of the model, represented by a graph with two parameters: True Positive Rate (TPR) and False Positive Rate (FPR).

After training the model, we achieved an AUC ROC of 0.83, indicating a higher probability of predicting the positive class over the negative class

The loss function LogLoss of the model is 0.29. However, we use the predicted probabilities of the positive class as the output of our model, as it help us rank the results of our future experiments.

Feature importance shows how much each feature contributes to the model prediction and determines a degree of usefullness of a specific variable for a model and prediction. The feature importance analysis revealed that the most relevant features in the model training are those representing CLIP embeddings. Text and its BERT embeddings play a supporting role in this model (See Table 3).

Table 3. The feature importance of the model

feature	clip l2	clip cs	img_clip_2	img_clip_1	img_bert_2	img_bert_1
importance	36.124	23.903	21.466	14.358	1.670	1.553

Thus, our model has the following features:

- This is the classification model, which predicts the binary target variable
- As we build a recommendation system, we use probability prediction of binary target variable
- The model is built on imbalanced and unmarked data
- Features of images play more important role than text features. This is due to the fact that dataset is unmarked and the text of the images is parsed from Yandex Images.

4   Experiment Settings

4.1   Experimental Datasets

For experiments we use two different datasets:

1. Yandex dataset. This is the same dataset as for model constructing in the previous chapter. However it is differently organized. Initially it contains 124695 images with text description.
2. Flickr Image Dataset. It is dataset of images with text description with capture people engaged in everyday activities and events [3]. It contains 28023 images with text description.

To prepare for the experiments with two datasets, we need to process them.

- We construct CLIP vectors from images and BERT vectors from the text description, using the same models of CLIP and BERT vectors as for the dataset for model [2,35].
- We divide them using the K-means algorithm from the sklearn library, which is a popular clustering method [5]. The algorithm selects initial cluster centroids using sampling based on an empirical probability distribution of the point's contribution to the overall inertia [11]. The datasets are divided by 1000 clusters based on CLIP and BERT vectors.

– We then obtain the cluster-distance for each image and determined the three nearest clusters for each image.
– For each pair, we calculate its cosine similarity and Euclidean distance for both CLIP and BERT vectors.

Therefore, all the datasets have the following structure:

– The CLIP vector for the related image.
– The CLIP vector for the candidate image.
– The BERT vector for the related image.
– The BERT vector for the candidate image.
– The cosine similarity between two CLIP vectors.
– The cosine similarity between two BERT vectors.
– The L2 distance between two CLIP vectors.
– The L2 distance between two BERT vectors.

By doing this, we create 57531198 pairs of images from the Yandex dataset and 3192694 from the Flickr dataset. We conduct three kinds of experiments: using our model with both CLIP and BERT embeddings, using only CLIP embeddings, using only BERT embeddings and using our model with both CLIP and BERT embeddings.

4.2   Experiments with Yandex Dataset

From the experimental Yandex dataset, we group a certain number of candidates with the highest (for cosine similarity) or the lowest (for Euclidean distance) metrics of CLIP-vectors, BERT-vectors, or probabilities of the positive class for each image. In this case, the number of candidates is equal to 5. This choice is optimal for a relatively small dataset, as it provides meaningful results. As a result, we obtain the related image and 5 candidate images for each case.

The algorithm is executed for seven images, and the results for all three metrics were combined and are now ready for comparison.

Our Catboost model achieved a remarkable accuracy of 1, selecting the correct images flawlessly. However, in the case of CLIP-vectors, the system made one mistake in the fifth image, which depicted a different person. Similarly, BERT-embeddings also produced one incorrect recommendation in the fourth image. Thus, the accuracy for both cases stood at 0.8 (See Fig. 2).

The overall system and CLIP-vectors demonstrated perfect accuracy, selecting the correct images without any errors. However, BERT-embeddings yielded a lower accuracy of 0.4 due to three mistaken recommendations, all of which depicted another pictures (See Fig. 3).

The images of Yandex dataset is various and unstructured. Text description of the images is not very detailed. For this reason recommendations by BERT embeddings had lower accuracy than others by Catboost model or CLIP vectors. However when we use the vectors together we get a better quality recommendation.

Fig. 2. Yandex Dataset: recommendation for the first random image

	Related image	Recommended image 1	Recommended image 2	Recommended image 3	Recommended image 4	Recommended image 5	Accuracy
Model							1
CLIP							1
BERT							0.4

Fig. 3. Yandex Dataset: recommendation for the second random image

4.3 Experiments with Flickr Image Dataset

In the Flickr Image Dataset we use the same algorithm, as for previous Yandex Dataset.

For the random image the model recommend images without mistakes. The accuracy here is 1. The recommendation using CLIP vectors made a mistake in the forth image with accuracy is equal to 0.8: the algorithm showed a person without boat. The BERT recommendation made mistakes in three last images with the accuracy 0.4: there were person without boat (See Fig. 4).

Unlike the previous, the Flickr dataset has better data structure. It is more convenient for experiments, and we obtained higher rate of accuracy than experiments with Yandex Dataset, but even for the prepared dataset our model outperforms CLIP and BERT used separately.

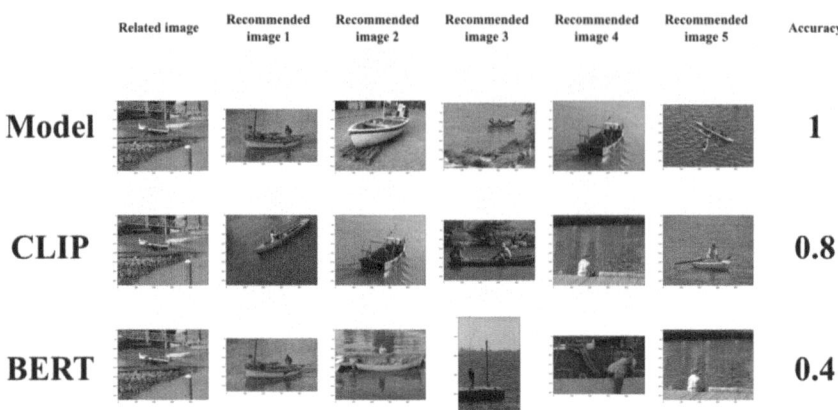

Fig. 4. Flickr Image Dataset: recommendation for the random image

5   Results

The outcomes of our multimodal recommender system for images, incorporating three models: multimodal algorithm with combination of vectors; CLIP only vectors; BERT only vectors. The performance is measured as the ratio of correctly recommended images to all recommended images.

225 random images were used for accuracy measure for three models: the multimodal recommender system shows accuracy equal to 0.814; the CLIP model has 0.792; the BERT model is equal to 0.478.

Also we measured 50 random images from the Flickr Image Dataset and compared these three models too: the Catboost model shows accuracy equal to 0.976; the CLIP model's accuracy is 0.952; the accuracy of BERT model is equal to 0.848.

6   Conclusion

Our multimodal recommender system for images has shown promising results with an average accuracy of 0.814 with Yandex dataset and 0.976 with Image Flickr Dataset. Compared to algorithms solely relying on CLIP embeddings or BERT-vectors, our system outperforms them. The system's strengths and novelty lie in its effective CLIP and BERT vectors recognition and efficient processing of all image features, including the ability to work smoothly with clusters, facilitating easy dataset splitting. Nevertheless, we acknowledge some limitations, particularly in the auxiliary role of text features and BERT embeddings. Further improvements may necessitate more detailed text or additional text features and vectors in the dataset. Our primary objective is to continually enhance the recommender system through diverse data and model parameter variations, ensuring sustained and improved performance over time.

References

1. beautifulsoup4 4.12.2 (2023). https://pypi.org/project/beautifulsoup4. Accessed 07 May 2023
2. clip-vit-base-patch (2023). https://huggingface.co/openai/clip-vit-base-patch32. Accessed 11 May 2023
3. Flickr Image dataset | Kaggle (2023). https://www.kaggle.com/datasets/hsankesara/flickr-image-dataset. Accessed 07 Sept 2023
4. Overview - CatBoostClassifier | Catboost (2023). https://catboost.ai/en/docs/concepts/python-reference_catboostclassifier. Accessed 07 Sept 2023
5. sklearn.cluster.KMeans - scikit-learn 1.3.0 documentation (2023). https://scikit-learn.org/stable/modules/generated/sklearn.cluster.KMeans.html. Accessed 07 Sept 2023
6. sklearn.ensemble.RandomForestClassifier (2023). https://scikit-learn.org/stable/modules/generated/sklearn.ensemble.RandomForestClassifier.html. Accessed 07 Sept 2023
7. sklearn.tree.DecisionTreeClassifier (2023). https://scikit-learn.org/stable/modules/generated/sklearn.tree.DecisionTreeClassifier.html. Accessed 07 Sept 2023
8. SMOTE - Version 0.11.0 (2023). https://imbalanced-learn.org/stable/references/generated/imblearn.over_sampling.SMOTE.html. Accessed 07 Sept 2023
9. XGBoost Documentation - xgboost 1.7.6 documentation (2023). https://xgboost.readthedocs.io/en/stable/. Accessed 07 Sept 2023
10. Yandex Pictures (2023). https://yandex.ru/images. Accessed 06 May 2023
11. Ahmed, M., Seraj, R., Islam, S.M.S.: The k-means algorithm: a comprehensive survey and performance evaluation. Electronics **9**(8), 1295 (2020)
12. Andreeva, E., Ignatov, D.I., Grachev, A., Savchenko, A.V.: Extraction of visual features for recommendation of products via deep learning. In: van der Aalst, W.M.P., et al. (eds.) AIST 2018. LNCS, vol. 11179, pp. 201–210. Springer, Cham (2018). https://doi.org/10.1007/978-3-030-11027-7_20
13. Anneroth, G., Batsakis, J., Luna, M.: Review of the literature and a recommended system of malignancy grading in oral squamous cell carcinomas. Eur. J. Oral Sci. **95**(3), 229–249 (1987)
14. Bisong, E., Bisong, E.: Introduction to scikit-learn. Building Machine Learning and Deep Learning Models on Google Cloud Platform: A Comprehensive Guide for Beginners, pp. 215–229 (2019)
15. Bobadilla, J., Ortega, F., Hernando, A., Gutiérrez, A.: Recommender systems survey. Knowl.-Based Syst. **46**, 109–132 (2013)
16. Dorogush, A.V., Ershov, V., Gulin, A.: CatBoost: gradient boosting with categorical features support. arXiv preprint arXiv:1810.11363 (2018)
17. Garson, J., Aggarwal, A., Sarkar, S.: ResNet manual. Ver **1**, 30 (2002)
18. Ge, T., et al.: Image matters: visually modeling user behaviors using advanced model server. In: Proceedings of the 27th ACM International Conference on Information and Knowledge Management, pp. 2087–2095 (2018)
19. Grechikhin, I., Savchenko, A.V.: User modeling on mobile device based on facial clustering and object detection in photos and videos. In: Morales, A., Fierrez, J., Sánchez, J.S., Ribeiro, B. (eds.) IbPRIA 2019, Part II. LNCS, vol. 11868, pp. 429–440. Springer, Cham (2019). https://doi.org/10.1007/978-3-030-31321-0_37
20. Kharchevnikova, A., Savchenko, A.: Neural networks in video-based age and gender recognition on mobile platforms. Opt. Memory Neural Netw. **27**, 246–259 (2018)

21. Kiela, D., Bottou, L.: Learning image embeddings using convolutional neural networks for improved multi-modal semantics. In: Proceedings of the 2014 Conference on Empirical Methods in Natural Language Processing (EMNLP), pp. 36–45 (2014)
22. Kim, P., Kim, P.: Convolutional neural network. MATLAB deep learning: with machine learning, neural networks and artificial intelligence, pp. 121–147 (2017)
23. Lazaridou, A., Pham, N.T., Baroni, M.: Combining language and vision with a multimodal skip-gram model. arXiv preprint arXiv:1501.02598 (2015)
24. Li, X., et al.: Adversarial multimodal representation learning for click-through rate prediction. In: Proceedings of The Web Conference 2020, pp. 827–836 (2020)
25. Lin, K.Y., Lu, H.P.: Why people use social networking sites: an empirical study integrating network externalities and motivation theory. Comput. Hum. Behav. **27**(3), 1152–1161 (2011)
26. Makarov, I., Bakhanova, M., Nikolenko, S., Gerasimova, O.: Self-supervised recurrent depth estimation with attention mechanisms. PeerJ Comput. Sci. **8**, e865 (2022)
27. Makarov, I., et al.: On reproducing semi-dense depth map reconstruction using deep convolutional neural networks with perceptual loss. In: Proceedings of the 27th ACM International Conference on Multimedia, pp. 1080–1084 (2019)
28. Makarov, I.: Temporal network embedding framework with causal anonymous walks representations. PeerJ Comput. Sci. **8**, e858 (2022)
29. Makarov, I., Veldyaykin, N., Chertkov, M., Pokoev, A.: Russian sign language dactyl recognition. In: 2019 42nd International Conference on Telecommunications and Signal Processing (TSP), pp. 726–729. IEEE (2019)
30. Malkiel, I., Ginzburg, D., Barkan, O., Caciularu, A., Weill, J., Koenigstein, N.: Interpreting BERT-based text similarity via activation and saliency maps. In: Proceedings of the ACM Web Conference 2022, pp. 3259–3268 (2022)
31. Monastyrev, V.V., Drobintsev, P.D.: Recommendation system based on user actions in the social network. Труды института системного программирования РАН **32**(3), 101–108 (2020)
32. Ngiam, J., Khosla, A., Kim, M., Nam, J., Lee, H., Ng, A.Y.: Multimodal deep learning. In: Proceedings of the 28th International Conference on Machine Learning (ICML 2011), pp. 689–696 (2011)
33. Resnick, P., Varian, H.R.: Recommender systems. Commun. ACM **40**(3), 56–58 (1997)
34. Salah, A., Truong, Q.T., Lauw, H.W.: Cornac: a comparative framework for multimodal recommender systems. J. Mach. Learn. Res. **21**(1), 3803–3807 (2020)
35. Sanh, V., Debut, L., Chaumond, J., Wolf, T.: DistilBERT, a distilled version of BERT: smaller, faster, cheaper and lighter. arXiv preprint arXiv:1910.01108 (2019)
36. Savchenko, A., Alekseev, A., Kwon, S., Tutubalina, E., Myasnikov, E., Nikolenko, S.: Ad lingua: text classification improves symbolism prediction in image advertisements. In: Proceedings of the 28th International Conference on Computational Linguistics, pp. 1886–1892 (2020)
37. Savchenko, A.V.: User preference prediction in visual data on mobile devices. In: Proceedings of International Joint Conference on Neural Networks (IJCNN), pp. 1–7. IEEE (2021)
38. Savchenko, A.V.: Recommending restaurants based on classification of photos from the gallery of mobile device. In: Proceedings of the 20th Jubilee International Symposium on Intelligent Systems and Informatics (SISY), pp. 431–436. IEEE (2022)

39. Savchenko, A.V., Demochkin, K.V., Grechikhin, I.S.: Preference prediction based on a photo gallery analysis with scene recognition and object detection. Pattern Recogn. **121**, 108248 (2022)
40. Savchenko, A.V., Savchenko, L.V., Makarov, I.: Fast search of face recognition model for a mobile device based on neural architecture comparator. IEEE Access **11**, 65977 65990 (2023)
41. Savchenko, A.: Deep neural networks and maximum likelihood search for approximate nearest neighbor in video-based image recognition. Opt. Memory Neural Netw. **26**, 129–136 (2017)
42. Savchenko, A., Khokhlova, Y.I.: About neural-network algorithms application in viseme classification problem with face video in audiovisual speech recognition systems. Opt. Memory Neural Netw. **23**, 34–42 (2014)
43. Savchenko, A., Savchenko, L.: Three-way classification for sequences of observations. Inf. Sci., 119540 (2023)
44. Savchenko, V.V., Savchenko, A.V.: Criterion of significance level for selection of order of spectral estimation of entropy maximum. Radioelectron. Commun. Syst. **62**(5), 223–231 (2019)
45. Schafer, J.B., Konstan, J., Riedl, J.: Recommender systems in e-commerce. In: Proceedings of the 1st ACM Conference on Electronic Commerce, pp. 158–166 (1999)
46. Sharma, K., Giannakos, M.: Multimodal data capabilities for learning: What can multimodal data tell us about learning? Br. J. Edu. Technol. **51**(5), 1450–1484 (2020)
47. Sharma, M., Mann, S.: A survey of recommender systems: approaches and limitations. Int. J. Innov. Eng. Technol. **2**(2), 8–14 (2013)
48. Smith, B., Linden, G.: Two decades of recommender systems at amazon. com. IEEE Internet Comput. **21**(3), 12–18 (2017)
49. Tikhomirova, K., Makarov, I.: Community detection based on the nodes role in a network: the telegram platform case. In: van der Aalst, W.M.P. (ed.) AIST 2020. LNCS, vol. 12602, pp. 294–302. Springer, Cham (2021). https://doi.org/10.1007/978-3-030-72610-2_22
50. Truong, Q.T., Salah, A., Lauw, H.: Multi-modal recommender systems: hands-on exploration. In: Proceedings of the 15th ACM Conference on Recommender Systems, pp. 834–837 (2021)
51. Walinder, L., Price, M., Lim, B., Smith, B.: Multimodal personalized recommender algorithm based on knowledge graph (2022)
52. Wirojwatanakul, P., Wangperawong, A.: Multi-label product categorization using multi-modal fusion models. arXiv preprint arXiv:1907.00420 (2019)
53. Yakovlev, K., et al.: Sinkhorn transformations for single-query postprocessing in text-video retrieval. In: Proceedings of the 46th International ACM SIGIR Conference on Research and Development in Information Retrieval, SIGIR 2023, pp. 2394–2398. Association for Computing Machinery, New York (2023). https://doi.org/10.1145/3539618.3592064
54. Zhang, W., Qin, J., Guo, W., Tang, R., He, X.: Deep learning for click-through rate estimation. arXiv preprint arXiv:2104.10584 (2021)

Date-Driven Approach for Identifying State of Hemodialysis Fistulas: Entropy-Complexity and Formal Concept Analysis

Vasilii A. Gromov[1]([⊠]) [iD], Ekaterina I. Zvorykina[1] [iD], Yurii N. Beschastnov[1] [iD], and Majid Sohrabi[1,2] [iD]

[1] School of Data Analysis and Artificial Intelligence, Faculty of Computer Science, HSE University, 109028 Moscow, Russian Federation
stroller@rambler.ru
[2] Laboratory for Models and Methods of Computational Pragmatics, HSE University, Moscow, Russian Federation
msohrabi@hse.ru

Abstract. The paper explores mathematical methods that differentiate regular and chaotic time series, specifically for identifying pathological fistulas. It proposes a noise-resistant method for classifying responding rows of normally and pathologically functioning fistulas. This approach is grounded in the hypothesis that laminar blood flow signifies normal function, while turbulent flow indicates pathology. The study explores two distinct methods for distinguishing chaotic from regular time series. The first method involves mapping the time series onto the entropy-complexity plane and subsequently comparing it to established clusters. The second method, introduced by the authors, constructs a concepts-objects graph using formal concept analysis. Both of these methods exhibit high efficiency in determining the state of the fistula.

Keywords: chaotic time series · clustering · medical diagnostics · entropy · complexity · arteriovenous fistula

1 Introduction

The escalating prevalence of chronic kidney disease (CKD) continues to grow annually and has already reached levels comparable to other socially significant conditions, such as hypertension, diabetes, obesity, and metabolic syndrome [1,2]. The conventional approach for patients requiring hemodialysis involves the establishment of permanent vascular access in the form of an arteriovenous fistula. However, this method is plagued by frequent occurrences of thrombosis, thereby jeopardizing patient safety. Currently, the global incidence of end-stage chronic kidney failure has risen to approximately 0.1% [2,3]. The advent of the SARS-CoV-2 pandemic has significantly impacted these statistics,

© The Author(s), under exclusive license to Springer Nature Switzerland AG 2024
D. I. Ignatov et al. (Eds.): AIST 2023, CCIS 1905, pp. 250–262, 2024.
https://doi.org/10.1007/978-3-031-67008-4_19

as acute kidney injury has emerged as one of the most common extrapulmonary manifestations of COVID-19, often associated with a considerable decline in patient prognosis and outcomes [1]. In a majority of cases [4], the sole recourse for preserving a patient's life lies in the diligent management of an arteriovenous fistula through hemodialysis procedures. Numerous studies underscore the advantages of native arteriovenous fistulas (AVF) over alternative approaches [5,6]. The timely detection of improper fistula function becomes challenging when reliant on patient self-monitoring, which has fueled an intensified interest in automated methods for identifying malfunctioning fistulas. Within this problem domain, the bruit series of the fistula [7] stands out as the most effective source of information. From a mathematical standpoint, this problem represents the resolution of an inverse problem employing some form of intelligent system [8–10]. In this context, the direct problem involves a complex system characterizing the dynamics of blood flow within the human body. This paper presents a methodology for automatically distinguishing between series corresponding to normally functioning fistulas and those exhibiting pathological behavior. The underlying hypothesis postulates laminar blood flow in normally functioning fistulas and turbulent flow in pathologically functioning ones [11–13], which can be attributed to stenosis and thrombosis of the blood vessels [14].

The rest of the paper is organized as follows. In the second section, we discuss related works. In the third section, we introduce the proposed algorithm in detail. In the fourth section, we analyze the results of algorithm evaluation on data. Finally, in the Conclusion section, we provide the conclusion and potential applications of the proposed methodology.

2 Related Works

The first documented instances of numerical analysis of AVF bruit date back to the mid-1980s and highlighted the correlation between sound amplitude variations and AVF functioning states. Subsequently, numerical analysis of phono-angiography signals has been primarily based on a two-class classification scheme, distinguishing AVF into "good" and "bad" categories, with only a few authors exploring a multi-class assessment approach, such as categorization into "good," "medium," and "bad" [15]. Recent strides have been made towards developing technical solutions for continuous AVF bruit monitoring and diagnostics [16,17]. Nevertheless, the demand for a cost-effective, automated diagnostic device for assessing AVF status remains unmet.

From a mathematical perspective, this task revolves around differentiating between regular and chaotic time series. Technically, implementing the methodology is straightforward, necessitating only the recording of fistula sounds with a mobile phone and their subsequent automatic processing using a dedicated application. Two methods for distinguishing chaotic series from regular ones have been considered. The first involves determining the series' position on the "entropy-complexity" plane and comparing it with identified clusters of values derived from the set of time series. The method proposed by the authors entails constructing an objects-concepts graph [18] within the framework of formal concept

analysis [19–21] based on the examined time series and comparing its characteristics with those of graphs definitively associated with specified types of time
series.

The methodology we propose hinges on distinguishing between chaotic and
regular time series. The problem at hand-differentiating between chaotic and
regular time series of blood flow in an arteriovenous fistula-possesses two features, one simplifying the task and the other complicating it. Firstly, during
the observation periods, the series are generated by the same dynamical system
with consistent parameter values [22–24]. Secondly, apart from the chaotic component inherent to the heartbeat and blood vessel motion, there exists a notable
random component introduced by measurement errors in the time series under
consideration.

One group of methods for distinguishing chaotic and regular series traces
its origins back to the classical Rosenstein method [25]. These methods rely on
estimating the rate of divergence of initially proximate segments of the series
and subsequently calculating the highest Lyapunov exponent, with positive values indicating randomness and negative values signifying regularity. Wernecke
introduces a binary test designed to differentiate between strong and partially
predictable chaos. Similar to the Lyapunov exponential procedure, this method
depends on the evolution of initially close trajectories over time [26]. The 0-
1 test, a contemporary tool for identifying chaos in time series, comes in two
forms: a regression test and a correlation test. The correlation test performs
exceptionally well with noise-free time series, but when noise contaminates the
data, a modified version of the original regression test minimizes sensitivity to
noise [27]. However, for real-world time series, these methods tend to exhibit a
slight bias towards chaos, often producing positive values even for entirely regular series [28,29]. Additionally, most methods in this category are sensitive to
noise. Another group of methods draws on the concept of ordinal structure and
the representation of a time series as a point on the entropy-complexity plane
[30,31]. This approach discerns between complex deterministic (chaotic), simple deterministic, and stochastic time series. It rests on the observation that,
unlike purely random series, deterministic series-no matter how complex and
chaotic-are characterized by numerous forbidden patterns [32,33]. These forbidden patterns are resilient to noise and decay at a rate contingent on the noise
level, making it feasible to establish a method for distinguishing deterministic
time series even in the presence of substantial noise [4,34].

Interestingly, the absence of ordinal patterns in deterministic dynamics can
indicate epileptic states [35]. Nonetheless, real-world time series typically exhibit
considerable noise levels, which curtail the applicability of this class in resolving the challenge of identifying pathological fistulas. In practical scenarios, these
methods should be complemented with approaches from other classes, as demonstrated in this paper. In the current paper, we made an attempt to address the
challenge of clustering a set of time series into a predefined number of clusters, precisely two clusters. Specialists who work with hemodialysis patients
have noted the need for categorizing the noises obtained from the fistula within

the scope of the examined subject area. Thus, this study considers the problem of clustering time series $T_i = (x_0, x_1, ..., x_N)$ into non-overlapping clusters $C_j : C_j \cap C_k = \emptyset, j \neq k$. The number of clusters is not predetermined. Another task is classification, which involves assigning a given series to one of the three categories mentioned above. This procedure is based on the results of the clustering performed.

3   Materials and Methods

Methods for Obtaining and Describing the Study Sample. During dialysis treatment, patients undergo the installation of an AVF, and the condition of the fistula is evaluated based on the continuous noise it produces. Changes in the fistula noise can indicate fistula thrombosis and the development of stenosis. The authors have developed a mobile phone application [36] that allows recording fistula sounds using an external microphone, along with an accumulating database of recordings [37]. The study involved 290 patients (131 women and 159 men) undergoing renal replacement therapy using peritoneal hemodialysis for chronic renal failure at the Dialysis Center of Surgut Regional Clinical Hospital and the Hemodialysis Department of the Samara State Medical University (Samara, Russia). The mean age of the patients was 62 (56.7 $\pm$ 5.3) years. The average duration of renal replacement therapy by renal hemodialysis was 43 (34.3 $\pm$ 8.7) months. The AVF bruit recordings were obtained from all patients for 20–30 s using the electronic stethoscope "Littmann Model 3200" at the auscultation point located 2 cm distal to the arteriovenous anastomosis. Each patient was auscultated 10 to 15 times at various times of the day before and after the hemodialysis session. In all patients, renal hemodialysis was performed three times a week through a native AVF formed in the distal third of the forearm. The research material consists of 678 records, collected from 290 patients on dialysis. The data collection and all experiments were performed in accordance with the relevant guidelines and regulations. All patients included in the study have been informed about its course and purpose, and they gave written informed consent to participate in the study. It's important to note that the information about fistula bruit for each patient was received already anonymized. The hemodialysis center doctors performed a clinical assessment of the fistulas, classifying 567 records as normal functioning, 50 as potentially problematic, and 61 as requiring imminent reconstruction. Therefore, the goal of the mathematical analysis of the spectrograms was to identify patterns that would allow for the categorization of recordings into the categories mentioned earlier.

Each patient had from one to five records, corresponding to various states of his or her fistula, recorded at different times. We select the last segment of the audio recording of the fistula bruit (5–10 s) for further analysis, thereby making a time series of 1000 observations. In the course of the research, we revealed that such a length was sufficient to solve the problem in question.

Methods for Analyzing Time Series. Calculation of the position on the "entropy-complexity" plane. In the work of O. A. Rosso et al. [29], a method is proposed to distinguish chaotic series from those generated by a simple deterministic system on one side, and from purely random series on the other. The approach involves computing two characteristics (entropy and complexity) for a given time series $T_i = \{x_i\}$. The position of these characteristics on the corresponding plane relative to the upper and lower theoretical bounds determines the type of the series. The algorithm is based on constructing a probability distribution for the given series. The concept of ordinal patterns is employed to construct this distribution [30]. For the constructed distribution, two characteristics are computed: entropy and complexity.

$$S[P] = -\Sigma_{i=1}^{n!} P_i.ln(P_i) \tag{1}$$

$$H[P] = \frac{S[P]}{S_{max}} \tag{2}$$

$$S_{max} = ln(n!) = S[P_e] \tag{3}$$

where P_e is the uniform distribution $P = \{P_i, i = 1, 2, ..., N\}$, (Entropy achieves its maximum at this very distribution.)

We will consider the sole limiting condition for P, representing the state of our system, as

$$\Sigma_{i=1}^{N} P_i = 1 \tag{4}$$

In the case when $S(P) = S_{min} = 0$, we can confidently predict which possible outcomes i with probability P_i will indeed occur. Thus, knowledge about the probability distribution is maximized in this case, while knowledge about the normal probability distribution P_e is minimized:

$$S_{max} = lnN = S(P_e), \tag{5}$$

where P_e is the uniform distribution and N is the number of degrees of freedom. The "entropy-complexity" method is the product of the information stored in the system and its disequilibrium $H(P)$, and is computed as follows:

$$C(P) = Q(P, P_e).H(P), \tag{6}$$

Here, Q is the Jenson-Shannon divergence between the considered distribution P and the uniform distribution P_e, which is calculated as follows:

$$Q(P, P_e) = Q_0.(S(\frac{P + P_e}{2}) - S(\frac{P}{2}) - S(\frac{P_e}{2})), \tag{7}$$

where Q_0 is the normalization constant equal to the reciprocal of the maximum possible value of $Q(P, P_e)$, computed by the formula:

$$Q_0 = -2\{(\frac{N+1}{N})ln(N+1) - 2ln(2N) + lnN\}^{-1} \tag{8}$$

It should be noted that the statistical complexity defined above is the product of two normalized entropies (Shannon entropy and Jenson-Shannon divergence). However, it is a non-trivial function of entropy since it depends on two different probability distributions, namely, the one corresponding to the system state P and the uniform distribution P_e. In the case of the analyzed series in this work, the complexity is always positive, $C > 0$. It is worth noting that the value of complexity varies depending on the nature of the observations and the size of the series used. Depending on the size of the series, each observation introduces a new set of available states with their corresponding probability distribution. The procedure described above maps each series to a point on the "entropy-complexity" plane, and the position of that point relative to the upper and lower theoretical bounds indicates the type of the observed series [29].

The Wishart Clustering Algorithm. To solve the given problem, it is necessary to perform clustering on the specified dataset. The primary requirement for selecting a clustering algorithm is the absence of prior information about the number of clusters [38]. Thus, in this study, the clustering method proposed by Wishart [39,40], modified by Lapko and Chentsov [41] for cluster vectors, was used. Some difficulties associated with applying the algorithm to time series are discussed in [42].

Construction of the "Objects-Concepts" Graph and Its Analysis. Within the framework of formal concept analysis [18], an alternative approach to distinguishing regular series from chaotic ones was proposed. It involves constructing an "objects-concepts" graph [19,20]. Such graphs comprise vertices of two types: vertices corresponding to the set of the objects, and vertices containing subsets of attributes that characterise these objects. In particular, community detection algorithms on these graphs make it possible to obtain not mere clusters, but clusters and sets of attributes that determine these clusters. Series with pronounced chaos yield graphs with an essentially more complex structure than those of graphs corresponding to regular time series. To evaluate the degree of structural complexity, we employed the following statistics: the average number of elements per cluster; and the share of non-cluster noise. To construct the clusters, we will use the asynchronous label propagation algorithm described earlier [21]. Figure 1b shows the result of applying this algorithm to the graph in Fig. 1a.

To assess the level of structural complexity of the graph, the following statistics were formulated:

- The number of clusters relative to the total number of elements.
- The proportion of non-cluster noise.
- In the course of a large-scale simulation, we found that the effective boundaries between regular and chaotic series are 10% for both the first statistics and the second. statistics.

4 Results

Figure 2a shows a typical 15-s sequence for a patient with a normally functioning fistula, while Fig. 2b shows a sequence for a patient with fistula dysfunction. For comparison, Fig. 2c presents a segment of the Lorenz series, which is a typical chaotic series.

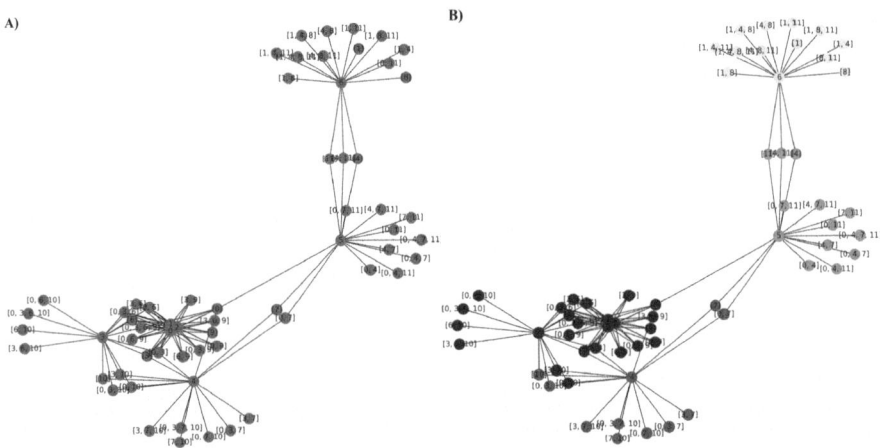

Fig. 1. Communities detected by the asynchronous label propagation (B) in the concepts-objects graph (A).

For each time series element in the examined dataset, the algorithm for constructing a point on the "entropy-complexity" plane was applied. Subsequently, the obtained dataset was subjected to the Wishart clustering algorithm, which does not require prior knowledge of the number of clusters and determines it autonomously during the clustering process (Fig. 3).

The clustering algorithm identified three clusters:

– The first one (green color in the figure) lies in the region of regular movements and, according to our hypothesis, corresponds to normally functioning fistulas.
– The second cluster (yellow color) is in the region of chaotic movements and movements close to chaos, corresponding to fistulas with dysfunction. It contains a small red subcluster, which represents samples with abnormal external noises.
– the third cluster (blue color) occupies an intermediate position and corresponds to uncertain patients. An alternative approach in analyzing the considered time series is the construction of an "objects-concepts" graph with the identification of clusters based on communities within the graph.

Figure 4a shows characteristic graphs corresponding to the case of a normally functioning fistula, while Fig. 4b shows the case of fistula dysfunction obtained

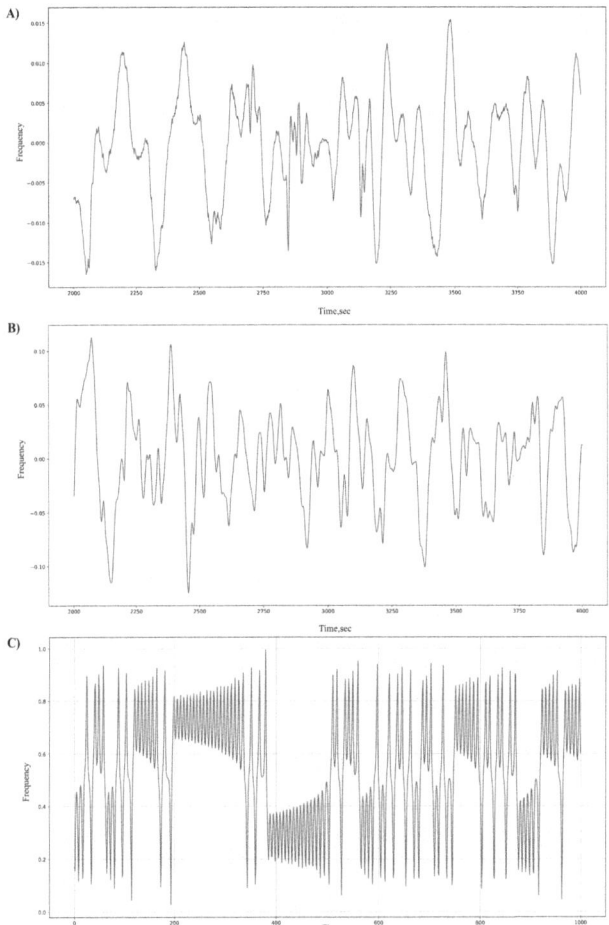

Fig. 2. 15-s recordings for a patient with a normally working fistula (A); with fistula dysfunction (B); sequence corresponding to the Lorenz series (C).

with the FCA method. To obtain integrated results using both the first and second approaches, a linear regression method was employed. Even without resorting to formal methods, we may observe that the graphs are distinctly different. For the first graph, the average number of elements per cluster amount is 0.091, the percentage of extra-cluster noise is 2.4%, maximal cluster size is 0.149. For the second graph, the average number of elements per cluster amount is 0.329, the extra-cluster noise is 28.7% and the maximal cluster size is 0.283 respectively.

The Wishart clustering algorithm was applied to the values of the metrics thus obtained. The objects-concepts algorithm distinguishes only two classes, corresponding to the states of - 'fistula works well', and 'there's a slight malfunction in the fistula'. In order to integrate results of the first and second

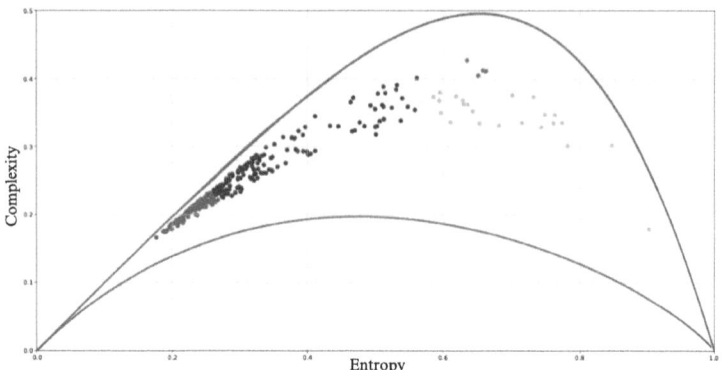

Fig. 3. Wishart algorithm clusterisation results.

approaches, we use linear regression. The integrated results appear to be more accurate in identifying fistula dysfunction than the methods alone. The integrated results were more accurate in determining fistula dysfunction compared to applying each method individually.

To verify the obtained results, a comparison was made with the solutions adopted by the attending physicians regarding the patients' data. The table presents the mentioned comparisons for the analysis conducted within the framework of the first and second approaches, as well as the integrated results.

We should stress that the diagnostic decisions of doctors (with which we compare our results) should not be considered as the ultimate truth: the overwhelming majority of cases of false positive and false negative diagnoses were related to situations when the doctors are not sure of their decision. They found the cases ambiguous and requiring additional analysis (Table 1).

5   Conclusions and Future Directions

The study employed a methodology to automatically distinguish between series representing normally and pathologically functioning arteriovenous fistulas (AVFs). This method relies on computing a pair of characteristics known as "entropy-complexity." Additionally, a novel approach was introduced, involving the creation of an objects-concepts graph for series classification and an analysis of its properties. Notably, these properties exhibited qualitative differences between regular and chaotic series. Both of these approaches, when used individually or in combination, were applied to assess the condition of AVFs and demonstrated high diagnostic effectiveness.

A promising avenue for future research in this field, particularly relevant to healthcare practitioners, could involve a thorough exploration of the structure of the objects-concepts graph to identify characteristic substructures associated with various types of fistula dysfunction.

A) B)

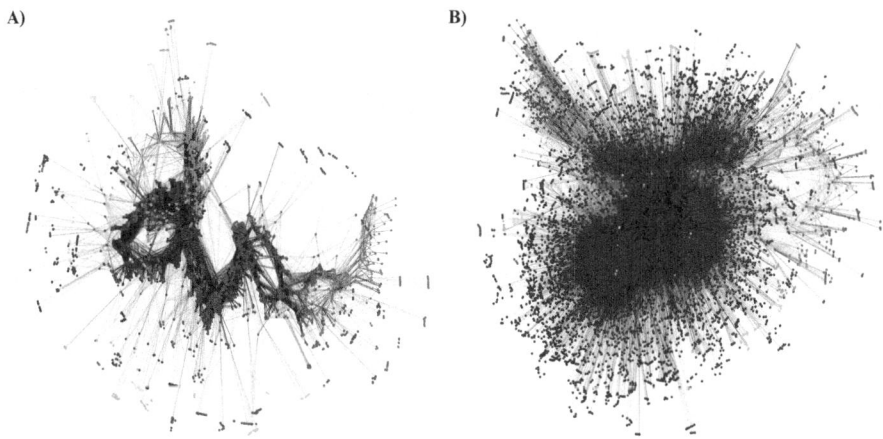

Fig. 4. (A)-Characteristic graph corresponding to a case of a normally functioning fistula (number of clusters - 45, proportion of noise - 0.024). (B)-Characteristic graph corresponding to a case of fistula dysfunction (number of clusters - 163, proportion of noise - 0.287).

Table 1. Relative efficiency of the approaches on the sample used (n = 290).

Method	True Positive	True Negative	False Positive	False Negative
Entropy-Complexity (Wishart): Precision = 0.76 Recall = 0.81 F-measure = 0.878 Accuracy = 72.3	130	90	40	30
objects-concepts graphs: Precision = 0.76 Recall = 0.78 F-measure = 0.77 Accuracy = 72.3	140	100	30	20
Linear regression based on both algorithm: Precision = 0.90 Recall = 0.77 F-measure = 0.83 Accuracy = 84.50	170	70	20	30

In previous research, various classification methods were employed to categorize arteriovenous fistulas (AVFs) into distinct subclasses. Some authors utilized acoustic frequency domain analysis, employing the Fast Fourier Transform (FFT) and subsequent clustering [15,43], resulting in the introduction of six subclasses. Simultaneously, prior studies introduced a deep learning method that defined five AVF subclasses [16,44,45]. However, these methods were found to be sensitive to the quality of the recorded acoustic data.

In contrast, our proposed method exhibits reduced dependence on noise extraction from the records. The first algorithm reveals three clusters, corresponding to healthy, malfunctioning, and intermediate fistulas. The second algorithm distinguishes four classes, including bad and extremely bad among malfunctioning fistulas. The Wishart clusterization method, which we employ, does not impose constraints on the number of clusters, allowing us to explore beyond the originally defined three states of the fistula. Moreover, we have observed a significant correlation between the classification results obtained using our methods and the outcomes of traditional diagnostic tests commonly employed to evaluate AVF function, such as Doppler ultrasound.

It is crucial to acknowledge the limitations of our study, including the relatively small sample size of patients studied and the subjective classification of AVFs into functional categories, based on the expertise of clinicians in the dialysis unit.

As a possible direction for future research, we may indicate that the structure of the concepts-objects graph allows identifying typical substructures corresponding to various variants of fistula dysfunction, which is important from the point of view of practitioners. We also plan to design algorithms for the early diagnosis of fistula health, which will assess the chances of a healthy functioning fistula within the first or second day after the operation.

Acknowledgements. This paper is an output of a research project implemented as part of the Basic Research Program at the National Research University Higher School of Economics (HSE University).

References

1. Chan, L., et al.: AKI in hospitalized patients with COVID-19. J. Am. Soc. Nephrol. **32**(1), 151–160 (2021)
2. Hill, N.R., et al.: Global prevalence of chronic kidney disease: a systematic review and meta-analysis. PLoS ONE **11**(7), 1–18 (2016)
3. Liyanage, T., et al.: Worldwide access to treatment for end-stage kidney disease: a systematic review. The Lancet **385**(9981), 1975–1982 (2015)
4. Burkhart, H.M., Cikrit, D.F.: Arteriovenous fistulae for hemodialysis. Semin. Vasc. Surg. **10**(3), 162–165 (1997). PMID: 9304733
5. Hasuike, Y., et al.: Imbalance of coagulation and fibrinolysis can predict vascular access failure in patients on hemodialysis after vascular access intervention. J. Vasc. Surg. **69**(1), 174-180.e2 (2019)
6. Ravani, P., et al.: Examining the association between hemodialysis access type and mortality: the role of access complications. Clin. J. Am. Soc. Nephrol. **12**(6), 955–964 (2017)
7. Salman, L., Beathard, G.: Interventional nephrology: physical examination as a tool for surveillance for the hemodialysis arteriovenous access. Clin. J. Am. Soc. Nephrol. **8**, 1220–1227 (2013). e onese ones
8. Sato, T.: New diagnostic method according to the acoustic analysis of the shunt blood vessel noise. Toin Univ. Yokohama Eng. Jpn. Soc. Dial. Ther. J. **2**, 332–341 (2005)

9. Kokorozashi, N.: Analysis of the shunt sound frequency characteristic changes associated with shunt stenosis. Jpn. Soc. Dial. Ther. J. **3**, 287–295 (2010)
10. Todo, A., Kadonaka, T., Yoshioka, M., Ueno, A., Mitani, M., Katsurao, H.: Frequency analysis of shunt sounds in the arteriovenous fistula on hemodialysis patients. In: Proceedings of the 6th International Conference on Soft Computing and Intelligent Systems, and the 13th International Symposium on Advanced Intelligence Systems (2012)
11. Remuzzi, A., Ene-Iordache, B.: Novel paradigms for dialysis vascular access: upstream hemodynamics and vascular remodeling in dialysis access stenosis. Clin. J. Am. Soc. Nephrol. **8**, 2186–2193 (2013)
12. Brahmbhatt, A., Remuzzi, A., Franzoni, M., Misra, S.: The molecular mechanisms of hemodialysis vascular access failure. Kidney Int. **89**, 303–316 (2016)
13. Badero, O.J., Salifu, M.O., Wasse, H., Work, J.: Frequency of swing-segment stenosis in referred dialysis patients with angiographically documented lesions. Am. J. Kidney Dis. **51**, 93–98 (2008)
14. Lee, T., Barker, J., Allon, M.: Needle infiltration of arteriovenous fistulae in hemodialysis: risk factors and consequences. Am. J. Kidney Dis. **47**, 1020–1026 (2006)
15. Du, Y.-C., Stephanus, A.: A novel classification technique of AVF stenosis evaluation using bilateral PPG analysis. Micromachines **7**, 147 (2016)
16. Grochowina, M., Leniowska, L.,x Gala-Błądzińska, L., The prototype device for noninvasive diagnosis of AVF condition using machine learning methods. Sci. Rep. **10**, 16387 (2020)
17. Lopes, I., Sousa, F., Moreira, E., Cardoso, J.: Smartphone-based remote monitoring solution for heart failure patients. Stud. Health Technol. Inform. **261**, 109–114 (2019)
18. Raghavan, U., Nandini, R.A., Soundar, K.: Near linear time algorithm to detect community structures in large-scale networks. Phys. Rev. E **76**(3), 036106 (2007)
19. Navarro, E., Prade, H., Gaume, B.: Clustering sets of objects using concepts-objects bipartite graphs. In: Húllermeier, E., Link, S., Fober, T., Seeger, B. (eds.) SUM 2012. LNCS (LNAI), vol. 7520, pp. 420–432. Springer, Heidelberg (2012). https://doi.org/10.1007/978-3-642-33362-0_32
20. Buzmakov, A., Egho, E., Jay, N., Kuznetsov, S., Napoli, A., Raïssi, C.: On mining complex sequential data by means of FCA and pattern structures. Int. J. Gener. Syst. **45**, 135–159 (2015)
21. Gromov, V.A., Lukyanchenko, P.P., Beschastnov, Y.N., Tomashchuk, K.K.: Time Ser. Struct. Anal. Number Law Cases. Proc. Cybernet. **4**(48), 37–48 (2022)
22. Le Guen, V., Thome, N.: Probabilistic time series forecasting with shape and temporal diversity. In: NeurIPS (2020)
23. Liu, P., Mahmood, T., Ali, Z.: The cross-entropy and improved distance measureas for complex q-rung orthopair hesitant fuzzy sets and their applications in multicriteria decision-making. Complex Intell. Syst. **8**, 1167–1186 (2022). https://doi.org/10.1007/s40747-021-00551-2
24. Sangma, J.W., Sarkar, M., Pal, V., et al.: Hierarchical clustering for multiple nominal data streams with evolving behaviour. Complex Intell. Syst. **8**, 1737–1761 (2022). https://doi.org/10.1007/s40747-021-00634-0
25. Rosenstein, M.T., Collins, J.J., De Luca, C.J.: Reconstruction expansion as a geometry-based framework for choosing proper delay times. Physica D **73**, 82–98 (1994)
26. Gottwald, G.A., Melbourne, I.: A new test for chaos in deterministic systems. Proc. R. Soc. Lond. Ser. A Math. Phys. Eng. Sci. **460**, 603–611 (2004)

27. Gottwald, G.A., Falconer, I.S., Wormnes, K.: Application of the 0–1 test for chaos to experimental data. SIAM J. Appl. Dyn. Syst. **6**(2), 395–402 (2007)

28. Gottwald, G.A., Melbourne, I.: The 0-1 test for chaos: a review. In: Skokos, C.H., Gottwald, G.A., Laskar, J. (eds.) Chaos Detection and Predictability. LNP, vol. 915, pp. 221–247. Springer, Heidelberg (2016). https://doi.org/10.1007/978-3-662-48410-4_7

29. Rosso, O.A., Carpi, L.C., Saco, P.M., Gómez Ravetti, M., Plastino, A., Larrondo, H.A.: Causality and the entropy-complexity plane: robustness and missing ordinal patterns. Physica A Stat. Mech. Appl. **391**(1), 42–55 (2012)

30. Bandt, C., Pompe, B.: Permutation entropy: a natural complexity measure for time series. Phys. Rev. Lett. **88**, 174102 (2002). https://doi.org/10.1103/PhysRevLett.88.174102

31. Zanin, M.: Forbidden patterns in financial time series. Chaos **18**(1), 013119 (2008). https://doi.org/10.1063/1.2841197. PMID: 18377070

32. Zunino, L., Zanin, M., Tabak, B.M., Pérez, D., Rosso, O.A.: Forbidden patterns, permutation entropy and stock market inefficiency. Phys. A **388**, 2854–2864 (2009)

33. Benettin, G., Galgani, L., Giorgilli, A., et al.: Lyapunov Characteristic Exponents for smooth dynamical systems and for hamiltonian systems; a method for computing all of them. Part 1: Theory. Meccanica **15**, 9–20 (1980). https://doi.org/10.1007/BF02128236

34. Wolf, A., Swift, J.B., Swinney, H.L., Vastano, JA.: Determining Lyapunov exponents from a time series. Physica D Nonlinear Phenomena **16**(3), 285–317 (1985). https://doi.org/10.1016/0167-2789(85)90011-9. ISSN 0167-2789

35. Ouyang, G., Li, X., Dang, C., Richards, D.A.: Deterministic dynamics of neural activity during absence seizures in rats. Phys. Rev. E **79**, 041146 (2009)

36. EA-56137: Application for registration of a database "Registry of data on the condition of vascular access in patients undergoing hemodialysis."

37. EA-56151: Application for registration of a computer program "Mobile application for collection, processing, and storage of data in the registry for the classification of vascular access status for hemodialysis"

38. Aggarwal, C.C., Reddy, C.K.: Data Clustering: Algorithms and Applications, 1st edn. Chapman and Hall/CRC, New York (2014)

39. Wishart, D.: A numerical classification methods for deriving natural classes. Nature **221**, 97–98 (1969)

40. Thrun, M.C., Ultsch, A.: Using projection-based clustering to find distance- and density-based clusters in high-dimensional data. J. Classif. **38**, 280–312 (2020)

41. Lapko, A.V., Chentsov, S.V.: Nonparametric Information Processing Systems. Nauka (2000)

42. Gromov, V.A., Borisenko, E.A.: Predictive clustering on non-successive observations for multi-step ahead chaotic time series prediction. Neural Comput. Appl. **2**, 1827–1838 (2015)

43. Malindretos, P., Liaskos, C., Bamidis, P., Chryssogonidis, I., Lasaridis, A., Nikolaidis, P.: Computer-assisted sound analysis of AVF in hemodialysis patients. Int. J. Artif. Organs **37** (2013). https://doi.org/10.5301/ijao.5000262

44. Ota, K., Nishiura, Y., Ishihara, S., Adachi, H., Yamamoto, T., Hamano, T.: Evaluation of hemodialysis arteriovenous bruit by deep learning. Sensors (Basel, Switzerland) **20**(17), 4852 (2020)

45. Kordzadeh, A., Esfahlani, S.S.: The role of artificial intelligence in the prediction of functional maturation of AVF. Ann. Vasc. Dis. **12**(1), 44–49 (2019). https://doi.org/10.3400/avd.oa.18-00129

Network Analysis

Approximate Density Computation for OA-Biclustering

Dmitry I. Ignatov[1]($\boxtimes$), Daria Komissarova[1], Kamila Usmanova[1],
Stefan Nikolić[1], Andrey Khrunin[2], and Gennady Khvorykh[2]

[1] National Research University Higher School of Economics, Moscow, Russia
dignatov@hse.ru
[2] National Research Centre "Kurchatov Institute", Moscow, Russia
gennady_khvorykh@gmail.com

Abstract. In this paper, we introduce an approach for approximate density computation of object-attribute biclusters (OA-biclusters) given in the form (m', g'), where $(.)'$ stands for Galois operators from Formal Concept Analysis. They serve as an approximate variant of formal concepts since not all object-attribute pairs for the original incidence relation are present in $m' \times g'$. Direct density computation of an OA-bicluster takes $O(|g'||m'|)$, which is laborious for large biclusters, while its high density serves as a high-quality guarantee (in the sense of interestingness measure). Our approach relies on the Monte Carlo sampling strategy and concentration inequalities known in statistics and machine learning. It is tested on different datasets including Southern Women from Social Networks Analysis domain, the Zoo dataset from the UCI ML repository, custom artificial genotype-phenotype data from genome association study, and web advertising data donated by Yahoo (formerly US Overture). The results of the experimental comparison show a reasonably good quality of approximation in terms of average error and demonstrate the possible applications of the approach in different domains.

Keywords: Formal Concept Analysis · Biclustering · Approximate Density · Monte Carlo approach

1 Introduction

One of the simple and natural representations of the structure of a complex system (in economics, biology, sociology, etc.) is a presentation of information in the form of a network [1]. The tendency of nodes to form communities or groups is a distinctive feature of most networks [2]; therefore, objects in a certain group will be similar to others under a certain criterion or feature, while at the same time, nodes in different communities will be different.

The identification of communities holds considerable interest in various fields. For instance, it can be used to identify groups of people with similar interests on social networks or with similar genetic variants, find web pages on related

---

The original version of the chapter has been revised. An affiliation had not been displayed correctly. This has been corrected. A correction to this chapter can be found at https://doi.org/10.1007/978-3-031-67008-4_23

© The Author(s), under exclusive license to Springer Nature Switzerland AG 2024, corrected publication 2024
D. I. Ignatov et al. (Eds.): AIST 2023, CCIS 1905, pp. 265–283, 2024.
https://doi.org/10.1007/978-3-031-67008-4_20

topics on the global Internet, and recognize typical representatives in biological networks.

In the case of social network analysis, methods for the analysis of single-mode networks (friend-to-friend), 2-mode networks, 3-mode networks and even multimodal dynamic networks are known [3,4]. The definition of multimodal networks is given as follows [5]: these are networks in which actors can be connected with other types of objects along the edges, for example, the relationship between users and their interests in the 2-mode case, or a hyperedge, which can be represented as the connection between users, tags and resources in the three-mode case [3,6]. Such networks are also called heterogeneous because the network involves different types of nodes. Various studies have been carried out in the field of 1-mode networks, where the data represent a certain set of objects that have connections within this set. For example, a social network for dating, where its members can have connections with some people, but do not know others.

By the 1980 s, the notion of clique required revisiting as the main tool for searching social communities due to its rather rigid nature (all vertices should be connected) and computational complexity. Thus, for example, the notion of k-plex was proposed, where k edges could be missing for a particular vertex (member) of the considered community [7]. Later on, the study of two-mode communities led to the notion of (k, l)-biplex [8] or k-biplex when $k = l$ [9].

Formal concepts, being analogues of maximal bicliques, and their lattices were extensively used for social network analysis of two-mode networks [10–12]. However, their usage is only possible for rather small networks both due to high computational complexity and an enormous number of possible communities. One way to cope with the difficulty is to use size constraints like in the so-called iceberg lattices, while the other is to propose suitable approximations where some incidence pairs are missing like in the case of k-biplexes. Among four possible candidates, approximate boxes [13], biplexes [9], fault-tolerant concepts [14], and dense OA-biclusters [15], we set our choice on the latter ones due to their simple definition and almost linear computational complexity (if we omit density constraint). The density computation time for a bicluster (X, Y) is proportional to its size $|X||Y|$, which is expensive for large biclusters.

The practical part of this work consists in applying approximate computation of OA-biclusters' density based on Chernoff-Hoefding inequality for various two-mode data sets: data about visits to various events by a group of women (Davis's Southern Club Women network), data with attributes of animals in a zoo (Zoo database) along with their classification to different classes, genotypes of human individuals over a range of genomic loci with classification of individuals into case and control groups, and data about advertising phrases on the US segment of Internet (Overture [16]). The main goal of the experimental part is to show the approximate density for a large bicluster being estimated in a reasonable time within acceptable error with the use of the proposed approach. Additionally, we outline some possible applications of our approach in different domains.

The paper is organized as follows. In Sect. 2, we discuss the notions of maximal biclique, formal concept, and OA-biclusters. Section 3 describes algorithms for dense OA-biclusters generation, while Sect. 4 describes an algorithm for the approximate density computation with a given accuracy in the case of large networks and biclusters. In Sect. 5, the problem of context coverage with various sets of biclusters under a given accuracy is analysed. Section 6 summarises our experiments. Section 7.

2 Basic Notions for Two-Mode Network Analysis

2.1 Basic Definition of FCA

The definition of a formal context [17] is as follows: It is a triple $\mathbb{K} = (G, M, I)$, where G is a set of objects, M set of attributes, and an incidence relation $I \subseteq G \times M$ shows which objects have which attributes.

For any $A \subseteq G$ and $B \subseteq M$ we can define the concept-forming or Galois operators as follows:

$$A' = \{m \mid gIm \text{ for all } g \in A\} \text{ and } B' = \{g \mid gIm \text{ for all } m \in B\}.$$

The operator $(\cdot)''$ is a closure operator (idempotent, monotone, extensive). A set of objects $A \subseteq G$ is called closed, if $A'' = A$. Similarly, for a set of attributes.

In the case of two-mode networks, formal contexts correspond to a bipartite graph having connections between two sets G, and M.

A formal concept of $\mathbb{K} = (G, M, I)$ is a pair (A, B):

$$A \subseteq G, B \subseteq M, A' = B \text{ and } B' = A.$$

In this case, the sets A and B are closed and are called the (formal) extent and the (formal) intent of the concept (A, B).

Thus, for a certain set of objects A, the common set of attributes A' will describe the objects in the set A, and the closure A'' is some community consisting of similar objects. Ordered by extent (dually, by intent) inclusion, formal concepts of a given context form its concept lattice. In the worst case, the number of formal concepts for the formal context will be $2^{\min(|G|,|M|)}$.

Searching for all formal concepts corresponds to that of maximal bicliques, a standard approach to community in a two-mode network.

Since we potentially have a rather large number of found "communities" (in the worst case, an exponential number of the size of the input data) and not necessarily all connections between objects have to be present in the real community, we need some "relaxation" of the formal concept conditions as an analogue of the notion of community in a two-mode network.

2.2 Basic Definition of Bicluster and Its Properties

Another approach in the detection of communities in a bimodal network could be the search of OA-biclusters (O stands for objects, A for attributes). Where a bicluster in a general sense, is understood, as a pair of sets which contains elements from the set of objects and the set of attributes, respectively. For the

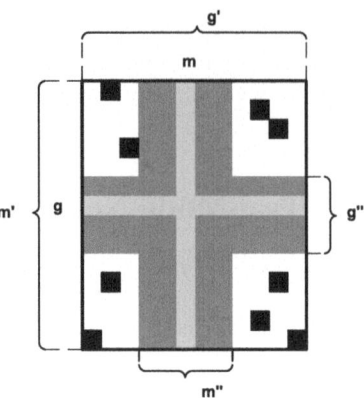

Fig. 1. OA-bicluster

formal context, $\mathbb{K} = (G, M, I)$ and any pair of object and attribute concepts $(g'', g') \in G$ and $(m', m'') \in M$, which are ordered as follows $(g'', g') \le (m', m'')$, a pair (m', g') is called an OA-bicluster [15].

The density of a bicluster (A, B) of the formal context $\mathbb{K} = (G, M, I)$ can be defined as the value $\rho(A, B)$, where

$$\rho(A, B) = \frac{|I \cap (A \times B)|}{|A||B|}.$$

Thus, for any bicluster, we have $0 \le \rho(A, B) \le 1$.

We should also note that any formal concept (A, B) has a density $\rho(A, B) = 1$. Let us take an OA-bicluster $(A, B) \in 2^G \times 2^M$ and a positive integer number $\rho_{min} \in [0, 1]$. Then the OA-bicluster is dense iff its density $\rho(A, B) \ge \rho_{min}$.

The main properties of OA-biclusters:

1. For all $(m', g') \subseteq 2^G \times 2^M$ we have $\frac{|m'| + |g'| - 1}{|g'||m'|} \le \rho(A, B) \le 1$;
2. OA-bicluster (m', g') is a formal concept iff $\rho(A, B) = 1$;
3. If (m', g') is a bicluster, then $(g'', g') \le (m', m'')$.

These properties show that the main difference between OA-biclusters and formal concepts is that the former does not necessarily have a density of 1. Geometrically, this means that not all OA-bicluster cells must be nonempty (see Fig. 1). The rectangle in Fig. 1 depicts the bicluster (m', g'). The horizontal grey line corresponds to the object g and contains only non-empty cells. The vertical grey line corresponds to the attribute m, and also contains only non-empty cells. White areas indicate empty cells and black areas indicate non-empty cells. In this figure, the formal concept will cover only the green and grey areas, the bicluster also covers white and black cells in the bounding rectangle $m' \times g'$.

3 Search Algorithm for Biclusters, Complexity Assessment of Algorithm

We can potentially consider two algorithms for OA-biclustering: the original algorithm [15] and its online version [18].

The original algorithm for finding OA-biclusters is as follows: it forms the bicluster (m', g') for each pair of $(g, m) \in I$. The prime operators are computed previously. Thus, for each object-attribute pair of an input context, we check that the generated bicluster has not been found previously. This can be done with the use of a hash function.

In the online version of the algorithm, new pairs come in a stream. Therefore, computing prime operators in advance is no longer possible and needed. Instead, they are updated on the fly with the help of pointers or references.

Input: J is a current collection of pairs (g, m), Υ is a current set of biclusters
Output: $\Upsilon = \{(X, Y)\}$
$PrimesObj := []$
$PrimesAttr := []$
for all $(g, m) \in J$ **do**
$\quad PrimesObj[g] := PrimesObj[g] \cup m$
$\quad PrimesAttr[m] := PrimesAttr[m] \cup g$
$\quad \Upsilon := \Upsilon \cup (\&PrimesObj[g], \&PrimesAttr[m])$
end for

The main idea of this algorithm is that, initially, there is no information about the complete data set, i.e. (G, M, I). Each pair $(g, m) \in I$ is processed sequentially.

This algorithm has some advantages:

1. The algorithm works fast enough in the case of sparse data, processing only available pairs $(g, m) \in I$.
2. It requires less memory because there is no need to store the entire incidence matrix of size $|G||M|$; the context is stored as a set of pairs.
3. If you need to analyse the context obtained from the previously existing one by the addition of new objects, attributes or their pairs, then the online version of the algorithm will allow supplementing the previously processed data set, without relaunching the algorithm.
4. It is insensitive to missing pairs in context (they may come next).

The algorithm runs in one pass. We can implement operation $(\cdot)'$ without information about the actual number of objects and attributes in the context by using references and pointers. Thus, the finding of all biclusters in the context takes $O(|I|)$ time. However, the following step of filtering dense biclusters takes $O(|\Upsilon||G||M|)$ in the worst case and our goal is to improve their selection via approximate density computation with probabilistic guarantees.

4   Approximate Density Estimation for the Large Biclusters

In general, we want to find an approximate density $\tilde{\rho}$ such that, for given parameters δ and ϵ, the probability of its discrepancy from the true value is bounded as following $P(|\rho - \tilde{\rho}| \le \epsilon) \ge \delta$.

For this purpose, we can use Chernoff-Hoefding's inequality [19]:

$$P(|X - E[X]| \le \epsilon) \ge 1 - -2e^{-2n\epsilon^2}.$$

If ϵ and δ are given, then we can express n as follows:

$$n = \frac{\ln \frac{2}{1-\delta}}{2\epsilon^2}.$$

In this case, n is equal to a threshold value of the bicluster size, above which it makes sense to count its density approximately because of its large size; n also equals the number of tests of n pairs from this bicluster for being in I. (See Fig. 2) for the behaviour of $n(\delta, \epsilon)$; it is unbounded in $(1,0)$, i.e. $\lim_{\substack{\delta \to 1- \\ \epsilon \to 0+}} n(\delta, \epsilon) \to +\infty$.

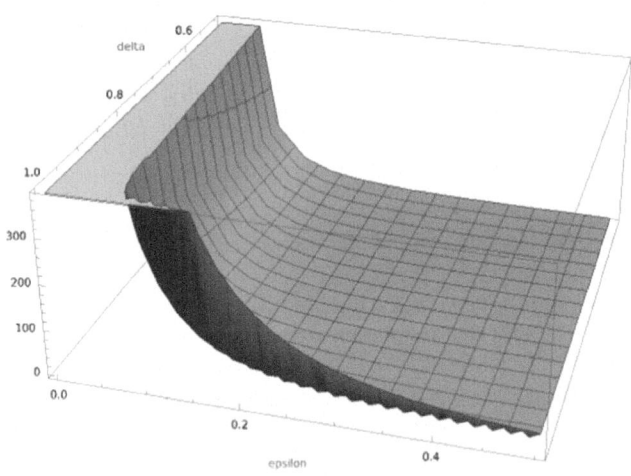

Fig. 2. n as a function of $\epsilon \in (0, 0.5]$ and $\delta \in [0.5, 1)$

However, for example, with $\epsilon = 0.3$ and $\delta = 0.7$, we obtain $n \approx 10$.

It is necessary to randomly select 10 pairs of elements $(k, l) \in A \times B$ without repetition in the modified subrelation $A \times B = m' \times g' \setminus (g \times g' \cup m' \times m)$. Among ten randomly selected pairs (highlighted in orange) we get 3 non-empty ones (Fig. 3).

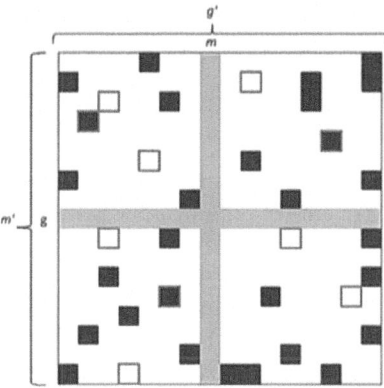

Fig. 3. Example of the approximate density calculation of a bicluster

Then the density of bicluster (m', g') for this example can be approximately computed as:

$$\rho(m', g') \approx \frac{|g'| + |m'| - 1}{|g'||m'|} + 0.3 \cdot \frac{(|g'| - 1)(|m'| - 1)}{|g'||m'|},$$

because we chose 10 cells not from the whole bicluster, but only from the rest part of size $(|g'| - 1)(|m'| - 1)$.

This method allows us to compute the approximate density of a bicluster with the given error ϵ and confidence δ parameters within $O(\min(|m'||g'|, n(\epsilon, \delta)))$.

5   Search for the Optimal Set of Biclusters in the Context Coverage Problem

The set of OA-biclusters for a formal context contains $|I|$ members (with duplicates) and can be large. However, in practice, we do not need them all: 1) this set most likely includes highly sparse biclusters, which are very difficult to interpret as an analogue of a community; 2) a large bicluster collection is difficult to analyze by human experts.

All this leads us to the following selection criteria:

1. biclusters should be reasonably dense (it is necessary for their interpretation in the community);
2. their size in terms of the number of objects and attributes should be large enough, because a bicluster containing only a single attribute or object is not appealing enough for research as a community (e.g., a group of people with only one common interest).

The given conditions above lead us to the problem of good coverage of the context (almost every pair of I is at least inside one selected bicluster) by choosing the most interesting set of biclusters (with respect to the selected criteria).

One of such criteria is proposed in [13] and deduced from the least square minimization principle: $f(X, Y) = \rho(X, Y)^2 |X||Y|$ where (X, Y) is a bicluster.

5.1 A Greedy Approach to Cover a Context

The main idea here is to cover a given fraction of object-attribute pairs in a given formal context (for example, 80% or 90%) with possibly few biclusters. To solve this problem, we consider a greedy search.

This algorithm allows us to solve several tasks: 1) we can select high-density biclusters and 2) simultaneously get rid of excessive "overlapping" biclusters.

First, we select a bicluster from the input set of biclusters that cover the maximum number of pairs $(g, m) \in I$. Then in the remaining biclusters, the "covered" part of incidence pairs will be excluded from consideration, the current bicluster is removed from the list of potential biclusters and the loop will continue. The main loop runs until the desired threshold for the share of coverage of the entire formal context is reached.

Input: $\Upsilon = \{(X, Y) \mid \rho(X, Y) \geq \rho_{min}\}, cover_{min}$ is a minimal data coverage
Output: $\tilde{\Upsilon}$ where $\tilde{\Upsilon} \subseteq \Upsilon, cover(\tilde{\Upsilon}) \geq cover_{min}$

while $cover(\Upsilon') \geq cover_{min}$ **do:**
   $\Sigma := sortMax_{cover(X,Y)}(\Upsilon)$
   $(X, Y) := \Sigma[1]$
   $\tilde{\Upsilon} := \tilde{\Upsilon} \cup \{(X, Y)\}$
   $\Upsilon := \Upsilon \setminus \{(X, Y)\}$
   for (U, V) **in** Υ **do:**
      $\Upsilon := \Upsilon \setminus \{(U, V)\} \cup \{(U, V) \setminus (X, Y)\}$
   end for
end while

6 Experiments

6.1 Description of Data

Our experiments are done with two-mode networks; the first three of them are summarised in Table 1.

a. Zoo dataset The dataset contains information about 101 animals, their characteristics, also their belonging to one out of the 7 classes of animals (mammals, birds, reptiles, etc.). Each of the representatives is described by 14 attributes: hair, feathers, the number of legs (scaled nominally), etc.

b. Internet advertising dataset The network describes companies that bought advertising terms to promote their goods or services via the sponsored search in Yahoo. Communities detection in this set is determined by the need to determine recommendations for market participants to purchase some advertising terms by identifying typical groups of competitors (Table 1).

Table 1. Datasets statistics and the processing time to obtain all their formal concepts and biclusters.

Name of dataset	$|I|$	$|G|$	$|M|$	# concepts	Computing biclusters, s	Computing formal concepts, s
Southern women	93	18	14	65	$1 \cdot 10^{-3}$	$8 \cdot 10^{-3}$
Zoo dataset	761	101	21	365	$2 \cdot 10^{-2}$	$11 \cdot 10^{-2}$
Advertising dataset	92 345	3000	2000	8 950 740	1.6 s	

c. Southern women dataset This dataset is fairly popular for identifying communities in social networks analysis [20]. It was collected in 1930 by a group of researchers, and describes the social activity of a group of 18 women during 9 months, each of whom attended a specific set of 14 social events (like daytime work at the store counter, women's club meeting, church supper, etc.).

6.2   Approximation of Formal Concepts by Biclusters

Computing all formal concepts for (the contexts of) large networks is a laborious task. For example, in the case of a contextual advertising dataset, counting a set of formal concepts would take more than a day. Even for rather small datasets, "Southern women" and "Zoo", the generation time of biclusters and that of formal concepts significantly differs.

We can impose constraints such as a minimum density threshold to ultimately reduce the total number of selected biclusters. Our experiments show that OA-biclusters cover almost all formal concepts under a reasonable density threshold. However, with an increase in the minimum density, we will lose part of the biclusters, which could cover some of the concepts. Let us consider the dynamics of changes in the number of biclusters for various density thresholds, and the fraction of covered formal concepts, and also evaluate the quality of the considered bicluster sets using the quality measures. The definitions of the following quality measures can be found in [2,4]: diversity, context coverage, and the averaged coefficient of local modularity (the larger the better for all).

We show the results on the Southern women dataset in terms of approximation (coverage) of formal concepts (FC) with the sets of biclusters under different threshold values of their minimum density (see Table 2).

Note that the transition from the complete approximation by biclusters of all the formal concepts also took place at a density threshold of 0.65 on the "Southern Women" dataset (the second line in bold), while the loss of object-attribute pairs happened at 0.75 (the last line in bold). A sharp drop was observed for the advertising data set contrary to the Southern Women dataset.

We also show the coverage dynamics with sets of biclusters under varied minimal density, and also the coverage of objects and attributes (see Fig. 4). For the large advertising dataset, it was not possible to find the set of formal concepts and therefore to cover them with a set of biclusters.

Table 2. Coverage of formal concepts by the biclusters under varied minimum density. "Southern women" Dataset.

ρ_{min}	# unique biclusters	# covered FC	Fraction of covered FC	Av. local modularity	Diversity	Coverage
0	83	65	1,00	0,68	0,88	1,00
0.05	83	65	1.00	0.68	0.88	1.00
0.1	83	65	1.00	0.68	0.88	1.00
0.15	83	65	1.00	0.68	0.88	1.00
0.2	83	65	1.00	0.68	0.88	1.00
0.25	83	65	1.00	0.68	0.88	1.00
0.3	83	65	1.00	0.68	0.88	1.00
0.35	82	65	1.00	0.68	0.89	1.00
0.4	81	65	1.00	0.68	0.89	1.00
0.45	77	65	1.00	0.69	0.88	1.00
0.5	71	65	1.00	0.70	0.87	1.00
0.55	63	65	1.00	0.73	0.84	1.00
0.6	60	65	1.00	0.74	0.83	1.00
0.65	**51**	**64**	**0.98**	**0.76**	**0.81**	**1.00**
0.7	40	63	0.97	0.79	0.71	1.00
0.75	**33**	**57**	**0.88**	**0.81**	**0.65**	**1.00**
0.8	22	51	0.78	0.85	0.62	0.97
0.85	13	35	0.54	0.86	0.50	0.86
0.9	7	20	0.31	0.91	0.62	0.71
0.95	0	0	0.00	0.00	0.00	0
1	0	0	0.00	0.00	0.00	0

6.3 Approximate Density Estimation

We know that computing biclusters' density for large networks takes a significant amount of time. However, we show that computing the approximate density with a given accuracy in much less time 3.

Since the Southern women dataset is the smallest among the considered ones (18×14), many experiments on calculating the approximate density (for various parameter values) included very few biclusters (often only 1–2), so the

Table 3. Estimation of the time for calculating the exact and approximate density values.

Dataset	Density computation, s	Approx. density computation, s	Error cases
Southern women	$4.4 \cdot 10^{-3}$	$2.9 \cdot 10^{-5}$	23%
Zoo	$3.7 \cdot 10^{-2}$	$6.4 \cdot 10^{-4}$	18%
Advertising	20.87	0.219	4%

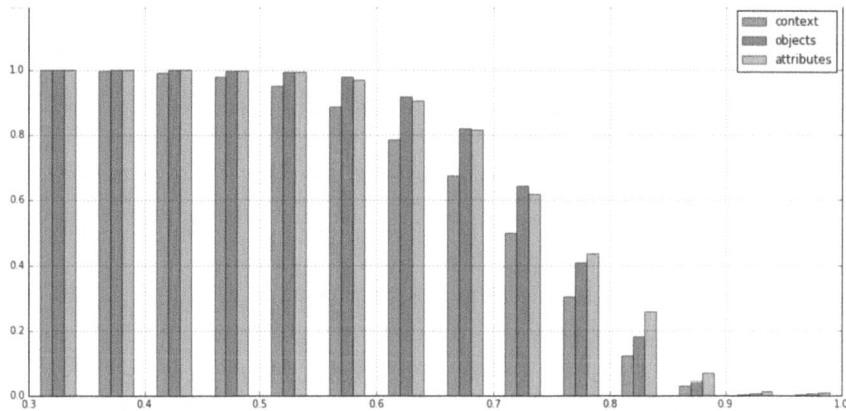

Fig. 4. Dynamics of coverage (vertical axis) vs. density of biclusters on the Internet advertising dataset

fraction of error cases were too high (Table 3). This proportion was lower for the Zoo dataset, even though in many experiments there were observed still a few biclusters (less than 10). At the same time, in the largest advertising dataset, the proportion of error cases sharply fell down and did not exceed 4%.

Here, we consider Zoo and Internet advertising datasets. In Table 4, one can see the results of the approximate density computation for the Zoo dataset with various error parameters. In the experiment, we set accuracy parameters $\epsilon \in [0.05, 0.5]$, where $|\rho - \tilde{\rho}| \leq \epsilon$, and the level of the maximum permissible fraction of error cases $\delta \in [0.5, 0.95]$. For example, in the case of density estimation for N biclusters, where N_{bad} is the number of biclusters with an erroneously estimated density ($|\rho - \tilde{\rho}| > \epsilon$), the fraction of errors should not exceed δ, i.e. $\frac{N_{bad}}{N} \leq \delta$.

In Table 4 below you can see cells marked in red, where the estimation exceeds the maximum permissible error level. For different values of the parameters, we obtain a different number n, and we also obtain a different number of biclusters to estimate the approximate density. Hence, due to the relatively small size of the dataset, and the small size of the biclusters, some experiments involved few biclusters, which exaggerated the fraction of erroneous results.

The main fraction of erroneous results refers to the level $c = 0.05$ or 0.1, because we allowed a very narrow range of deviations of the estimated density, and also we obtained a value of n in the range 500–700 (for example, at $\epsilon = 0.05$ and $\delta = 0.9, n = 599$), for some combinations of parameters.

In this experiment, we used an approximate calculation method for biclusters having sizes $|g'||m'| > 599$ (their total number is 34), which is relatively small for an effective statistical estimation.

Table 4. Accuracy of approximate density calculation for Zoo dataset with $\epsilon \in (0, 0.5]$ and $\delta \in [0.5, 1)$.

ϵ \ δ	0,5	0,55	0,6	0,65	0,7	0,75	0,8	0,85	0,9	0,95
0,05	0,44	0,40	0,45	0,43	0,37	0,38	0,37	0,34	0,27	1,00
0,1	0,67	0,67	0,70	0,69	0,70	0,72	0,71	0,71	0,73	0,76
0,15	0,81	0,82	0,78	0,83	0,84	0,84	0,84	0,84	0,86	0,84
0,2	0,84	0,88	0,88	0,86	0,88	0,92	0,90	0,91	0,92	0,94
0,25	0,91	0,88	0,90	0,87	0,92	0,94	0,93	0,93	0,94	0,96
0,3	0,89	0,93	0,93	0,93	0,94	0,95	0,93	0,94	0,95	0,97
0,35	0,93	0,95	0,93	0,94	0,94	0,95	0,97	0,97	0,98	0,98
0,4	0,95	0,96	0,97	0,97	0,97	0,97	0,96	0,97	0,98	0,99
0,45	0,97	0,97	0,96	0,98	0,98	0,99	0,98	0,99	0,99	0,99
0,5	0,98	0,98	0,99	1,00	0,99	0,99	0,99	1,00	1,00	1,00

Let us consider the results of estimating the approximate density value for the Internet advertising dataset.

Table 5. Accuracy of approximate density calculation for the Internet advertising dataset with $\epsilon \in (0, 0.5]$ and $\delta \in [0.5, 1)$

ϵ \ δ	0,5	0,55	0,6	0,65	0,7	0,75	0,8	0,85	0,9	0,95
0,05	0,74	0,75	0,75	0,76	0,77	0,78	0,79	0,81	0,83	0,86
0,1	0,89	0,90	0,91	0,92	0,93	0,94	0,95	0,96	0,97	0,98
0,15	0,92	0,93	0,94	0,95	0,95	0,96	0,97	0,98	0,98	0,99
0,2	0,93	0,94	0,95	0,96	0,96	0,97	0,98	0,98	0,99	1,00
0,25	0,94	0,94	0,95	0,96	0,97	0,97	0,98	0,99	0,99	1,00
0,3	0,94	0,95	0,95	0,96	0,97	0,98	0,98	0,99	0,99	1,00
0,35	0,94	0,96	0,96	0,97	0,97	0,98	0,99	0,99	0,99	1,00
0,4	0,95	0,96	0,97	0,97	0,97	0,98	0,99	0,99	0,99	1,00
0,45	0,96	0,96	0,96	0,98	0,98	0,99	0,99	0,99	1,00	1,00
0,5	0,95	0,95	0,98	0,98	0,98	0,99	0,99	1,00	1,00	1,00

In Table 5, the method showed very good results in estimating the approximate density of biclusters for an advertising dataset.

Out of 100 experiments (combinations of ϵ and δ) there were only 4 error results, and even in the "erroneous experiments", the accuracy values were very close to the required ones: at the level of $0.8 - 0.85$, and this with the permissible error of $\epsilon = 0.05$. It is clear to see that the fraction of correctly determined densities never dropped below 0.74, and in most cases was within $[0.95, 1]$. It is worth noting that due to the large size of the entire dataset, in each of the experiments, there were a lot of biclusters. For example, with the maximum value of the parameter among the other experiments $n = 738$ (for $\epsilon = 0.05$ and $\delta = 0.95$, i.e., the most stringent constraint from below on the minimum size), there were 78.6 thousand biclusters with size greater than n, (the total number of unique biclusters is 92 thousand).

6.4   Genome Association Study

The codes and experiment protocols for this section are available in iPython on GitHub: https://github.com/dimachine/EpsilonDeltaApproxBicl.

Data Description and Preprocessing. The original data was downloaded from GitHub repository https://github.com/inzilico/synthetic-data. They contained genotypes of 5581 artificial samples simulated from the haplotypes of 99 unrelated individuals from CEO population of 1000 Genomes project (Phase I) [21]. The data included 32028 single-nucleotide polymorphisms (SNPs) located on chromosome 22 and coded as $\{0, 1, 2\}$, where the number was the count of the minor allele in the given genotype. The minor allele was the less frequent allele in the whole group of individuals. It was denoted via underscore in the notations of SNPs mentioned in this article. The phenotype was an onset of a disease. The sizes of case and control groups equaled to 4929 and 652, respectively. Five SNPs were made to be associated with phenotype with odds ratios of 1.5 and 2.25 for heterozygous and homozygous loci, respectively. The casual SNPs are rs4823464, rs137425, rs3761422, rs9606478 and rs7292279. The SNPs were ordered with respect of their position on a chromosome under GRCh37 genome assembly. The file size was about 105 Mb. The simulation of this data and validation of SNPs associated with phenotype were described previously [22] (Section Data Simulation).

Using this data, we generated three data sets that represented formal contexts $\mathbb{K}_v = (P, S, I_v)$ induced by each genotype given as code value $v \in \{0, 1, 2\}$, where individuals are related to SNPs via pI_vs, iff the individual p has the code value v for SNP s. From a biological sense, the formal context $\mathbb{K}_v$ corresponds to one of the following inheritance models: dominant ($v = 0$), overdominant ($v = 1$), and recessive ($v = 2$). Since the inheritance model is not known in advance and specific for each genomic loci, all of them are worth to be considered.

Experiments with OA-Biclusters. The number of biclusters having causal SNPs for each context $\mathbb{K}_v = (P, S, I_v)$ is given in Table 6. As seen, it varied with v, demonstrating the minimum 4700 biclusters for $v = 2$ and the maximum 10154 biclusters for $v = 1$. The average bicluster densities varied with v as well, and we observed the largest one for $v = 0$, while the highest one for $v = 2$.

Table 6. Genome data sets and time elapsed on processing them

v	Model	$\|G\|$	$\|M\|$	$\|I\|$	ρ	# biclusters	Computing biclusters, HH:MM:SS	Average (ϵ, δ) density estimation, s
0	Dominant	5581	32028	131375034	≈ 0.735	10154	00:27:30	235.12
1	Overdominant	5581	32028	38596588	≈ 0.216	13051	00:08:13	260.83
2	Recessive	5581	32028	8776646	≈ 0.049	4700	00:01:42	32.43

In addition, for a given bicluster (A, B), we also considered $Purity = \frac{|A \cap G_-|}{|A|}$ with respect to the control group G_-. The distributions of density, purity, intent and extent for $v = 0$ are given in Fig. 5.

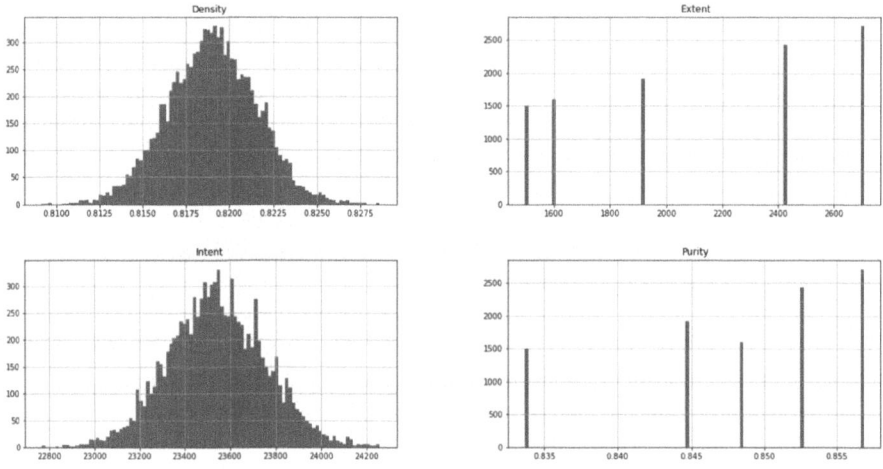

Fig. 5. Distributions for biclusters with $v = 0$

For example, we take one of the associated SNPs, s (rs9606478), and build biclusters for each of three contexts as follows (s'^v, p'^v), where the individual p is from the respective context $pI_v s$.

Context $\mathbb{K}_0$. We considered an OA-bicluster $(A, B) = (s'^0_{22327}, p'^0_{722})$ generated by individual p_{722} and causal SNP s_{22327} (rs9606478_G). Its main characteristics are $\rho(A, B) \approx 0.826$, $|A| = 2712$ (the number of individuals), $|B| = 23201$ (the number of SNPs), and $Purity(A, B) \approx 0.857$.

Ordering SNPs by their support (then in lectic order), we obtained 5386 SNPs with relative support 1 (or 2712 in terms of individuals), so, they are closed in the subcontext (A, B, I_0): rs5771007_C, rs6151427_C, rs8137073_C, rs2269383_T, rs8141871_T, etc.

Another associated SNP, rs4823464_G, was also presented in the bicluster but with a smaller support ≈ 0.432 (1171 individuals).

Context $\mathbb{K}_1$. Let us consider bicluster (C, D) generated by (p_{1339}, s_{22327}) for $v = 1$, i.e. SNP is rs9606478_G.

Its density is relatively low, i.e. $\rho(C, D) \approx 0.388$, $|C| = 2341$ (the number of individuals), $|D| = 7083$ (the number of SNPs), and $Purity(C, D) \approx 0.904$.

SNPs' supports are not that high:

- $sup = 2341$, $rsup = 1.0$ for rs9606478_G,
- $sup = 1459$, $rsup \approx 0.623$ for rs874836_A,
- $sup = 1405$, $rsup \approx 0.600$ for rs9606481_C,
- $sup = 1405$, $rsup \approx 0.600$ for chr22:17312790_T,
- $sup = 1310$, $rsup \approx 0.600$ for chr22:17306270_A,
- $sup = 1292$, $rsup \approx 0.552$ for chr22:17307742_A,
- ...

Two associated SNPs are also in D but with lower supports: rs137425_A has $sup = 1117$ and $rsup \approx 0.477$, while rs4823464_G has $sup = 1014$ and $rsup \approx 0.433$.

Context $\mathbb{K}_2$. The last bicluster, (E, F) is generated by (p_{3400}, s_{22327}) with the lowest density, $\rho(E, F) \approx 0.162$, $|E| = 528$ (the number of individuals), $|F| = 1276$ (the number of SNPs), and $Purity(E, F) \approx 0.926$.

Other associated SNPs are absent inside the bicluster.

The last example might be the most interesting, since it is sparse and its cross-like structure can be visible with a naked eye (Fig. 6).

Experiments with (ϵ, δ)-estimation We performed three experiments with contexts for the three various vs similar to those of in Subsect. 6.3.

Table 7. Accuracy of approximate density calculation with $\epsilon \in (0, 0.5]$ and $\delta \in [0.5, 1)$ for $v = 0$.

ϵ	δ									
	0.5	0.55	0.6	0.65	0.7	0.75	0.8	0.85	0.9	0.95
0.05	0.969	0.978	0.981	0.985	0.989	0.992	0.994	0.998	0.998	1.0
0.1	0.972	0.979	0.979	0.983	0.988	0.993	0.994	0.996	0.998	0.999
0.15	0.977	0.979	0.977	0.982	0.992	0.992	0.996	0.997	0.998	0.999
0.2	0.967	0.987	0.987	0.990	0.995	0.990	0.995	0.996	0.999	0.999
0.25	0.987	0.986	0.979	0.991	0.983	0.994	0.996	0.997	0.998	0.999
0.3	0.957	0.989	0.986	0.976	0.991	0.988	0.996	0.999	0.999	1.0
0.35	0.988	0.979	0.977	0.993	0.995	0.988	0.996	0.994	0.996	0.999
0.4	0.957	0.957	0.989	0.987	0.985	0.997	0.994	0.998	0.996	0.998
0.45	0.980	0.979	0.980	0.994	0.994	0.987	0.986	0.997	0.999	0.999
0.5	0.995	0.994	0.979	0.979	0.979	0.997	0.996	0.999	1.0	0.999

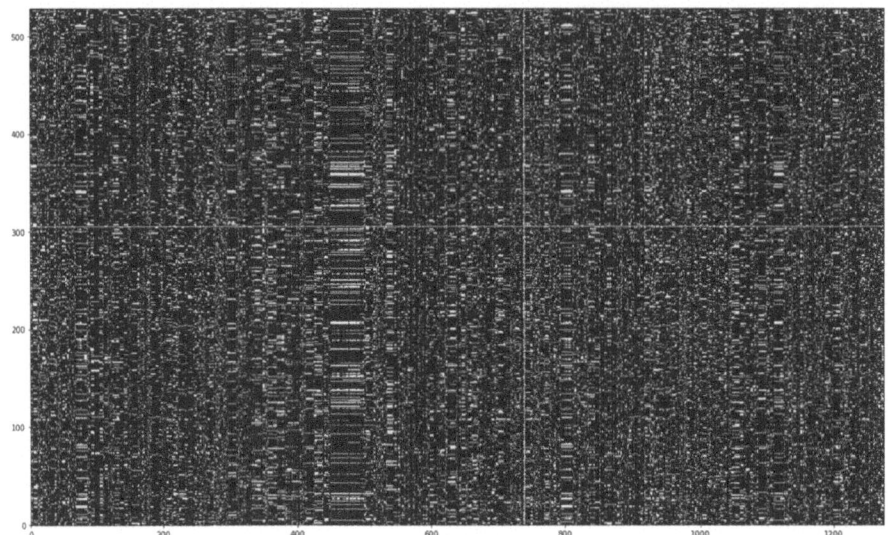

Fig. 6. A visual plot of the bicluster generated by (p_{3400}, s_{22327})

For $v = 0$ the accuracy values lie in $[0.95676581, 0.99960607]$, while for $v = 1$ and $v = 2$ and the same $(\epsilon\text{-}\delta)$-grid, the accuracy values vary within $[0.90805302, 0.99923378]$ and $[0.94361702, 1.0]$, respectively (Table 7).

These values were always higher than the theoretical δ for all three experiments.

6.5 Community Detection Cases

We also showcase that properly biclusters out of the whole of their set can be good analogues to communities in bimodal networks. In the experiment, several biclusters were selected, firstly, they were dense, and secondly, which have enough objects and attributes. It is possible to visually assess the quality of the partition based on the fact that how tight will be the connections within the community in comparison with their density outside the given sets.

Let us show a few biclusters of the Zoo dataset. In this set, it is much easier to interpret the results, because we know the names of objects and attributes in the network, and each object belongs to one of 7 classes (mammals, birds, etc.). To search for the example bicluster, density constraints ($\rho_{min} = 0.7$) and objects cardinality (over 15) were applied.

In Fig. 7, we can see 3 classes of animals: mammals, birds, and fishes, as well as their attributes. Additionally, there is a group of objects and attributes that were not assigned to any of these 3 classes. It is also noticeable that the overwhelming majority of connections are located within the identified communities, which indicates the high quality of the set partition into communities. For example, it is clear to see in the class of mammals such kinds of animals as a lion, an

Community division, Zoo dataset

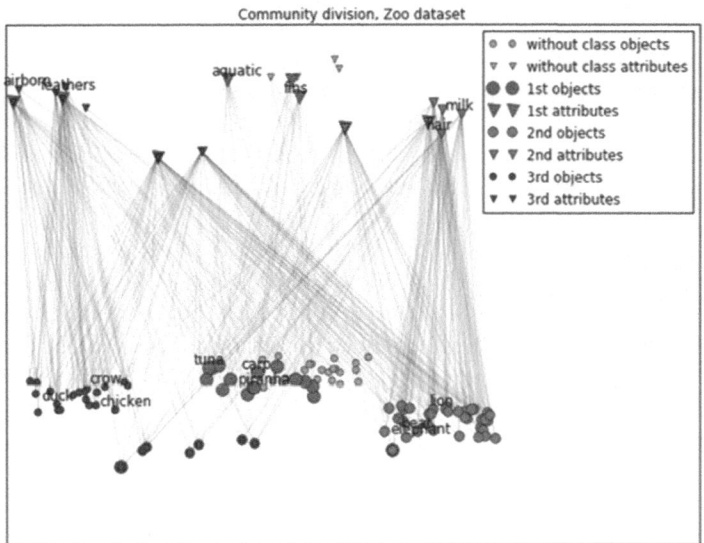

Fig. 7. Visualization of the found communities for the Zoo dataset.

elephant, and a bear, and their corresponding features such as "feeding offspring with milk" and "presence of hairline". As shown in Fig. 7, the representatives of the fish class are tuna, piranha and carp, and their corresponding features are the ability to swim and the presence of fins.

7 Conclusion

This paper departs from the variants of the community notions, such as biplexes and formal concepts, to the dense OA-biclusters. In particular, a practical comparison of formal concepts with OA-biclusters was conducted. More to that, the possibility of approximating formal concepts by a subset of biclusters was investigated. The experiments conducted during the work include an examination of the community detection biclustering process for a bimodal network. Since the search for formal concepts is a computationally difficult task, this paper considers the adequacy of their approximation by a certain collection of biclusters.

Finding an approximate density estimate results in a significant waste of time in searching for the exact density value while working with large networks. For small and medium datasets, the gain in computation time is approximately 100 times, while for large networks the difference in time is bigger. The method of approximate estimation, which was carried out in the experiment, showed that it has a sufficiently high accuracy of the approximation. However, the accuracy is 80% for small datasets, which is low, while for a large data set the estimation accuracy was 96%, which is a very good indicator. We can see the difference in results, which is explained by the following: for different input parameters, there was a different number of biclusters in the experiment, and since an insufficient number of biclusters participated in small datasets, this significantly reduced the accuracy of estimation. Also, when using the input parameters, the results for small datasets showed that the error is $\epsilon = 20\%$ and the minimum proportion of accurately estimated densities in the collection is $\delta = 75\%$; it is worth noting that the accuracy of the approximate density estimate was stably not less than 80% even for small networks.

In conclusion, several OA-biclusters from four different domains generated within a reasonable time with the proposed approach had a good interpretation as communities. This allows a broad practical applicability for our approach. For example, the biclusters identified in genotype-phenotype data can be potentially applied for the detection of new genetic loci associate with the multifactorial diseases.

Finally, we can conclude that the use of biclusters can be successful in the problems of approximating formal concepts and finding communities, for the following reasons: 1) algorithms of finding bicluster are computationally efficient; 2) with high accuracy, it is possible to approximate formal concepts with a small set of biclusters; 3) since we have weakened restrictions on bicluster in comparison with formal concepts, we are able to identify large communities with a well-defined structure even for large datasets.

Acknowledgements. The authors would like to thank Santo Fortunato and Vladimir Batagelj for a piece of advice related to network analysis. The research was funded by the Russian Science Foundation, grant number 23-14-00131 (the application of OA-biclustering to genotype-phenotype data for identification of SNPs associated with a disease). The contribution of the first author is also an output of a research project implemented as part of the Basic Research Program at the National Research University Higher School of Economics (HSE University). This research was also supported in part through computational resources of HPC facilities at HSE University.

References

1. Cerinšek, M., Batagelj, V.: Sources of Network Data, pp. 2843–2851. Springer, New York, New York, NY (2018)
2. Fortunato, S., Hric, D.: Community detection in networks: a user guide. Phys. Rep. **659**, 1–44 (2016)
3. Jelassi, M.N., Ben Yahia, S., Mephu Nguifo, E.: Towards more targeted recommendations in folksonomies. Soc. Netw. Anal. Min. **5**(1), 1–18 (2015). https://doi.org/10.1007/s13278-015-0307-8
4. Ignatov, D.I., Semenov, A., Komissarova, D., Gnatyshak, D.V.: Multimodal clustering for community detection. In: Missaoui, R., Kuznetsov, S.O., Obiedkov, S. (eds.) Formal Concept Analysis of Social Networks, pp. 59–96. Springer International Publishing, Cham (2017). https://doi.org/10.1007/978-3-319-64167-6_4
5. Missaoui, R., Kuznetsov, S.O., Obiedkov, S. (eds.): Formal Concept Analysis of Social Networks. LNSN, Springer, Cham (2017). https://doi.org/10.1007/978-3-319-64167-6

6. Jäschke, R., Hotho, A., Schmitz, C., Ganter, B., Stumme, G.: TRIAS - an algorithm for mining iceberg tri-lattices. In: Proceedings of the 6th IEEE International Conference on Data Mining (ICDM 2006), 18-22 December 2006, Hong Kong, China, IEEE Computer Society, pp. 907–911 (2006)
7. Seidman, S., Foster, B.: A graph-theoretic generalization of the clique concept. J. Math. Sociol. **6**(1), 139–154 (1978)
8. Borgatti, S., Everett, M.: Network analysis of 2-mode data. Soc. Netw. **19**(3), 243–269 (1997)
9. Luo, W., Li, K., Zhou, X., Gao, Y., Li, K.: Maximum biplex search over bipartite graphs. In: 2022 IEEE 38th International Conference on Data Engineering (ICDE), pp. 898–910 (2022)
10. Freeman, L.C., White, D.R.: Using Galois lattices to represent network data. Sociol. Methodol. **23**(1), 127–146 (1993)
11. Freeman, L.C.: Cliques, Galois lattices, and the structure of human social groups. Soc. Netw. **18**, 173–187 (1996)
12. Duquenne, V.: Lattice analysis and the representation of handicap associations. Soc. Netw. **18**(3), 217–230 (1996)
13. Mirkin, B.G., Kramarenko, A.V.: Approximate bicluster and tricluster boxes in the analysis of binary data. In: Proceedings of the 13th international conferenceon Rough sets, fuzzy sets, data mining and granular computing. RSFDGrC'11, Berlin, Heidelberg, Springer-Verlag, pp. 248–256 (2011). https://doi.org/10.1007/978-3-642-21881-1_40
14. Besson, J., Robardet, C., Boulicaut, J.F.: Mining a new fault-tolerant pattern type as an alternative to formal concept discovery. In: Scharfe, H., Hitzler, P., Ohrstrom, P. (eds.) Conceptual Structures: Inspiration and Application. Lecture Notes in Computer Science, vol. 4068, pp. 144–157. Springer, Berlin / Heidelberg (2006)
15. Ignatov, D.I., Kuznetsov, S.O., Poelmans, J.: Concept-based biclustering for internet advertisement. In: 2012 IEEE 12th International Conference on Data Mining Workshops, IEEE, 123–130 (2012)
16. Zhukov, L.: Spectral Clustering of Large Advertiser Datasets (2004)
17. Ganter, B., Wille, R.: Formal Concept Analysis. Springer, Heidelberg (1999). https://doi.org/10.1007/978-3-642-59830-2
18. Gnatyshak, D., Ignatov, D.I., Kuznetsov, S.O., Nourine, L.: A one-pass triclustering approach: is there any room for big data? In: CLA, pp. 231–242 (2014)
19. Phillips, J.M.: Chernoff-hoeffding inequality and applications. arXiv preprint arXiv:1209.6396 (2012)
20. Freeman, L.C.: Finding social groups: a meta-analysis of the southern women data (1994)
21. Auton, A., et al.: A global reference for human genetic variation. Nature **526**(7571), 68–74 (2015)
22. Khvorykh, G.V., Sapozhnikov, N.A., Limborska, S.A., Khrunin, A.V.: Evaluation of density-based spatial clustering for identifying genomic loci associated with ischemic stroke in genome-wide data. Int. J. Mol. Sci. **24**(20), 15355 (2023)

Theoretical Machine Learning
and Optimization

Distributed Bayesian Coresets

Vladimir Omelyusik[1] and Maxim Panov[2,3(✉)]

[1] University of Missouri, Columbia, MO, USA
[2] Technology Innovation Institute, Abu Dhabi, UAE
[3] Mohamed bin Zayed University of Artificial Intelligence and Technology Innovation Institute, Abu Dhabi, UAE
maxim.panov@mbzuai.ac.ae

Abstract. Bayesian coresets are small weighted subsamples of the original data, which aim to preserve the full posterior. Many algorithms for Bayesian coreset construction approximate likelihood functions as vectors in a normed space. This approximation often requires iterative recomputation of the full dataset's likelihood, making it very time-consuming, especially for high-dimensional data. One way to reduce computation time is to construct parts of the coreset on separate processors. This approach is well-studied in the frequentist coreset construction for K-Means clustering, which provides guarantees on random partitioning by using cluster centers as global anchors. However, since likelihood can describe more sophisticated data geometry, distributed Bayesian coreset construction is still challenging. We use relations between K-Means and EM algorithms and propose a partitioning strategy that uses maximum likelihood estimates as anchors of the dataset's global properties. We compare this strategy with random partitioning. Experiments show that our method outperforms the latter on datasets with complex geometry and still provides sufficient time benefit compared to the coreset construction on the full dataset.

Keywords: Bayesian Inference · Coreset · Distributed Computation

1 Introduction

Bayesian statistics provides flexibility which empowers it to solve problems where frequentist methods fail. However, since widely used posterior approximation algorithms scale at least linearly with data size, this flexibility can be overshadowed by poor scalability [17]. A recent branch of the literature suggests dealing with this problem by constructing a small weighted subsample of the original data such that the posterior estimated on this subsample would closely approximate the posterior estimated on the full dataset. Such a subsample is called a Bayesian coreset [6,10].

Many algorithms for Bayesian coreset construction rely on an idea of choosing those points whose likelihood is best-aligned with the full dataset's likelihood. Alignment is formalized by representing the likelihood function of each

© The Author(s), under exclusive license to Springer Nature Switzerland AG 2024
D. I. Ignatov et al. (Eds.): AIST 2023, CCIS 1905, pp. 287–298, 2024.
https://doi.org/10.1007/978-3-031-67008-4_21

data point as a vector in a normed space using Monte-Carlo methods or ad-hoc approximations [6]. This requires at least one iteration of evaluating the likelihood function for the entire dataset times the desired vector length, which can be time-consuming, especially for high-dimensional datasets. Moreover, it was shown that iterative coreset construction results in a better quality, which poses a challenge to scalability of the coreset construction methods themselves [4].

One approach to this problem is to construct the coreset in a distributed setting. The procedure goes by assigning parts of the original dataset to individual processors, running a construction algorithm on each part concurrently and combining the results. Fundamental similarities between coreset construction methods make this approach appealing, since given the right procedure it can be applied to any construction method.

Distributed coreset construction is well-studied in the frequentist literature for the problem of preserving K-Means clustering [7,11]. Since the cluster centers are used as indicators of the dataset's global structure, it holds that the union of coresets built on processors is the coreset for the full dataset, which makes it possible to distribute data points randomly and obtain significant time benefit [8]. Although similar guarantees were formulated for the Bayesian case [6], our experiments show that the partitioning strategy has a clear influence on the quality of approximation.

In this paper, we propose a partitioning strategy applicable for Bayesian coreset construction methods. We rely on the fact that EM-based clustering can be viewed as generalized K-Means clustering and use the maximum likelihood estimates as global anchors. We compare our strategy with random partitioning and show that it outperforms the latter on datasets with complex geometry.

2   Background

2.1   Bayesian Coresets

Let $\mathcal{D} = \{X_i\}_{i=1}^N$ be a sample of random i.i.d. variables from a distribution $p(x \mid \theta)$ parameterized by a vector $\theta = (\theta_1, \ldots, \theta_m)$ with prior $\pi_0(\theta)$. Denote a single-observation log-likelihood function as $\ell_i(\theta) := \ln p(X_i \mid \theta)$ and the sample log-likelihood function as $\ell(\mathcal{D} \mid \theta) = \sum_{i=1}^N \ell_i(\theta)$. The posterior of θ is then expressed as

$$\pi(\theta \mid \mathcal{D}) = \frac{\exp\big(\ell(\mathcal{D} \mid \theta)\big)\,\pi_0(\theta)}{\int_{\theta \in \Theta} \exp\big(\ell(\mathcal{D} \mid \theta)\big)\,\pi_0(\theta)d\theta}.$$

As computation of the denominator is often infeasible in practical problems, the posterior is usually approximated by variational inference (VI) or Markov chain Monte Carlo (MCMC) methods [2,3]. However, since these algorithms scale at least linearly with the dataset size, further approximations are required for large samples [6]. One solution is to construct a small weighted subsample for which VI and MCMC's approximations are close to those obtained with the full dataset.

Formally, we are searching for a vector $w \in \mathbb{R}_+^N$ of non-negative weights, among which at most k are positive, such that the sample log-likelihood $\ell(\mathcal{D} \mid \theta)$ is well-approximated by a weighted sum of single-observation log-likelihoods $\ell_w(\mathcal{D} \mid \theta) = \sum_{i=1}^N w_i \ell_i(\theta)$ with respect to some distance D,

$$\begin{cases} w^* = \arg\min_{w \in \mathbb{R}_+^N} D\big(\ell(\mathcal{D} \mid \theta), \ell_w(\mathcal{D} \mid \theta)\big), \\ \sum_{i=1}^N \mathbb{I}\{w_i \neq 0\} \leq k, \ w_i \geq 0, \end{cases} \tag{1}$$

where $\mathbb{I}(a) = 1$ if a is true and 0 otherwise.

The weight-observation pairs $\{(w_i, X_i)\}_{i=1}^N$ are called the *Bayesian coreset* of $\{X_i\}_{i=1}^N$ [6,10]. The posterior estimated on a coreset, also called the weighing distribution, is defined as

$$\pi_w(\theta \mid \mathcal{D}) = \frac{\exp\big(\ell_w(\mathcal{D} \mid \theta)\big) \pi_0(\theta)}{\int_{\theta \in \Theta} \exp\big(\ell_w(\mathcal{D} \mid \theta)\big) \pi_0(\theta) d\theta},$$

and is expected to closely approximate π.

2.2   Sensitivity

A common approach to making problem (1) solvable with standard optimization tools is to consider likelihood functions $\ell_i(\theta)$ as vectors in a space with the norm $\| \cdot \|$ that induces D. To simplify notation, further on we write ℓ_i, ℓ and ℓ_w when referring to $\ell_i(\theta)$, $\ell(\mathcal{D} \mid \theta)$ and $\ell_w(\mathcal{D} \mid \theta)$ as vectors. If the norm $\| \cdot \|$ is induced by the scalar product $\langle \ell_i, \ell_j \rangle_{\|\cdot\|}$ (e.g., the L2 $\|v_i\|_2$ norm induced by $v_i^T v_j$), then ℓ_i can be approximated by a vector of arbitrary length S by the Bohner's theorem [6,15].

The estimation procedure goes by sampling parameter vectors $(\theta_1, \dots, \theta_S)$ from the weighing distribution π_w and computing an approximation v_i in a way to have

$$\langle \ell_i, \ell_j \rangle_{\|\cdot\|} \approx v_i^T v_j.$$

For example, for the $\| \cdot \|_2$ norm one gets

$$v_i \propto \big[\ell_i(\theta_1), \dots, \ell_i(\theta_S)\big]^T.$$

Since the ratio $\|\ell_i\| / \sum_i \|\ell_i\|$ can be interpreted as a relative weight of a data point X_i, it can be used as an anchor for coreset construction. Mirroring the non-Bayesian coreset theory, the entity

$$\sigma_i := \|\ell_i\|$$

is called the *sensitivity* of a point X_i. The sample sensitivity is defined as $\sigma := \sum_{i=1}^N \sigma_i$.

It is convenient to express all coreset construction methods in terms of sensitivities, because if no scalar product that induces the norm $\| \cdot \|$ can be found, one can take sensitivity approximations available in the non-Bayesian theory. For example, the estimate

$$\hat{\sigma}_i = \frac{N}{1 + \sum_{j=1}^{P} |X_{-i}^j| \exp\left(R\|X_i - \bar{X}_{-i}^j\|_2\right)}. \tag{2}$$

can approximate the uniform norm sensitivity $\sup_{\theta \in \Theta} |\ell_i(\theta)/\ell(\theta)|$; see [6] for details. This estimate is directly adapted from the problem of constructing a non-Bayesian coreset which preserves K-Means clustering performed for the full dataset. Here P is the number of clusters, X_i is the point for which σ_i is computed, $|X_{-i}^j|$ is the number of points in cluster j excluding X_i, $\bar{X}_{-i}^j$ is the mean of those points, and R is a scale constant [8,10].

2.3   Related Work

The literature features several ways of approaching problem (1) which differ in the choice of the metric D and in the way of dealing with the sparsity constraint $\sum_{i=1}^{N} \mathbb{I}\{w_i \neq 0\} \leq k$.

Importance sampling methods use ratios σ_i/σ as probabilities for the multinomial distribution $\text{Multi}\left(k, (\sigma_i/\sigma)_{i=1}^N\right)$ to draw an auxiliary vector $(k_1, \ldots, k_N)$. The weights are then set proportional to the drawn values, $w_i \leftarrow (\sigma k_i)/(\sigma_i k)$ which makes the coreset size no bigger than k. This procedure ensures $\mathbb{E}(w_i) = 1$, so that $\mathbb{E}(\ell_w) = \ell$ as $k \to N$.

Huggins et al. (2016) were the first ones to apply this method for the Bayesian logistic regression problem [10]. They used sensitivity estimates from Eq. (2). These estimates favor points which are distant from all cluster centers, and assign low probabilities to the points near the cluster centers (the centers themselves are selected with some baseline probability). However, since the most different points are not necessarily the most error-reducing ones, Campbell et al. (2019) proposed to use estimates based on L2 and Fisher norms, which favor points whose log-likelihood is most aligned with the sample log-likelihood [6].

Iterative methods use ratios σ_i/σ to determine the direction of the next update. Aiming to develop a more accurate approach than importance sampling, Campbell et al. (2019) utilize the directionality property of Hilbert norms and propose a Frank-Wolfe optimization scheme which replaces the binary constraint with a polytope one, and at each iteration extends the coreset with the observation that is best aligned with the residual error $\ell - \ell_w$; see [6]. Greedy Iterative Geodesic Approach [5] takes advantage of the fact that the coreset weights can be rescaled by an arbitrary positive number, which under a Hilbert norm makes it possible to restate the optimization problem such that it only depends on the alignment of log-likelihood vectors irrespective of their norms. Both algorithms run for k iterations ensuring the specified coreset size.

Iterative Hard Thresholding method [17] applies two-step gradient descent to minimize the euclidean distance between ℓ and ℓ_w. The first step updates the weights in the same way as in the classical Ordinary Least Squares, and the second step projects the updated vector into the k-sparsity subspace, ensuring it contains no more than k positive elements.

Sparse Variational Inference method [4] minimizes Kullback-Leibler divergence $D_{KL}(\pi_w, \pi)$ between the weighted and true posterior. Although not a true distance, KL divergence produces analogous expressions for sensitivities,

$$\sigma_i = \|\ell_i\|_{KL} = \mathrm{Cov}_{\pi_w}(\ell_i, \ell_i),$$

which makes the approach similar to Frank-Wolfe optimization. The method is shown to produce better estimates than those obtained with the Hilbert norms, but at the expense of running time, since it requires to re-estimate sensitivities at each of k iterations.

3 Distributed Setting

Although algorithms described above have different time efficiency in terms of dataset and coreset size (e.g., $\mathcal{O}(N)$ for importance sampling, $\mathcal{O}(Nk)$ for Frank-Wolfe and SparseVI and $\mathcal{O}(N \log k)$ for IHT), they all require evaluating the likelihood function at least NS times for sensitivity computation. Moreover, iterative methods, shown to produce more accurate approximations, typically require several rounds of this evaluation. One way to reduce coreset construction time without changing fundamental properties of the algorithms is to build the coreset in a distributed setting, i.e., separate original data points across individual processors, build a coreset on each processor concurrently and unite the results.

Non-Bayesian coreset construction in a distributed setting is a well-studied problem for K-Means and K-Medians clustering with Euclidean distance. It is based on a proposition that if the cluster centers are fixed, one can arbitrarily partition the dataset into r chunks and build a coreset using the estimates from Eq. (2) for each chunk. The union of the resulting coresets is the coreset for the full dataset [8,11]. In this strategy, cluster centers can be viewed as anchors which are sufficient to preserve the dataset's global properties under estimates (2) even in the absence of the full data.

For the Bayesian case, Campbell et al. (2019) formulated error upper bounds for distributed coreset construction with importance sampling and Frank-Wolfe under the Hilbert norm and any data partition [6]. However, these error bounds describe the worst-case scenario (the processors have no information of the dataset's global properties), and our initial experiments demonstrated that in practice approximation quality clearly depends on the partitioning strategy.

We propose to adapt the anchoring approach from the frequentist setting to improve the quality of partitioning. Firstly, we note that due to the relation between the uniform norm and estimates (2), we can automatically solve a Bayesian problem equivalent to preserving K-Means or K-Medians clustering in the distributed setting. However, since likelihood can describe more sophisticated data geometry than K-Means centers, Hilbert norms might require different anchors of global properties.

It was shown that EM-clustering can be considered as generalized K-Means clustering with probabilistic distances [12]. Particularly, EM algorithm approximates maximum likelihood estimates which describe cluster partitioning, providing more likelihood-relevant information on the data geometry. Consequently, we propose to use maximum likelihood estimates $\hat{\theta}_{ML}$ as anchors of the full dataset's global properties and assign a point X_i to a processor w with probability

$$p(X_i \in w) = \frac{\left| \ell(X_i \mid \hat{\theta}_{ML}) \right|}{\sum_{j=1}^{N} \left| \ell(X_j \mid \hat{\theta}_{ML}) \right|} = \frac{\left| \ell_i(\hat{\theta}_{ML}) \right|}{\sum_{j=1}^{N} \left| \ell_j(\hat{\theta}_{ML}) \right|}.$$

The assignment is done without replacement which ensures the processors have non-intersecting subsets. A coreset construction algorithm is set to output a coreset of size k/W, where W is the number of workers. It is then run on each processor concurrently, and produced coresets are combined.

Since $\ell(X_i \mid \hat{\theta}_{ML}) \geq \ell(X_i \mid \theta)$ for any θ, we have

$$\sigma_i \approx \left\| [\ell_i(\theta_1), \dots, \ell_i(\theta_S)] \right\| \leq C \times \left| \ell_i(\hat{\theta}_{ML}) \right|,$$

where C is a constant (e.g., for the L2 norm $C = \sqrt{S}$), i.e., the numerator of $p(X_i \in w)$ is an upper bound on σ_i. This forces points which are close in terms of sensitivity to be distributed to the same processor, much like each worker has information about the cluster centers in case of K-Means clustering.

Maximum likelihood estimates $\hat{\theta}_{ML}$ can be efficiently approximated by EM algorithm or computed explicitly if the distribution is common. Although it is impossible to formally quantify the exact time difference of EM estimation in comparison to MCMC in the general case [16], larger datasets may require larger S to better capture the likelihood's properties, so that EM algorithm may converge faster in practice. The speed of coreset construction itself will be reduced by at least W.

Since frequentist guarantees are applicable to arbitrary partitioning strategy, we expect our approach to be comparable in quality with random partitioning on datasets which are well-clustered by K-Means and superior to the latter on data with more sophisticated geometry on which K-Means fails.

4   Experiments

Our preliminary experiments showed that Hilbert norm-based coreset construction methods described above are close in quality without extensive tuning [13]. Thus, we test our parallelization strategy on a state-of-the-art algorithm based on the Euclidean norm, the Campbell's Frank-Wolfe [6]. We compare a problem-specific quality metric and running time of this algorithm in the sequential setting and in the parallel setting based on a random split and our strategy.

Experiments are done on a Macbook Pro M1 with 8 cores, one of which acts as a manager, so that the coresets are concurrently constructed on 7 cores. EM estimates are obtained with `sklearn` [14]. The code is available in our GitHub repository https://github.com/stat-ml/efficient_bayesian_coresets.

4.1 Multivariate Gaussian

We begin with a problem for which Kullback-Leibler divergence between π and π_w can be computed explicitly: estimation of the mean of a single multivariate Gaussian. Let us assume that

$$X_i \sim \mathcal{N}(\mu, I), \quad \mu \sim \mathcal{N}(\mu_0, I), \quad i = 1, \ldots, N,$$

for which we choose $N = 5000$ and $\mu_0 = \{0\}^5$. It can be shown that the weighing distribution is normal [4] with

$$\Sigma_w = \left(1 + \sum_{i=1}^{N} w_i\right)^{-1} I, \quad \mu_w = \Sigma_w \left(\mu_0 + \sum_{i=1}^{N} w_i X_i\right).$$

The full posterior π is also normal. For two normal distributions p and q the Kullback-Leibler divergence can be computed [18] with

$$D_{KL}(p, q) = \frac{1}{2}\left[\log\frac{|\Sigma_q|}{|\Sigma_p|} - k + (\mu_p - \mu_q)^T \Sigma_q^{-1}(\mu_p - \mu_q) + \mathrm{tr}\left(\Sigma_q^{-1}\Sigma_p\right)\right],$$

which we use to calculate $D_{KL}(\pi, \pi_w)$ and $D_{KL}(\pi_w, \pi)$, often referred to as forward and backward KL divergence respectively. Our final metric is the symmetric divergence computed as the halved sum of the two.

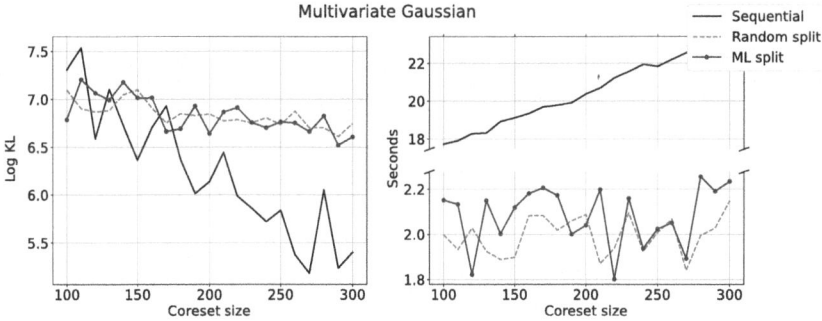

Fig. 1. Campbell's Frank-Wolfe algorithm on multivariate Gaussian data ($N = 5000$, $d = 5$, median over 20 runs). MCMC subsample size is set to 0.7 of the node's data, S is set to $0.02N$.

As Fig. 1 shows, on average, our parallelization strategy is similar in quality to random partitioning. This result is expected, since the problem is equivalent to 1-Means clustering, so that guarantees on arbitrary partitioning apply [8]. Average running time of our strategy exceeds that of random partitioning by 5% while being about 9 times smaller than that of the sequential run.

4.2   Gaussian Mixture

Next, we compare partitioning results on data with several intersecting clusters modeled as a Gaussian Mixture,

$$X_i \sim \sum_{j=1}^{J} \pi_j \mathcal{N}(\mu_j, \sigma_j), \quad \pi_0(\mu_j), \ \pi_0(\sigma_j) \propto FP, \quad i = 1, \dots, N,$$

where FP is a flat prior. We consider a univariate problem for which $N = 5000$, $\mu_1 = 6$, $\mu_2 = 8$, $\mu_3 = 10$, $\sigma_1 = 0.8$, $\sigma_2 = 0.5$, $\sigma_3 = 0.2$, $\pi_1 = 0.2$, $\pi_2 = 0.5$ and a multivariate problem with $N = 3000$, $\mu_1 = \begin{bmatrix} 0 & 0 \end{bmatrix}$, $\mu_2 = \begin{bmatrix} -1 & 1 \end{bmatrix}$, $\sigma_{j,(m,m)} = 1$, $\sigma_{j,(m,n)} = (-1)^{j+1} 0.9$, $\pi_1 = \pi_2 = 0.5$.

Similar to a single Gaussian, posterior distribution of each mixture component is normal. We approximate forward and backward Kullback-Leibler divergence with the following upper bound [9]:

$$D_{KL}(p_A, q_{B+C}) \leq w_B D_{KL}(p_A, q_B) + w_B D_{KL}(p_A, q_C),$$

and use the halved sum of the two as the final metric.

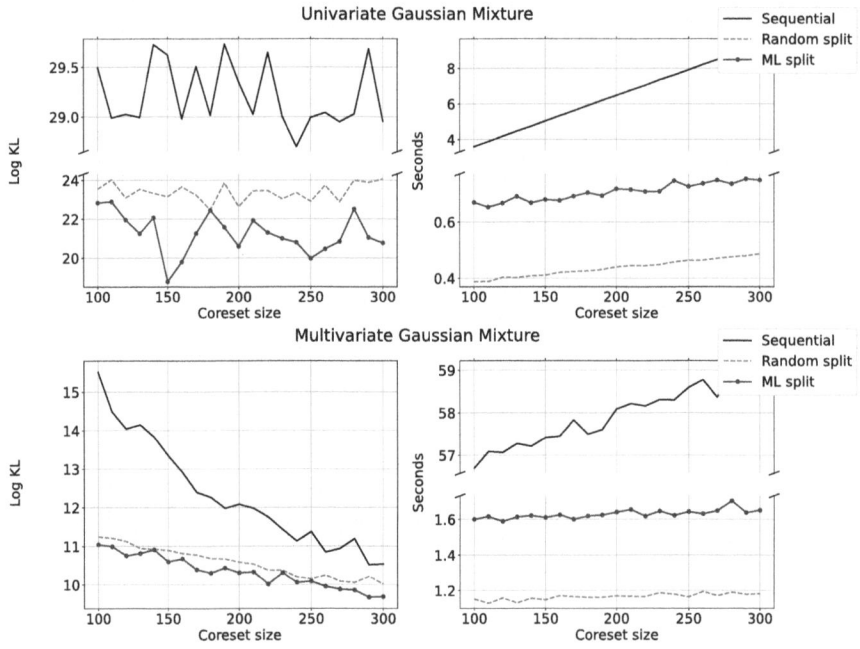

Fig. 2. Campbell's Frank-Wolfe algorithm on Gaussian Mixture data (median over 20 (univariate) and 10 (multivariate) runs). Univariate mixture: $N = 5000$, $J = 3$. Multivariate mixture: $N = 3000$, $J = 2$. MCMC subsample size is set to 0.7 of the node's data, S is set to 0.3 of the MCMC subsample size for both cases.

As Fig. 2 demonstrates, on average, the KL divergence of our strategy is bounded from above by the divergence of random partitioning in both cases. These results are consistent with Sect. 3: since K-Means algorithm fails to correctly classify mixed clusters, random partitioning describes the worst-case scenario, whereas maximum likelihood estimates provide additional information on the data geometry. Therefore, given the likelihood is specified correctly, our strategy allows for more accurate distributed coreset construction. In addition, random partitioning is on average at most 1.4 times faster, while sequential construction is at least 8 times slower than our approach.

4.3 Classification

To assess practical applicability of our partitioning strategy, we consider a classification problem with 10 classes on the UCI ML Optical Recognition of Handwritten Digits dataset ($N = 1797$, $d = 64$) which has similar geometry to the Gaussian mixture example [1]. We use logistic regression as a classification model to allow for explicit formulation of log-likelihood,

$$p(y_i = j) = \frac{\exp(X_i^T \beta^j)}{\sum_{c=1}^{10} \exp(X_i^T \beta^c)}, \quad \pi_0(\beta) \propto FP, \quad i = 1, \ldots, N.$$

Since true posterior is unknown, we use test accuracy of a model trained on the coreset as quality metric. The train-test ratio is $7 : 3$.

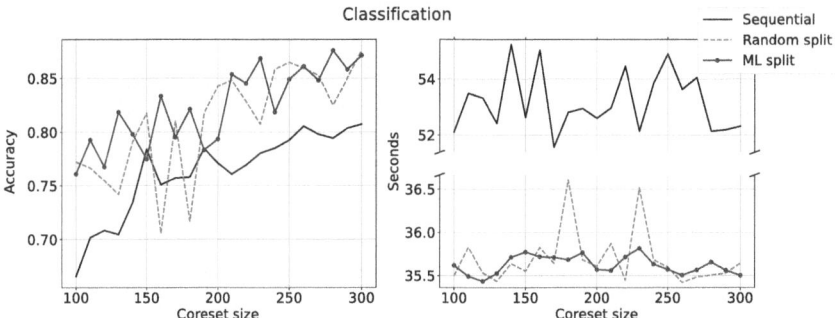

Fig. 3. Campbell's Frank-Wolfe algorithm (median over 10 runs) on Digits data ($N = 1797$, $d = 64$) [1]. The data is randomly split into a train (1257 obs.) and a test (540 obs.) sample. In the sequential case, multiclass logistic regression is estimated on the whole train sample. In the parallel cases, the train sample is split across workers. MCMC subsample size is set to 0.7 of the node's data, S is set to 0.3 of MCMC subsample size. Multinomial distribution is approximated by normal distribution for efficiency.

As Fig. 3 shows, although the upper bound effect from Sect. 4.2 is not entirely present, possibly due to approximating the true likelihood, on average, our strategy allows for higher accuracy compared to random partitioning. The two are comparable in average running time, while being approximately 1.5 times faster than the sequential run.

4.4   Number of Processors

Since the number of points on each worker depends on the total number of processors, the latter might affect the quality of coreset construction. To study this effect, we fix the coreset size at $k = 300$ for the Gaussian mixture examples above and vary the number of processors from 2 to 7.

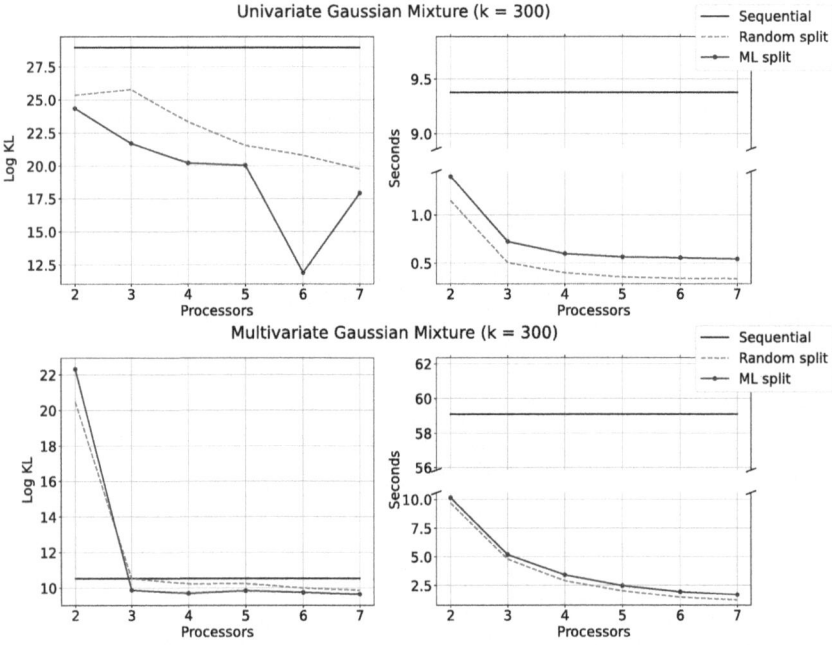

Fig. 4. Campbell's Frank-Wolfe algorithm on Gaussian mixture data with respect to the number of processors (median over 20 runs). MCMC subsample size is set to 0.7 of the node's data, S is set to 0.3 of the MCMC subsample size.

As Fig. 4 shows, on average, the upper bound effect from Sect. 4.2 holds irrespective of the number of processors. We attribute the only exception, Multivariate Gaussian Mixture case for $W = 2$, to random effects, and we expect the strict upper bound to hold for all cases as the number of runs increases. Average running time of the two methods scales almost identically up to a shift.

5   Conclusion

In this paper, we present a partitioning strategy for simultaneous Bayesian coreset construction on several processors. The strategy is based on using maximum likelihood estimates as anchors of the dataset's geometry, so that the points with close likelihood are more likely to be distributed to the same processor. We compare our strategy with random partitioning.

Experiments on synthetic data demonstrated that if true likelihood is known, our strategy is comparable in quality to random partitioning on data with simple geometry (one cluster), and better on data with complex geometry (several mixed clusters). This effect is on average applicable to real cases with unknown likelihood. The time difference between the two strategies is small if the maximum likelihood estimates are approximated with the EM algorithm, and both are substantially faster than the sequential run.

References

1. Alpaydin, E., Kaynak, C.: Optical Recognition of Handwritten Digits. UCI Machine Learning Repository (1998). https://doi.org/10.24432/C50P49
2. Andrieu, C., De Freitas, N., Doucet, A., Jordan, M.I.: An introduction to MCMC for machine learning. Mach. Learn. **50**, 5–43 (2003)
3. Blei, D.M., Kucukelbir, A., McAuliffe, J.D.: Variational inference: a review for statisticians. J. Am. Stat. Assoc. **112**(518), 859–877 (2017)
4. Campbell, T., Beronov, B.: Sparse variational inference: Bayesian coresets from scratch. Adv. Neural. Inf. Process. Syst. **32**, 11461–11472 (2019)
5. Campbell, T., Broderick, T.: Bayesian coreset construction via greedy iterative geodesic ascent. In: International Conference on Machine Learning, pp. 698–706. PMLR (2018)
6. Campbell, T., Broderick, T.: Automated scalable Bayesian inference via Hilbert coresets. J. Mach. Learn. Res. **20**(1), 551–588 (2019)
7. Fan, Y., Li, H.: Communication efficient coreset sampling for distributed learning. In: 2018 IEEE 19th International Workshop on Signal Processing Advances in Wireless Communications (SPAWC), pp. 1–5. IEEE (2018)
8. Har-Peled, S., Mazumdar, S.: On coresets for K-means and K-median clustering. In: Proceedings of the Thirty-Sixth Annual ACM Symposium on Theory of Computing, pp. 291–300 (2004)
9. Hershey, J.R., Olsen, P.A.: Approximating the Kullback Leibler divergence between gaussian mixture models. In: 2007 IEEE International Conference on Acoustics, Speech and Signal Processing - ICASSP 2007, vol. 4, pp. IV–317–IV–320 (2007)
10. Huggins, J., Campbell, T., Broderick, T.: Coresets for scalable Bayesian logistic regression. In: Advances in Neural Information Processing Systems, vol. 29, pp. 4080–4088 (2016)
11. Lu, H., Li, M.J., He, T., Wang, S., Narayanan, V., Chan, K.S.: Robust coreset construction for distributed machine learning. IEEE J. Sel. Areas Commun. **38**(10), 2400–2417 (2020)
12. Lücke, J., Forster, D.: K-means as a variational EM approximation of gaussian mixture models. Pattern Recogn. Lett. **125**, 349–356 (2019)

13. Omelyusik, V.: Effective methods of sample size reduction in application to Bayesian neural networks. Master's thesis, Skolkovo Institute of Science and Technology (2022)
14. Pedregosa, F., et al.: Scikit-learn: machine learning in Python. J. Mach. Learn. Res. **12**, 2825–2830 (2011)
15. Rahimi, A., Recht, B.: Random features for large-scale kernel machines. In: Advances in Neural Information Processing Systems, vol. 20 (2007)
16. Rydén, T.: EM versus Markov chain Monte Carlo for estimation of hidden Markov models: a computational perspective. Bayesian Anal. **3**(4), 659–688 (2008)
17. Zhang, J., Khanna, R., Kyrillidis, A., Koyejo, S.: Bayesian coresets: revisiting the nonconvex optimization perspective. In: International Conference on Artificial Intelligence and Statistics, pp. 2782–2790. PMLR (2021)
18. Zhang, Y., Liu, W., Chen, Z., Wang, J., Li, K.: On the properties of Kullback-Leibler divergence between multivariate gaussian distributions. arXiv preprint arXiv:2102.05485 (2021)

Demo Paper

Development of a Visualization Tool for the Occurrence of Life Events on the Demographic Lexis Grid

Ekaterina Mitrofanova, Dmitry I. Ignatov, Tatyana Maximova[✉],
Anna Podshivalova, and Anna Muratova

National Research University Higher School of Economics, Moscow, Russia
dignatov@hse.ru,
tmaksimova@hse.ru
http://www.hse.ru

Abstract. We present an interactive web-based tool that allows researchers to visualize the occurrence of life events on the demographic Lexis grid. The tool can be used to compare different generations, find the most important patterns for the occurrence of certain chains of events, and predict next the most likely events. The current version of the tool is already available by the link (https://tanyammm-visualization-v-2--data-31ue11.streamlit.app/).

Keywords: Demography · Sequence Analysis · Lexis Grid · Visualisation · Human-Computer Interaction · UX/UI Design

1 Introduction

Demographers work with datasets containing information about demographic and socioeconomic events and their sequences. These events occur along the life courses of people who differ in such features as age, gender, education, place of living, etc. Scientists compare, for example, different generations, find the most important patterns for the occurrence of certain chains of events and predict next the most likely events (see, for example, our previous study [1,2,10]).

So far, there is no convenient automated way to visualise the onset of several events at once, and especially for sequences. Scientists have to do calculations and some basic visualisation first, and then to draw graphs using such non-scientific programs as Adobe Photoshop. Our team has developed an interactive web-based tool that allows a researcher to access advanced data visualisations via several clicks.

The current version of our tool is already available by the link[1]. It is not a final version, but it already allows a user to fully work with a preloaded dataset and to build chronograms for the occurrence of sequences of events, including

---

[1] https://tanyammm-visualization-v-2--data-31ue11.streamlit.app/.

© The Author(s), under exclusive license to Springer Nature Switzerland AG 2024
D. I. Ignatov et al. (Eds.): AIST 2023, CCIS 1905, pp. 301–308, 2024.
https://doi.org/10.1007/978-3-031-67008-4_22

the ones on the demographic Lexis grid. It also allows to depict median ages of events on the Lexis grid and, in the future, to explore the data using Machine Learning algorithms. The tool is bilingual which is suitable for extracting graphs for both Russian and English papers.

2   Data

We used dataset of the first waves of the Generations and Gender Survey (GGS) for France, Estonia, and Russia (2004–2005). The numbers of respondents are following: France – 8493, Estonia – 6797, and Russia – 9433. The gender proportions are: 63.9% of women in France, 55.7% in Estonia, 63.1% in Russia. The respondents were born between 1930 and 1979.

We studied the following starting life course events which were presented in GGS: completing professional education, leaving parental home, first partnership, first marriage, and first childbirth.

The preview in Fig. 1 displays information for the first 10 respondents. First five columns contain the sociodemographic characteristics of the respondents, and other variables are the statuses of respondents at each month of life from the age of 15 until the age of 35.

	RESPID	country	sex	generations	y15m1	y15m2	y15m3	y15m4	y15m5	y15m6	y15m7
0	185,321	1	2	2	SCON	SCON	SCON	SCON	SCON	SCON	SCOE
1	185,322	1	1	2	SCOE	SCOE	SCOE	SCOE	SCOE	SCOE	SCOE
2	185,324	1	2	3	SCON	SCON	SCON	SCON	SCON	SCON	SCON
3	185,325	1	1	4	SCON	SCON	SCON	SCON	SCON	SCON	SCON
4	185,326	1	1	2	SCON	SCON	SCON	SCON	SCON	SCON	SCON
5	185,327	1	1	4	SCON	SCON	SCON	SCON	SCON	SCON	SCON
6	185,328	1	2	1	SCON	SCON	SCON	SCON	SCON	SCON	SCON
7	185,329	1	2	4	SCON	SCOL	SCOL	SCOL	SCOL	SCOL	SCOL
8	185,330	1	2	3	SCON	SCON	SCON	SCON	SCON	SCON	SCON
9	185,331	1	2	1	SCON	SCON	SCON	SCON	SCON	SCON	SCON

Fig. 1. Example of the dataset

3   Events' Occurrence

To visualise the occurrence of events, we used the median ages (they work well with distributions containing "outliers" and asymmetry) and the demographic

Lexis grid [8,9]. The Lexis grid comprises three temporal axes: calendar time (X-axis), age (Y-axis), and generation (diagonal).

In Fig. 2, our events of interest are displayed at the Lexis grid as pictograms. Their labels are represented in the legend. The legend is formed automatically according to the set of events of the first presented generation. In the current version, the colour of the legend will be blue if you choose either men or both genders and red if you choose only women.

Each event has its own position within a parallelogram for a particular age and generation. Horizontally, each event has its own position regarding other events; vertically, events are ordered according to gender: pictograms for men are displayed at the top, and for women at the bottom of each parallelogram. The position of the pictograms inside one parallelogram always shows the exact age and does not mean that men obtain events later than women.

At the presented example of the graph, we compared Russian men and women of 1950–59 and 1970–79 years of birth. Men obtain their demographic events later than women in both generations, while the ages of socioeconomic events are almost equal.

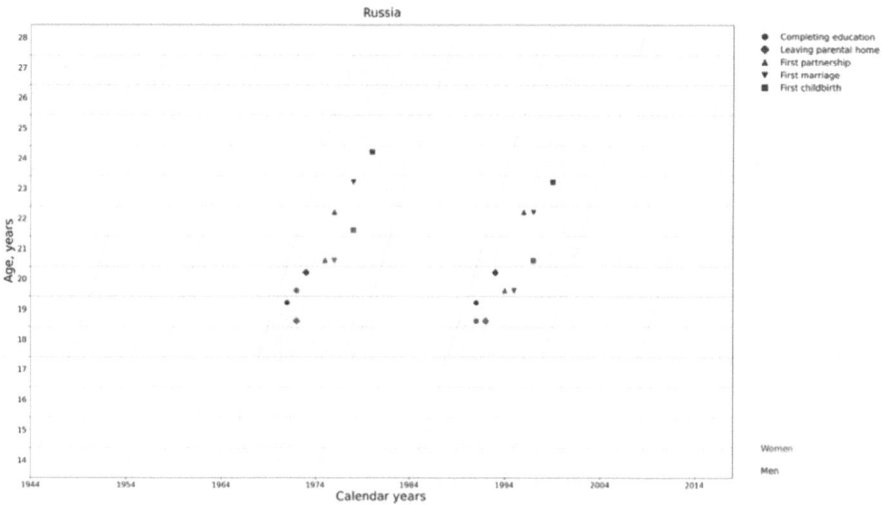

Fig. 2. Lexis grid with median ages of events occurrence

4   Setting Menu for Events' Occurrence

To build graphs, a user should select countries, events, genders and generations in the menu on the left. A user can change the labels and boundaries of the axes. In the menu "Selection of "corridors"", a user can customize the colour "corridors" for any axis. To remove such a corridor, one needs to merge two points of a line (Fig. 3).

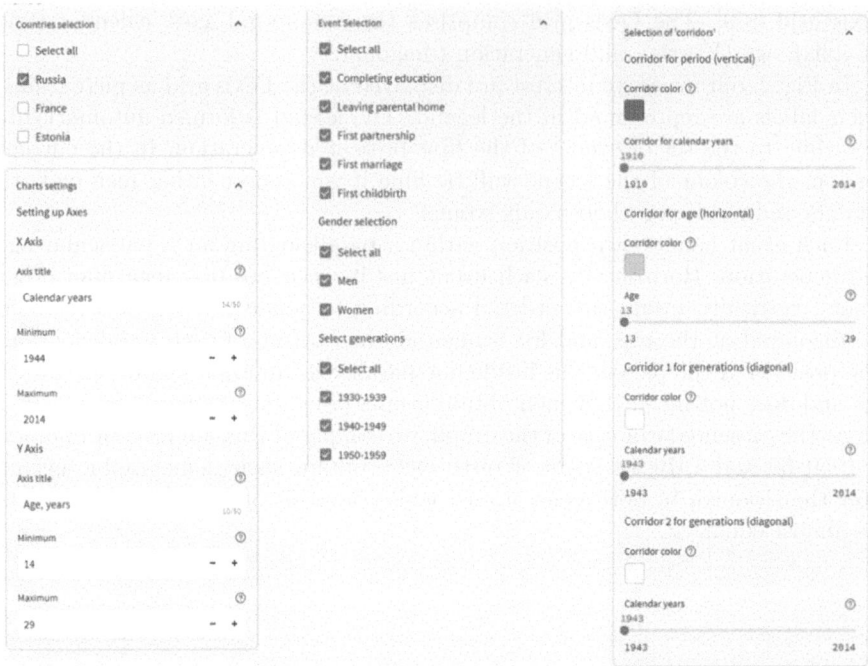

Fig. 3. Setting menu for analysis of events' occurrence

5 Sequence Analysis

For sequence analysis, we aggregated events into statuses representing respondents' states at each point in time from the age of 15 until the age of 35. Each status was assigned with an "alphabetic" code and colour (check the table in the menu "Status Selection").

The status codes along with the ascribed colours are presented in Fig. 4. Gray colour indicates "censoring", which is a proportion of respondents who were younger than a specific age at the time of the survey.

Statuses are visualised using chronograms [3–7]. A chronogram (Fig. 3) illustrates how 100% of the representatives of a certain generation are distributed by certain statuses at each age (Fig. 5). Moving from left to right along the X-axis, we observe the transition of respondents from the less "saturated" statuses (when no one has almost any events) to more "saturated" statuses (when respondents have nearly all events). The legend for the graph is on the right. It is built automatically according to the selected statuses.

Socioeconomic events	Demographic events					
	No children (C0)			1ˢᵗ child (C1)		
	Single (S)	1ˢᵗ partner (P)	1ˢᵗ marriage (M)	Single (S)	1ˢᵗ partner (P)	1ˢᵗ marriage (M)
No events (N)	SC0N	PC0N	MC0N	SC1N	PC1N	MC1N
Completing education (E)	SC0E	PC0E	MC0E	SC1E	PC1E	MC1E
Leaving parental home (L)	SC0L	PC0L	MC0L	SC1L	PC1L	MC1L
Education -> Leaving (EL)	SC0EL	PC0EL	MC0EL	SC1EL	PC1EL	MC1EL
Leaving -> Education (LE)	SC0LE	PC0LE	MC0LE	SC1LE	PC1LE	MC1LE
Education and Leaving simultaneously (E&L)	SC0E&L	PC0E&L	MC0E&L	SC1E&L	PC1E&L	MC1E&L
Censoring events						

Fig. 4. Colour coding for Sequence Analysis (Color figure online)

Fig. 5. Sequences of events occurrence on the chronogram

Another way of visualisation is to put chronograms on the Lexis grid. To do so, we rotated the chronogram for 90° and tilted it according to the diagonal axis (Fig. 6). X axis (calendar time) works here to indicate the boundaries of birth years of the selected generations. X axis should be interpreted not as a set of calendar years, but as shares of the statuses in which the selected generations were at specific ages.

At the provided example, we observe that red colours appear in life courses of new generations: it means partnerships become more popular than marriages for youngsters.

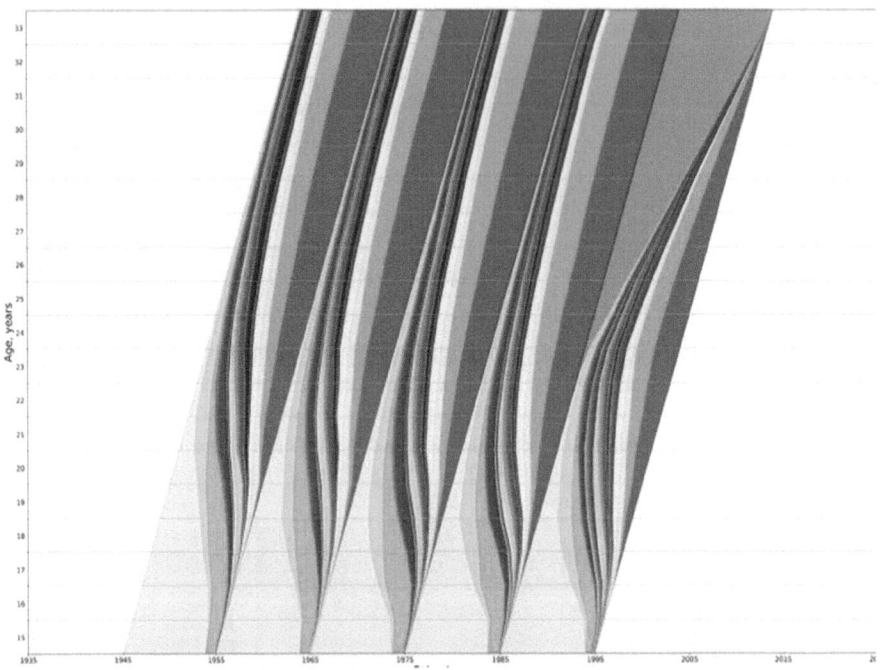

Fig. 6. Lexis grid with sequences of events occurrence on the chronograms

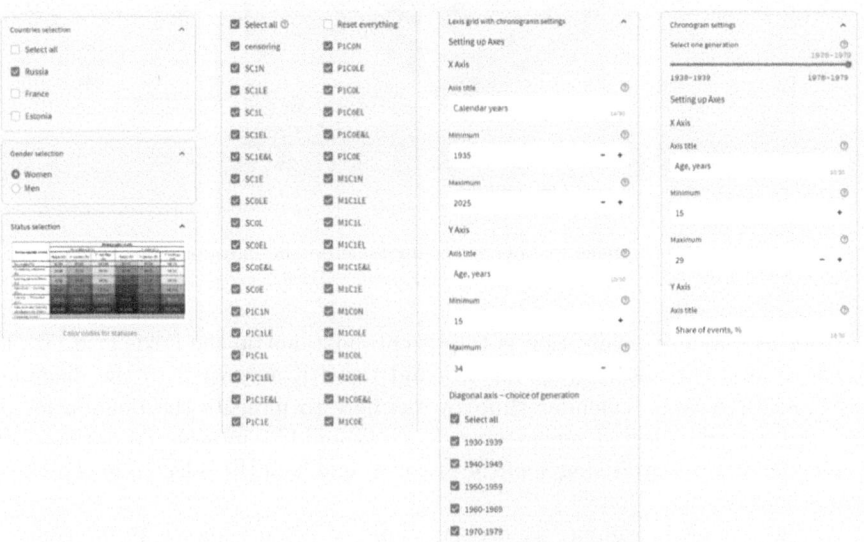

Fig. 7. Setting menu for Sequence analysis

6   Setting Menu for Sequence Analysis

To build graphs, a user should select countries, genders, statuses and generations in the menu on the left. A user can change the labels and boundaries of the axes (Fig. 7).

7   Conclusion

The presented visulaisation tool can already work with the following tasks: 1) to depict median ages of events occurrence on the Lexis grid, 2) to build chronograms for statuses of respondents throughout the whole period of observation, and 3) to rotate chronograms in order to put them on the Lexis grid and make comparisons.

The current version of the preloaded dataset contains the life courses derived from the Generations and Gender Survey (GGS). We prepared median ages and statuses of the respondents from France, Estonia, and Russia at the moment of the first waves of GGS (2004–2005).

We also prepared the closed versions of our web-tool for panel respondents of the Russian GGS (2004–2011). The next step is to prepare the datasets of European Social Survey for our tools usage and to release the version which will be able to work with users datasets.

Acknowledgements. This work is an output of a research project implemented as a part of the Basic Research Program at the National Research University Higher School of Economics (HSE University). This research was also supported in part through computational resources of HPC facilities at HSE University.

References

1. Ignatov, D.I., Mitrofanova, E., Muratova, A., Gizdatullin, D.: Pattern mining and machine learning for demographic sequences. In: Klinov, P., Mouromtsev, D. (eds.) KESW 2015. CCIS, vol. 518, pp. 225–239. Springer, Cham (2015). https://doi.org/10.1007/978-3-319-24543-0_17
2. Gizdatullin, D., Baixeries, J., Ignatov, D.I., Mitrofanova, E., Muratova, A., Espy, T.H.: Learning interpretable prefix based patterns from demographic sequences. In: Strijov, V.V., Ignatov, D.I., Vorontsov, K.V. (eds.) IDP 2016. CCIS, vol. 794, pp. 74–91. Springer, Cham (2019). https://doi.org/10.1007/978-3-030-35400-8_6
3. Abbott, A., Tsay, A.: Sequence analysis and optimal matching methods in sociology: review and prospect. Sociol. Methods Res. **29**(1), 3–33 (2000)
4. Aisenbrey, S., Fasang, A.: Beyond optimal matching: the second wave' of sequence analysis (2007)
5. Aisenbrey, S., Fasang, A.E.: New life for old ideas: the second wave' of sequence analysis bringing the course' back into the life course. Sociol. Methods Res. **38**(3), 420–462 (2010)
6. Billari, F.C.: Sequence analysis in demographic research. Can. Stud. Popul. **28**(2), 439–458 (2001)

7. Billari, F.C., Piccarreta, R.: Analyzing demographic life courses through sequence analysis. Math. Popul. Stud. **12**(2), 81–106 (2005)
8. Calot, G.: Demographic Techniques: LEXIS Diagram. Elsevier (2001)
9. Smith, P.C.: Demographic analysis: Methods, Results, Applications by Roland Pressat. Philippine Sociological Review (1973)
10. Muratova, A., Mitrofanova, E., Islam, R.: Comparison of machine learning methods for life trajectory analysis in demography. In: Nguyen, N.T., Chittayasothorn, S., Niyato, D., Trawiński, B. (eds.) ACIIDS 2021. LNCS (LNAI), vol. 12672, pp. 630–642. Springer, Cham (2021). https://doi.org/10.1007/978-3-030-73280-6_50

Correction to: Recent Trends in Analysis of Images, Social Networks and Texts

Dmitry I. Ignatov⬭, Michael Khachay⬭, Andrey Kutuzov⬭, Habet Madoyan,
Ilya Makarov⬭, Irina Nikishina⬭, Alexander Panchenko⬭, Maxim Panov⬭,
Panos M. Pardalos⬭, Andrey V. Savchenko⬭, Evgenii Tsymbalov⬭,
Elena Tutubalina⬭, and Sergey Zagoruyko⬭

Correction to:
D. I. Ignatov et al. (Eds.): *Recent Trends in Analysis of Images, Social Networks and Texts*, **CCIS 1905,**
https://doi.org/10.1007/978-3-031-67008-4

Chapter 4:
 The original published version of this chapter, an acknowledgement before references was generated. This now has been added.
Chapter 15:
 The original published version of this chapter, Author name displayed incorrectly. This now has been added.
Chapter 20:
 The original published version of this chapter, affiliation displayed incorrectly. This now has been added.

The updated version of these chapters can be found at
https://doi.org/10.1007/978-3-031-67008-4_4
https://doi.org/10.1007/978-3-031-67008-4_15
https://doi.org/10.1007/978-3-031-67008-4_20

© The Author(s), under exclusive license to Springer Nature Switzerland AG 2024
D. I. Ignatov et al. (Eds.): AIST 2023, CCIS 1905, p. C1, 2024.
https://doi.org/10.1007/978-3-031-67008-4_23

Author Index

© The Editor(s) (if applicable) and The Author(s), under exclusive license
to Springer Nature Switzerland AG 2024
D. I. Ignatov et al. (Eds.): AIST 2023, CCIS 1905, pp. 309–310, 2024.
https://doi.org/10.1007/978-3-031-67008-4

SPRINGER NATURE

GPSR Compliance

The European Union's (EU) General Product Safety Regulation (GPSR) is a set of rules that requires consumer products to be safe and our obligations to ensure this.

If you have any concerns about our products, you can contact us on ProductSafety@springernature.com

In case Publisher is established outside the EU, the EU authorized representative is:

Springer Nature Customer Service Center GmbH
Europaplatz 3
69115 Heidelberg, Germany

The manufacturer's authorised representative in the EU is Springer
Nature Customer Service Centre GmbH, Europaplatz 3, 69115 Heidelberg,
Germany. If you have any concerns regarding our products, please
contact ProductSafety@springernature.com

Printed and bound by CPI Group (UK) Ltd, Croydon, CR0 4YY

07/05/2026

02104559-0002